Essential Mathematics for GCSE

by Ronald L. Bolt, M.Sc.

Formerly Senior Mathematics Master
Woodhouse Grove School

Unwin Hyman

Published in 1986 by
UNWIN HYMAN LTD
15–17 Broadwick Street
London W1V 1FP

British Library Cataloguing in Publication Data
    Bolt, Ronald
        Essential mathematics for GCSE.
        1. Mathematics—1961–
        I. Title
        510   QA39.2
    ISBN 0-7135-2677-7

Printed and bound in Great Britain
at the University Press, Cambridge

Preface

This is a course for students working for the higher level papers of the GCSE. It is intended for use in the two years prior to the examination but it is also suitable for more mature students on one year courses. The book is an adaptation of my *Essential Mathematics for O Level*. Some sections have been removed, some others have been modified and some new ones have been added so that most of the topics in the syllabuses of the various GCSE examining groups are covered. There is emphasis on basic work in each topic and the exercises contain a wide selection of graded questions in order to build students' confidence. Many sections also include some GCE questions of the types which are likely to appear in GCSE papers.

I am grateful to the following Examining Boards for permission to reproduce questions from their recent GCE and 16+ examination papers:

The University of Cambridge Local Examinations Syndicate (C)
The East Anglian Examination Board (EA)
The University of London Entrance and School Examination Council (L)
The Joint Matriculation Board (JMB)
The Welsh Joint Education Committee (W)

The answers to the examination questions are entirely the responsibility of the author and have been neither provided nor approved by the Examining Boards.

R. L. Bolt

Contents

Preface   iii

1   Sets

Basic ideas, large and infinite sets, the empty set, subsets, intersection, union, the universal set, complement, shading sets in Venn diagrams, problems using Venn Diagrams   1

2   Numbers

Natural numbers, integers, rational and irrational numbers, factors, prime numbers, HCF, multiples, LCM, order of operations   16

3   Fractions

Equivalent fractions, comparing fractions, reducing to lowest terms, addition and subtraction, multiplication and division   20

4   Decimals

Fractions to decimals, addition and subtraction, multiplication, division, approximations, significant figures, decimal places, fractions to decimals   26

5   Powers, roots, reciprocals and standard form

Squares and square roots, cubes and cube roots, reciprocals, approximate values, standard form, surds   34

6   Percentages

Percentages to fractions or decimals, fractions or decimals to percentages, one quantity as percentage of another, percentage increases and decreases, profit and loss, simple interest, compound interest, depreciation   41

7   Ratio and proportion

Increasing and decreasing in a given ratio, direct and inverse proportion, foreign exchange, speeds   52

8   Length, area and volume

Rectangles and squares, parallelograms, triangles, trapezia, circles, length of arc, area of sector, volumes and surface areas of prisms, cylinders, pyramids, cones, spheres   58

9   Household and personal finance   80

10   Basic processes of algebra

Addition and subtraction, powers, directed numbers, evaluating expressions, brackets, binomial products, $(a + b)^2, (a - b)^2, (a + b)(a - b)$, completing the square, maximum and minimum values of $ax^2 + bx + c$ 83

11   Indices: negative, fractional and zero   97

12   Factors

$ax \pm ay, a^2 - b^2$, trinomials, four-term expressions   100

13 Simple algebraic fractions
    Reduction, addition and subtraction, multiplication and
    division   107
14 Linear equations and inequalities   112
15 Formulae
    Construction, rearranging   121
16 Simultaneous linear equations
    Methods of substitution and elimination   128
17 Quadratic equations
    Solution by factorisation, forming equations with given
    roots, checking roots, formula, simultaneous linear and
    quadratic equations   133
18 Functions
    Domain and range, inverse, composite functions   143
19 Graphs
    Coordinates, straight line graphs, quadratic functions,
    intersection of graphs, inequalities in x–y plane, travel graphs
20 Variation
    Direct, inverse and joint   170
21 Gradients
    Gradient of a straight line, gradient of a curve,
    distance–time graphs, velocity–time graphs   177
22 Areas under graphs
    Counting squares, trapezium method, velocity–time graphs   183
23 Matrices
    Storing information, order of a matrix, addition and
    multiplication, zero matrix, unit matrix, inverse matrices,
    singular matrices   188
24 Vectors
    Multiplication by a scalar, addition, magnitude and
    direction, unit base vectors, position vectors, proving lines
    parallel   197
25 Transformations
    Translations, reflections, rotations, enlargements, stretches   208
26 Matrices and transformations
    Combining transformations, inverse mappings   220
27 Angles
    Angle properties of parallel lines, angle properties of
    triangles and polygons   228
28 Similarity and congruence
    Similar triangles, ratio of areas of similar figures, ratio of
    volumes, congruent triangles   241
29 Symmetry
    Line symmetry, point symmetry, rotational symmetry,
    properties of quadrilaterals   251
30 Pythagoras' theorem
    The theorem, special right-angled triangles, converse   258

Contents

31 Sine, cosine and tangent
    Finding a side, finding an angle, using an isosceles triangle,
    finding a hypotenuse, angles of elevation and depression,
    bearings, ratios of angles of any size and their graphs,
    finding obtuse angles   263
32 Chord, angle and tangent properties of circles
    Properties of chords, intersecting chords, angles in a circle,
    cyclic quadrilaterals, tangent properties, alternate segment
    theorem   280
33 Constructions and loci
    Standard constructions, standard loci   297
34 Sine and cosine rules
    Problems using the sine and cosine rules, area of a triangle as
    $\frac{1}{2} bc \sin A$ 304
35 Bar charts and pie charts   310
36 Mean, median and mode   313
37 Frequency distributions
    Frequency tables and histograms, means of frequency
    distributions   317
38 Dispersion
    Range, interquartile range, cumulative frequency   325
39 Probability
    Basic idea, mutually exclusive events, independent events,
    tree diagrams   329
    Answers   339
    Index   373

1 Sets

Basic ideas

We speak of a set of saucepans, a chess set, a set of stamps. A set is a collection of things or people such that we can definitely decide whether or not a particular object belongs to that set.

To define a set we can make a list of the elements (or members) of the set, or we can describe the set in words. Here are some examples:

List of elements	*Description of set*
{a, e, i, o, u}	{vowels}
{spring, summer, autumn, winter}	{seasons of the year}
{2, 4, 6, 8}	{even numbers less than 10}

Notice that we place curly brackets (or braces) round a set.

Often we use a capital letter for the name of a set. For example, we might write V = {a, e, i, o, u} and S = {seasons of the year}.

The number of elements in set V is 5. We write this as $n(V) = 5$. Likewise $n(S) = 4$.

$\in$ means *is a member of* and $\notin$ means *is not a member of*. So we can write:

Tuesday $\in$ {days of the week}

h $\notin$ V.

Equal sets

The elements of a set can be listed in any order. {5, 2, 3, 1, 4} is the same as {4, 2, 1, 5, 3}. However it is helpful when working with sets to place the elements in logical order, in this case {1, 2, 3, 4, 5}.

Large sets and infinite sets

When a set is large and its elements can be placed in an obvious order, we often list just the first few and last few elements, with dots in between. For example, {a, b, c, d, ..., y, z}. Of course, we must be able to fill in the other elements if asked to do so.

Some sets never end. We say they are *infinite*. An example is the set {positive integers} which can also be written as {1, 2, 3, 4, ...}.

The empty set

Sometimes a set has no elements. Examples are {triangles with four sides} and {odd numbers ending in 0}. The set with no elements is called the *empty set* or *null set* and is denoted by ϕ.

Exercise 1.1

1. Describe the following sets:

    (i) {a, b, c, d}   (ii) {spades, clubs, hearts, diamonds}

    (iii) {knife, fork, spoon}   (iv) {1, 3, 5, 7, 9}.

2. List the elements of the following sets:

    (i) {last five letters of the English alphabet}

    (ii) {months of the year beginning with J}

    (iii) {even numbers between 11 and 19}.

3. List three elements of each of the following sets:

    (i) metals   (ii) wild animals   (iii) means of travel.

4. Write a name for the set which includes the following elements, and give another element of the set:

    (i) Belgium, Italy, West Germany

    (ii) thrush, starling, owl

    (iii) Jupiter, Venus, Pluto.

5. D = {days of the week}, M = {months of the year}, V = {vowels}.

    (i) State $n(D)$, $n(M)$ and $n(V)$.

    (ii) State whether the following are true or false:

    (a) Saturday $\in$ D (b) Tuesday $\in$ M (c) u $\notin$ V (d) e $\in$ V

6. Copy the following sets, filling in the missing elements:

    (i) {21, 22, 23,......, 29}   (ii) {1, 3, 5, 7,......, 19}

    (iii) {3, 6, 9,......, 21}   (iv) {h, i, j, k,......, r}.

7. Write down the number of elements for each set in question 6.

    What can you say about the set {1, 2, 3, 4,......}?

Subsets

Suppose that P = {a, b, c} and Q = {b, c}. Q is part of P and is called a *proper subset* of P. We write Q $\subset$ P.

    The complete list of subsets of P is:

                 {a, b, c}, {a, b}, {b, c}, {c, a}, {a}, {b}, {c}, ϕ

    Notice that we include the empty set and P itself as subsets of P, but they are not proper subsets.

The two sets P and Q are shown in Fig. 1.1. This type of diagram is called a *Venn diagram*.

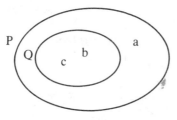

Fig. 1.1

Intersection

Consider the sets R = {d, e, f, g, h} and S = {e, g, k, n}. Notice that e and g are elements of both sets, as shown in the Venn diagram below. If T = {e, g} then T is a subset of both R and S. T is called the *intersection set* of R and S. The symbol ∩ is used for intersection. We write T = R ∩ S.

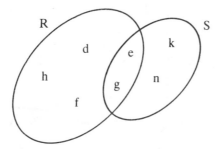

Fig. 1.2

The sets E = {2, 4, 6} and F = {3, 5} have no common elements. We say that the intersection of E and F is the empty set, and write E ∩ F = φ. Fig. 1.3 shows E and F.

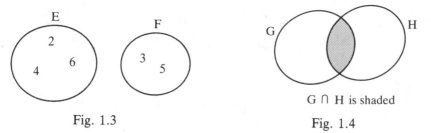

G ∩ H is shaded

Fig. 1.3   Fig. 1.4

If G = {people who wear spectacles} and H = {people who wear hats}, then G ∩ H is {people who wear spectacles and hats}. It is shown by shading in Fig. 1.4.

Union

The set W = {d, e, f, g, h, k, n} is formed by putting together the elements of the sets R and S. It is the *union* of R and S. The symbol ∪ is used for union. We write W = R ∪ S.

G ∪ H is {people who wear spectacles or hats or both}. It is shown by shading in Fig. 1.5.

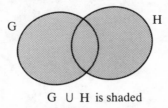

G ∪ H is shaded

Fig. 1.5

To form the union of the two sets A = {m, x, p, r} and B = {t, y, p, m, w}, we write down all the elements of A and then add those elements of B which are not down already. Thus A ∪ B = {m, x, p, r, t, y, w}.

Exercise 1.2

1. Draw a Venn diagram to show the sets {p, q, r, s, t} and {q, r}.

2. F = {foods}, V = {vegetables}.

   (i) Draw a Venn diagram to show F and V.

   (ii) Enter in your diagram the elements butter, carrots, bread and potatoes.

   (iii) Which is true, V ⊂ F or F ⊂ V?

3. Write down all the subsets of {x, y}.

4. Write down all the subsets of {p, q, r}.

5. (i) Copy Fig. 1.6 and shade A ∩ B.

   (ii) Make another copy and shade A ∪ B.

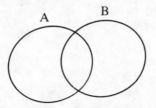

Fig. 1.6

6. C = {1, 3, 5, 7, 9} and D = {3, 7, 11, 15}. List the elements of C ∩ D and show the sets in a diagram like Fig. 1.2.

7. E = {k, m, n, p} and F = {n, p, r, k, t}. List the elements of E ∩ F and show the sets in a Venn diagram.

8. Using the sets of question 6, write down the elements of C ∪ D.

9. Using the sets of question 7, write down the elements of E ∪ F.

10. Write down a subset of {countries of Europe} which has three elements.

11. Fig. 1.7 shows three sets A, B and C. Examine each of the following statements and say whether it is true or false:

   (i) The shaded part represents B ∩ C

   (ii) B ⊂ C   (iii) B ⊂ A

   (iv) B ∩ C = φ.

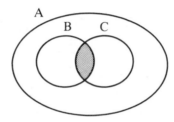

Fig. 1.7

12. From Fig. 1.8, list the elements of: (i) P (ii) Q (iii) P ∩ Q (iv) P ∪ Q.

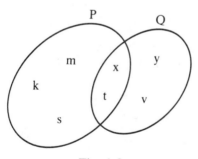

Fig. 1.8

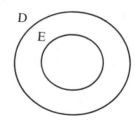

Fig. 1.9

13. Using Fig. 1.9, examine each of the following statements and say whether it is true or false:

   (i) E ⊂ D (ii) D ⊂ E (iii) D ∩ E = φ

   (iv) D ∩ E = E (v) D ∩ E = D.

14. A = {k, m, p, q, r, t, v}, B = {m, p, r}, C = {p, k, t}.
    List the elements of: (i) B ∩ C (ii) B ∪ C (iii) A ∩ B.

Draw a Venn diagram to show the sets A, B and C.
Make two statements about the sets using the symbol ⊂.

15. D = {1, 2, 3,......, 9},   E = {4, 6, 8},   F = {3, 5, 7}.
Comment on E ∩ F. Draw a diagram to show the sets D, E and F.

16. G = {1, 3, 5, 7, 9, 11},   H = {5, 7, 13, 15},
K = {7, 9, 11, 13, 17, 19}.
Write down the elements of G ∩ H, H ∩ K and K ∩ G. Which
element is in all three sets? The set containing this element is
G ∩ H ∩ K. Copy Fig. 1.10 and enter the elements in the appro-
priate places.

17. Make four copies of Fig. 1.10 and use them to show by shading:
(i) G ∩ H   (ii) H ∩ K   (iii) K ∩ G   (iv) G ∩ H ∩ K.

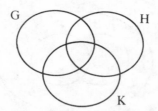

Fig. 1.10

18. Make four more copies of Fig. 1.10 and use them to show by
shading:
(i) G ∪ H   (ii) H ∪ K   (iii) K ∪ G   (iv) G ∪ H ∪ K.

19. M = {a, b, c},   N = {b, d, e},   P = {a, f, g}.
Write down the elements of:
(i) M ∪ N   (ii) N ∪ P   (iii) P ∪ M   (iv) M ∪ N ∪ P.
(M ∪ N ∪ P must contain all the elements of M, N and P.)

20. Copy the following diagrams and in each case shade the intersection
of the two sets.

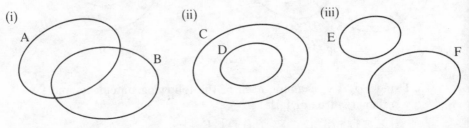

Fig. 1.11

21. Copy the above three diagrams and in each case shade the union
of the two sets.

22. The elements of sets W and Y are shown by crosses, in Fig. 1.12. Find $n(W)$, $n(Y)$, $n(W \cap Y)$ and $n(W \cup Y)$.

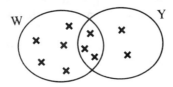

Fig. 1.12

23. (i) Set K has 5 elements and set M has 6. If K ∩ M has 2 elements, use a diagram like Fig. 1.12 to find $n(K \cup M)$.

(ii) $n(P) = 12$ and $n(Q) = 18$. If $n(P \cup Q) = 22$, use a diagram to find $n(P \cap Q)$.

24. Draw a Venn diagram to illustrate each of the following:

(i) two sets A and B such that $A \subset B$.

(ii) two sets C and D such that $C \cap D = \phi$.

(iii) two sets E and F such that $E \cap F = F$.

(iv) two sets G and H such that $G \cup H = H$.

The universal set

The *universal set* is the set of all elements which are being considered. It is denoted by the symbol ξ. For example, if the whole numbers from 1 to 9 are being considered, then $\xi = \{1, 2, 3, 4, \ldots, 9\}$. In a Venn diagram the universal set is usually represented by a rectangle.

Complement

The *complement* of set A contains all the elements of the universal set which are not in A. The complement of A is denoted by A'.

So if $\xi = \{1, 2, 3, 4, \ldots, 9\}$ and $A = \{3, 6, 9\}$, then
A' = {1, 2, 4, 5, 7, 8}.

In the Venn diagram of Fig. 1.13, A' is the part outside A.

Notice that $A \cup A' = \xi$ and $A \cap A' = \phi$.

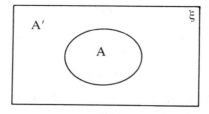

Fig. 1.13

Shading sets in Venn diagrams

Union. To shade the area representing $P \cup Q$, first shade the area representing P and then any part of the area of Q which is not shaded already.

Intersection. To shade the area representing $P \cap Q$, look for the area which is in both P and Q.

Examples

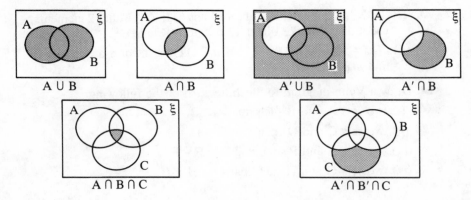

Fig. 1.14

Exercise 1.3

1. If $\xi = \{a, b, c, d, e, f\}$ and $P = \{a, c, e, f\}$, list the elements of P′.

2. If $\xi = \{1, 2, 3, 4, 5, 6, 7\}$ and $Q = \{4, 6, 7\}$, list the elements of Q′.

3. State the complement of each of the following sets:

    (i) E = {even numbers} where ξ = {whole numbers}

    (ii) M = {males} where ξ = {people}

    (iii) V = {vowels} where ξ = {letters of the English alphabet}.

4. Draw a Venn diagram to show ξ = {all triangles}, I = {isosceles triangles} and E = {equilateral triangles}.

In which set would you place a triangle with sides of:

    (i) 6 cm, 4 cm, 6 cm   (ii) 5 cm, 7 cm, 10 cm?

5. From Fig. 1.15, list the elements of:

    (i) ξ (ii) R   (iii) R′   (iv) T   (v) T′.

6. From Fig. 1.16, list the elements of:

    (i) ξ (ii) A   (iii) A′   (iv) B   (v) B′.

7. Make four copies of Fig. 1.17 and use them to show by shading:

    (i) $P \cup Q'$ (ii) $P \cap Q'$ (iii) $P' \cup Q'$ (iv) $P' \cap Q'$.

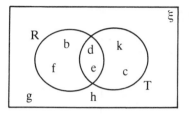

Fig. 1.15

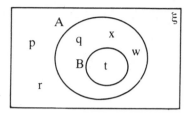

Fig. 1.16

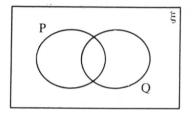

Fig. 1.17

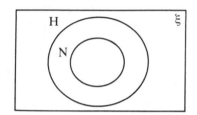

Fig. 1.18

8. Make six copies of Fig. 1.18 and use them to show by shading:
(i) H′ (ii) N′ (iii) H′ ∩ N′
(iv) H ∩ N′ (v) H′ ∪ N′ (vi) H ∪ N′.

9. Make four copies of Fig. 1.19 and use them to show by shading:
(i) W′ (ii) W′ ∩ R′ (iii) W′ ∩ R (iv) W′ ∪ R′.

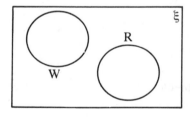

Fig. 1.19

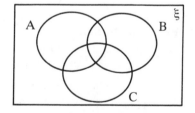

Fig. 1.20

10. Make two copies of Fig. 1.20 and use them to show by shading:
(i) A ∪ B ∪ C (ii) A′ ∪ B′ ∪ C′.

Your finished diagram for (ii) should be the complement of one of the diagrams of Fig. 1.14. Which one?

11. Make three copies of Fig. 1.20 and use shading on them to show:
(i) A ∩ B ∩ C (ii) A ∩ B ∩ C′ (iii) A ∩ B′ ∩ C′.

12. For Fig. 1.21 list the elements of:

(i) P   (ii) P ∩ Q   (iii) P ∩ Q ∩ R

(iv) R'   (v) R' ∩ P   (vi) P' ∩ Q' ∩ R'.

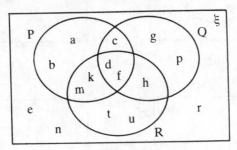

Fig. 1.21

13. Make two copies of Fig. 1.22 and use shading on them to show:

(i) P ∩ Q' ∩ R   (ii) P ∩ Q' ∩ R'.

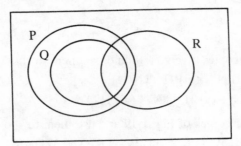

Fig. 1.22

14. ξ = {a, b, c, d,......, 1},   T = {a, e, f, i, k, l},   W = {a, b, e, f, g},
X = {a, g, h, k, l}. List the elements of:

(i) T'   (ii) W'   (iii) T' ∩ W'   (iv) T ∪ W

(v) (T ∪ W)', which is the complement of T ∪ W

(vi) T ∪ W ∪ X   (vii) (T ∪ W ∪ X)'.

15.   (i) Make two copies of Fig. 1.23. On one show A ∪ B and on the other show (A ∪ B)'.

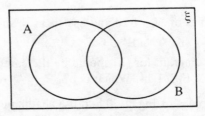

Fig. 1.23

(ii) Make two more copies of Fig. 1.23. On one show A ∩ B and on the other show (A ∩ B)'.

(iii) Copy Fig. 1.23 again and shade A' ∩ B'. Which of the sets illustrated in (i) and (ii) is the same as A' ∩ B'?

16. Using letters and symbols, describe the shaded parts in Fig. 1.24:

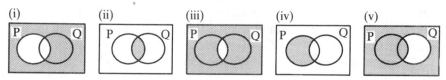

Fig. 1.24

17. ξ = {houses}, C = {houses with central heating}, D = {houses with double glazing}. Use letters and symbols to represent:

(i) {houses without central heating}

(ii) {houses with both central heating and double glazing}

(iii) {houses which have central heating but not double glazing}

(iv) {houses with neither central heating nor double glazing}.

18. ξ = {men leaving a train}, N = {men carrying newspapers}, H = {men wearing hats}. Describe:

(i) H'   (ii) N ∩ H   (iii) N ∩ H'   (iv) N' ∩ H'.

19. ξ = {1, 2, 3,...... 12},   E = {2, 4, 6, 8, 10, 12},   F = {4, 8, 12}, T = {3, 6, 9, 12}. List the members of:

(i) E'   (ii) T ∩ E'   (iii) T ∪ E

(iv) (T ∪ E)'   (v) F'   (vi) E ∩ F'.

Which is true: E ⊂ F or F ⊂ E?

Draw a Venn diagram to show ξ, E, F and T.

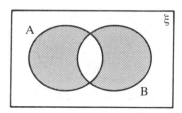

Fig. 1.25

20. The shading in Fig. 1.25 shows A △ B which is the set of elements in either A or B but not in both. It is called the *symmetric difference* of A and B.

(i) If A = {k, m, t, v, w} and B = {k, v, x, y}, list the elements of A △ B.

(ii) If C = {s, v, w, y, z}, list the elements of (A △ B) △ C.

(iii) List the elements of B △ C and A △ (B △ C).

(iv) Is A △ (B △ C) = (A △ B) △ C?

(v) Draw a Venn diagram and enter the elements of A, B and C. Hence shade the area representing A △ B △ C.

21. The universal set ξ has non-empty subsets P, Q and R such that P is not a subset of Q and Q is not a subset of P. It is also given that P ∩ Q ≠ φ, R ⊂ P and Q ∩ R = φ. Complete the Venn diagram to show the relationship between ξ, P, Q and R.   (JMB)

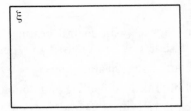

Fig. 1.26

Problems using Venn diagrams

In Fig. 1.27, ξ = {people who got off a certain bus}, B = {those carrying bags}, N = {those carrying newspapers}. The number of people in each set is shown in the area representing that set. The number of people carrying bags = $n(B) = 6 + 3 = 9$. The number carrying both bags and newspapers = $n(B \cap N) = 3$. The number outside set B and set N (i.e. the number carrying neither bags nor newspapers) = 7. The total number leaving the bus = $n(\xi) = 6 + 3 + 4 + 7 = 20$.

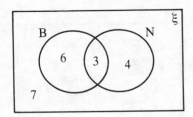

Fig. 1.27

Exercise 1.4

1. 40 people were asked whether they had cats or dogs. The resulting information is shown in Fig. 1.28 where C = {those with cats} and D = {those with dogs}.

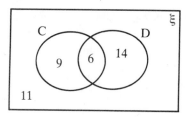

Fig. 1.28

How many people had:

  (i) cats   (ii) dogs   (iii) both

  (iv) neither   (v) cats but not dogs?

Use words, and then symbols, to describe the set which had 14 people.

2. From Fig. 1.29, state:

  (i) $n(A)$ (ii) $n(B)$ (iii) $n(A')$ (iv) $n(B')$

  (v) $n(A \cap B)$ (vi) $n(A \cup B)$ (vii) $n(\xi)$.

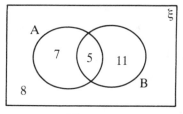

Fig. 1.29

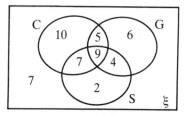

Fig. 1.30

3. In a class of 28 pupils, 22 passed a mathematics examination, 19 passed an English examination and 17 passed both. Show the information in a Venn diagram. How many passed neither examination?

4. Draw a Venn diagram showing sets F and G, such that $n(\xi) = 24$, $n(F) = 11$, $n(G) = 7$ and $n(F \cap G) = 3$. What is $n(F \cup G)$ and what is $n(F \cup G)'$?

5. On a cold day, 50 people were observed in a street. Some wore coats, some wore gloves, some wore scarves, and some wore none of these. The results are shown in Fig. 1.30. How many wore:

  (i) coats

  (ii) coats and gloves

  (iii) coats, gloves and scarves

  (iv) scarves but not coats or gloves

  (v) coats and scarves but not gloves?

Use words, and then symbols, to describe the set which had 10 people.

6. Some swimming tests were held at a swimming club. There were
   tests for the crawl (C), the backstroke (B) and diving (D). 60
   members took the tests. The numbers of passes were: C, 39; B, 30;
   D, 30; C and B but not D, 9; B and D but not C, 2; C and D but
   not B, 6; all three, 18. Draw a Venn diagram to show this information, and write the appropriate number in each section. From
   your diagram, find:

   (i) the number who passed the crawl test only

   (ii) the number who passed the diving test only

   (iii) the number who passed no test.

7. In a survey of 80 college students, the numbers studying various
   subjects were: mathematics 32, physics 27, chemistry 24, physics and
   chemistry 17, mathematics and physics 20, mathematics and chemistry 16, all three subjects 14. Show this information on a Venn
   diagram. How many studied:

   (i) mathematics only

   (ii) mathematics and physics but not chemistry

   (iii) none of the three subjects?

8. (i) For the sets P and Q in Fig. 1.31, $n(P) = 15$, $n(Q) = 19$ and
       $n(P \cap Q) = 7$.
       What is the value of y?
       $x + y = 15$.
       What is the value of x?
       What is the value of z?
       What is $n(P \cup Q)$?

   (ii) If $n(R) = 38$, $n(S) = 22$ and $n(R \cap S) = 10$, use a Venn diagram
        to find $n(R \cup S)$.

   (iii) If $n(T) = 20$, $n(V) = 24$ and $n(T \cup V) = 29$, find $n(T \cap V)$.

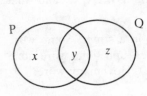

Fig. 1.31

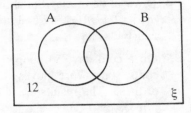

Fig. 1.32

9. (i) Copy Fig. 1.32. If $n(\xi) = 28$ and $n(A) = 9$, find $n(A')$.

   (ii) There are 12 elements outside both A and B. How many are
        in B but not in A? Is it correct to describe this as $n(A' \cap B)$?

   (iii) If $n(B) = 13$, find $n(B')$ and $n(A \cap B')$.

   (iv) Enter all the numbers in your diagram and find $n(A \cap B)$.

10. Out of 40 people questioned, 22 had watched 'Magic and Mystery' on TV, 14 had watched 'Fact or Fantasy' and 10 had watched neither. Use a Venn diagram to find how many watched:

(i) 'Magic and Mystery' only

(ii) 'Fact or Fantasy' only

(iii) both programmes.

11. In Fig. 1.33,

ξ = {pupils questioned in a survey}
P = {pupils who went to a pop-concert}
C = {pupils who went to a cinema}
Y = {pupils who went to a youth club}.

Find the number of pupils:

(i) who went to the youth club

(ii) who went to the pop-concert and to the cinema

(iii) in C ∪ Y

(iv) in (C ∪ Y) ∩ P

(v) in (P ∪ C ∪ Y)', given that $n(\xi) = 90$.

Explain the meaning of your answer to (v).

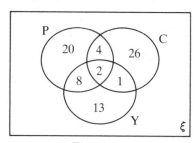

Fig. 1.33

12. In a survey carried out at a sports centre, men were asked about their sporting activities. Of the men questioned, 12 played rugby, 16 played squash, 13 played tennis. 8 played none of these games. 3 played both rugby and squash, 5 played both rugby and tennis, 2 men played tennis only.

Let R, S and T be the sets of rugby, squash and tennis players respectively. Let the number of men playing all three games be x. Draw a Venn diagram and show, in terms of x, the number of men in each region of the diagram in set R. Also show the number in each of the other four regions.

Find the total number of men questioned and state the possible values of x. (W)

2 Numbers

The different sets of numbers

Numbers can be classified as follows:

Natural numbers. These are the counting numbers. The set of natural numbers is $\{1, 2, 3, 4, 5,......\}$.

Integers. The set of integers is $\{......, -3, -2, -1, 0, 1, 2, 3,......\}$.

Rational numbers. A rational number has the form $\frac{a}{b}$ where a and b are integers. (b cannot be zero. We write this as $b \neq 0$.)

Any decimal number can be put in this form.

For example $0.56 = \frac{56}{100} = \frac{14}{25}$ and $7.5 = \frac{75}{10} = \frac{15}{2}$.

Any integer can be written in the form $\frac{a}{b}$ where $b = 1$.

For example $33 = \frac{33}{1}$. Hence the set of integers is a subset of the set of rational numbers.

Irrational numbers. These cannot be expressed in the form $\frac{a}{b}$ where a and b are integers. Examples are π, $\sqrt{5}$, $\sqrt[3]{6}$, $\cos 43°$. For calculations we must use approximations such as 3.142 for π and 2.236 for $\sqrt{5}$.

Factors

The number 12 can be divided exactly by 1, 2, 3, 4, 6 and 12. These are the factors of 12.

Example List the factors of 42.

We try the numbers 2, 3, 4, etc. in turn, and find that $42 = 1 \times 42$, 2×21, 3×14, 6×7. We stop here since 7 gives 7×6 which is the same as 6×7.

The factors of 42 are therefore 1, 2, 3, 6, 7, 14, 21, 42.

Prime numbers. A natural number which has no factors other than 1 and itself is called a prime number. The first few prime numbers are 2, 3, 5, 7, 11, 13. Note that 1 is not regarded as a prime number.

Even numbers. These have 2 as a factor. They end in 0, 2, 4, 6 or 8.

Odd numbers. These do not have 2 as a factor. They end in 1, 3, 5, 7 or 9.

Highest common factor (HCF)

The factors of 42 are 1, 2, 3, 6, 7, 14, 21, 42.
The factors of 56 are 1, 2, 4, 7, 8, 14, 28, 56.
The common factors of 42 and 56 are 1, 2, 7, 14.
The largest common factor, 14, is called the *highest common factor* (HCF) of 42 and 56.

A number as the product of primes

The factors of 40 are 1, 2, 4, 5, 8, 10, 20, 40. Of these, 2 and 5 are prime numbers. They are the only *prime factors* of 40. We can write:
$40 = 2 \times 20 = 2 \times 2 \times 10 = 2 \times 2 \times 2 \times 5$.

Here 40 is expressed as the product of prime numbers.
The working can be set out in division form as shown below.

$$
\begin{array}{c|c}
2 & 40 \\ \hline
2 & 20 \\ \hline
2 & 10 \\ \hline
5 & 5 \\ \hline
& 1
\end{array}
$$

Using index form, we can write $40 = 2^3 \times 5$.

Example Express 126 as the product of prime numbers in ascending order (smallest to largest).

$$126 = 2 \times 63 = 2 \times 3 \times 21 = 2 \times 3 \times 3 \times 7 = 2 \times 3^2 \times 7$$

Multiples

If we multiply the natural numbers 1, 2, 3, 4, ... by 7 we get 7, 14, 21, 28, ... These numbers are multiples of 7.

Lowest common multiple (LCM)

The multiples of 6 are 6, 12, 18, 24, 30, 36, 42, 48, 54, ...
The multiples of 9 are 9, 18, 27, 36, 45, 54, 63, 72, 81, ...
The common multiples of 6 and 9 are 18, 36, 54, ...
The smallest, 18, is called the *lowest common multiple* (LCM) of 6 and 9.

Exercise 2.1

1. List all the prime numbers up to 37.

2. Which of the following are prime numbers: 41, 43, 49, 51, 53, 59, 61, 63, 69?

3. List the factors of:

   (i) 6   (ii) 7   (iii) 8   (iv) 9   (v) 10.

4. Write down the set of factors of:

   (i) 14   (ii) 16   (iii) 20   (iv) 31   (v) 34.

5. (i) List the factors of 12.

   (ii) List the factors of 18.

   (iii) List the common factors of 12 and 18.

   (iv) State the highest common factor of 12 and 18.

6. Repeat question 5 for the numbers 16 and 28.

7. Repeat question 5 for the numbers 42 and 63.

8. P = {factors of 24} and Q = {factors of 30}.
   List the elements of: (i) P   (ii) Q   (iii) P ∩ Q.
   Show the sets P and Q in a Venn diagram. State the HCF of 24 and 30.

9. Write down the HCF of: (i) 9 and 12   (ii) 16 and 24.

10. Write down the HCF of: (i) 15, 25 and 30   (ii) 18, 27 and 36.

11. Express as the product of prime numbers: (i) 18   (ii) 30   (iii) 45.

12. Express as the product of prime numbers: (i) 39   (ii) 42   (iii) 60.

13. Write down:

    (i) the multiples of 5 up to 45

    (ii) the multiples of 3 up to 45

    (iii) the common multiples of 3 and 5 up to 45

    (iv) the lowest common multiple of 3 and 5.

14. Repeat question 13 for 4 and 6 up to 48.

15. Repeat question 13 for 9 and 15 up to 135.

16. Write down the LCM of: (i) 2 and 3   (ii) 6 and 9   (iii) 4 and 10.

17. One lighthouse flashes every 12 seconds and another flashes every 15 seconds. How often do they flash together?

18. Three bells toll at intervals of 12, 15 and 16 seconds. If they sound together at a certain instant, how long will it be before they sound together again?

Order of operations

Multiplication and division must be done before addition and subtraction.

Examples $5 + 3 \times 4 = 5 + 12 = 17$ (Not 8×4)
$30 \div 6 - 2 = 5 - 2 = 3$ (Not $30 \div 4$)
$6 \times 3 + 8 \div 2 = 18 + 4 = 22$ (Not $6 \times 11 \div 2$)

Computers work in this way, but many electronic calculators do not. Many calculators take each step in turn, working from left to right. So for $6 \times 3 + 8 \div 2$ such calculators give the answer 13. Try $5 + 3 \times 4$ and $30 \div 6 - 2$ on a calculator. These examples show that you must be very careful when using a calculator.

If in doubt, brackets should be used. Calculations inside brackets must be worked before any other calculations.

Examples $(3 + 5) \div 2 = 8 \div 2 = 4$
$16 - (9 - 4) = 16 - 5 = 11$
$3 \times (5 + 7) \div 4 = 3 \times 12 \div 4 = 36 \div 4 = 9$

Exercise 2.2

Find the values of the following expressions.

1. $(3 + 2) \times 4$ **2.** $7 + (2 \times 3)$ **3.** $(18 - 8) \div 2$
4. $24 \div (3 \times 4)$ **5.** $(24 \div 3) \times 4$ **6.** $15 - (7 - 2)$
7. $(15 - 7) - 2$ **8.** $9 - (4 + 3)$ **9.** $(9 - 4) + 3$

Find the values of the following expressions. (Remember that $\times$ and $\div$ must be used before $+$ and $-$.)

10. $10 - 2 \times 4$ **11.** $9 + 6 \div 3$ **12.** $5 \times 2 - 3 \times 3$
13. $20 \div 4 + 18 \div 3$ **14.** $20 - 12 \div 4$ **15.** $5 + 7 \times 2$
16. $(5 + 7) \times 2 - 1$ **17.** $13 - 3 \times (6 - 4)$ **18.** $(10 - 4 \times 2) - 2$

Insert brackets in each of the following so that the required answer is obtained.

19. $11 - 5 - 2$ to give 8   **20.** $15 - 3 + 7$ to give 5
21. $8 - 3 \times 2$ to give 10   **22.** $40 \div 4 \times 2$ to give 5

3   Fractions

In a fraction such as $\dfrac{3}{5}$ the top number, 3, is called the *numerator* and the bottom number, 5, is called the *denominator*.

In a *proper fraction* the numerator is smaller than the denominator, for example $\dfrac{4}{9}$.

In an *improper fraction* the numerator is larger than the denominator, for example $\dfrac{12}{7}$.

A *mixed number* is an integer plus a proper fraction, for example $2\dfrac{1}{4}$, which means $2 + \dfrac{1}{4}$.

Changing mixed numbers to improper fractions

Examples
$$2\frac{3}{4} = 2 + \frac{3}{4} = \frac{8}{4} + \frac{3}{4} = \frac{11}{4}$$
$$5\frac{4}{7} = 5 + \frac{4}{7} = \frac{35}{7} + \frac{4}{7} = \frac{39}{7}$$

Changing improper fractions to mixed numbers

Examples
$$\frac{17}{5} = \frac{15}{5} + \frac{2}{5} = 3 + \frac{2}{5} = 3\frac{2}{5}$$
$$\frac{13}{3} = \frac{12}{3} + \frac{1}{3} = 4 + \frac{1}{3} = 4\frac{1}{3}$$

Equivalent fractions

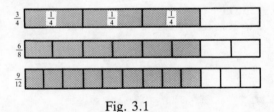

Fig. 3.1

Figure 3.1 shows that $\dfrac{3}{4}, \dfrac{6}{8}$ and $\dfrac{9}{12}$ are the same size. They are *equivalent fractions*. A fraction remains the same size if its numerator and denominator are both multiplied by the same number.

For example $\dfrac{2}{7} = \dfrac{2 \times 5}{7 \times 5} = \dfrac{10}{35}$.

Comparing fractions

To compare two or more fractions, they must first be given the same denominator.

Example   Arrange $\dfrac{5}{7}, \dfrac{11}{14}$ and $\dfrac{3}{4}$ in order of size with the smallest first.

The LCM of 7, 14 and 4 is 28.

$$\dfrac{5}{7} = \dfrac{5 \times 4}{7 \times 4} = \dfrac{20}{28}, \qquad \dfrac{11}{14} = \dfrac{11 \times 2}{14 \times 2} = \dfrac{22}{28}, \qquad \dfrac{3}{4} = \dfrac{3 \times 7}{4 \times 7} = \dfrac{21}{28}$$

Since $\dfrac{20}{28} < \dfrac{21}{28} < \dfrac{22}{28}$, the order of size is $\dfrac{5}{7}, \dfrac{3}{4}, \dfrac{11}{14}$.

Reducing to lowest terms

This means finding the equivalent fraction with the smallest possible numerator and denominator.

Example 1 $\dfrac{12}{18} = \dfrac{6 \times 2}{6 \times 3} = \dfrac{2}{3} \qquad \dfrac{20}{35} = \dfrac{5 \times 4}{5 \times 7} = \dfrac{4}{7}$

Example 2   Express $\dfrac{3}{5}$ hour in minutes.

1 hour is 60 minutes

$\dfrac{1}{5}$ hour is $60 \div 5 = 12$ minutes

$\dfrac{3}{5}$ hour is $12 \times 3 = 36$ minutes

Example 3   Express 60p as a fraction of £1.05.

$$\dfrac{60p}{£1.05} = \dfrac{60p}{105p} = \dfrac{60}{105} = \dfrac{12}{21} = \dfrac{4}{7}$$

Exercise 3.1

1. Copy and complete: $1 = \dfrac{}{5}$, $2 = \dfrac{}{3}$, $5 = \dfrac{}{7}$, $6 = \dfrac{}{10}$.

2. Change to improper fractions: $1\dfrac{1}{2}, 2\dfrac{1}{3}, 3\dfrac{2}{5}, 1\dfrac{5}{8}, 3\dfrac{7}{10}$.

3. Express as mixed numbers: $\dfrac{5}{4}, \dfrac{4}{3}, \dfrac{12}{5}, \dfrac{19}{4}, \dfrac{43}{10}$.

4. Express as twelfths: $\dfrac{1}{2}, \dfrac{2}{3}, \dfrac{3}{4}, \dfrac{1}{6}, \dfrac{5}{6}$.

5. Express as twentieths: $\dfrac{3}{4}, \dfrac{4}{5}, \dfrac{7}{10}$. Now write the original fractions in order of size with the smallest first.

6. Express as eighteenths: $\dfrac{2}{3}, \dfrac{5}{6}, \dfrac{7}{9}$. Now write the original fractions in order of size with the largest first.

7. Arrange the following in order of size with the smallest first: $\dfrac{3}{7}, \dfrac{5}{14}, \dfrac{1}{2}$.

8. Which of the following are equivalent to $\dfrac{3}{4}$: $\dfrac{6}{8}, \dfrac{9}{16}, \dfrac{12}{16}, \dfrac{15}{24}, \dfrac{21}{28}$?

9. Reduce to their lowest terms: $\dfrac{4}{6}, \dfrac{6}{10}, \dfrac{3}{12}, \dfrac{15}{25}, \dfrac{30}{40}$.

10. Reduce to their lowest terms: $\dfrac{9}{12}, \dfrac{16}{18}, \dfrac{45}{55}, \dfrac{9}{36}, \dfrac{44}{77}$.

11. Express $\dfrac{1}{2}$ minute in seconds.   **12.** Express $\dfrac{1}{5}$ cm in mm.

13. Express £$\dfrac{1}{4}$ in pence.   **14.** Express $\dfrac{2}{5}$ hour in minutes.

15. Express £$\dfrac{4}{5}$ in pence.   **16.** Express $\dfrac{5}{6}$ day in hours.

Express the first quantity as a fraction of the second, and reduce the fraction to its lowest terms:

17. £12, £15   **18.** 16p, 24p   **19.** 60p, £1
20. 35 minutes, 1 hour   **21.** 70 cm, 1 metre   **22.** £1.40, £3.50

23. $\dfrac{3}{4}$ of a sum of money is £18. What is $\dfrac{1}{4}$ of the sum? What is the whole sum?

24. $\dfrac{4}{5}$ of a length is 20 cm. What is $\dfrac{1}{5}$ of it? What is the whole length?

25. $\dfrac{5}{8}$ of a sum of money is £160. What is the whole sum?

26. When 19.8 litres of water are poured into an empty tank, it is $\dfrac{3}{8}$ full. How many more litres are needed to fill the tank?

Addition of fractions

If the denominators are the same, just add the numerators.

Example $\dfrac{4}{11} + \dfrac{3}{11} + \dfrac{9}{11} = \dfrac{4+3+9}{11} = \dfrac{16}{11} = 1\dfrac{5}{11}$

If the denominators are different, first find equivalent fractions having the same denominators.

Example $\dfrac{2}{3} + \dfrac{3}{5} = \dfrac{10}{15} + \dfrac{9}{15} = \dfrac{19}{15} = 1\dfrac{4}{15}$

To add mixed numbers, first turn them into improper fractions.

Example $2\dfrac{1}{2} + 1\dfrac{1}{3} = \dfrac{5}{2} + \dfrac{4}{3} = \dfrac{15}{6} + \dfrac{8}{6} = \dfrac{23}{6} = 3\dfrac{5}{6}$

Subtraction of fractions

This is similar to addition of fractions.

Examples $\dfrac{6}{7} - \dfrac{2}{7} = \dfrac{4}{7}$

$\dfrac{3}{4} - \dfrac{1}{6} = \dfrac{9}{12} - \dfrac{2}{12} = \dfrac{7}{12}$

$3\dfrac{1}{4} - 2\dfrac{5}{8} = \dfrac{13}{4} - \dfrac{21}{8} = \dfrac{26}{8} - \dfrac{21}{8} = \dfrac{5}{8}$

Exercise 3.2

Simplify, giving your answers in their lowest terms:

1. $\dfrac{5}{9} + \dfrac{2}{9}$

2. $\dfrac{2}{11} + \dfrac{4}{11}$

3. $\dfrac{1}{2} + \dfrac{1}{3}$

4. $\dfrac{1}{4} + \dfrac{2}{3}$

5. $\dfrac{9}{11} - \dfrac{2}{11}$

6. $\dfrac{1}{3} - \dfrac{1}{4}$

7. $\dfrac{3}{4} - \dfrac{1}{3}$

8. $\dfrac{4}{5} - \dfrac{3}{10}$

9. $\dfrac{7}{12} + \dfrac{1}{6}$

10. $\dfrac{3}{5} + \dfrac{7}{10}$

11. $\dfrac{5}{6} - \dfrac{2}{9}$

12. $\dfrac{11}{12} - \dfrac{7}{8}$

13. $\dfrac{1}{5} + \dfrac{1}{2} + \dfrac{1}{10}$

14. $\dfrac{1}{3} + \dfrac{1}{6} + \dfrac{1}{4}$

15. $\dfrac{5}{8} + \dfrac{1}{8} + \dfrac{3}{4}$

16. $\dfrac{4}{9} + \dfrac{1}{6} + \dfrac{2}{3}$

17. $\dfrac{1}{2} + \dfrac{1}{3} - \dfrac{1}{4}$

18. $\dfrac{4}{5} + \dfrac{1}{2} - \dfrac{3}{10}$

19. $1\dfrac{1}{3} + 1\dfrac{1}{4}$

20. $1\dfrac{1}{6} + 1\dfrac{1}{2}$

21. $2\dfrac{3}{5} + 2\dfrac{1}{2}$

22. $3\dfrac{1}{5} + 4\dfrac{2}{5}$

23. $1 - \dfrac{2}{7}$

24. $2 - 1\dfrac{1}{2}$

25. $3\frac{1}{3} - \frac{2}{3}$ **26.** $4\frac{4}{5} - 1\frac{4}{5}$ **27.** $2\frac{3}{4} - \frac{2}{3}$

28. $2\frac{1}{4} - 1\frac{2}{3}$ **29.** $2\frac{1}{6} - 1\frac{3}{4}$ **30.** $3\frac{1}{8} - 2\frac{1}{2}$

Multiplication of fractions

Multiply the numerators together and multiply the denominators together.

Examples $\dfrac{2}{7} \times \dfrac{3}{5} = \dfrac{2 \times 3}{7 \times 5} = \dfrac{6}{35}$ and $\dfrac{6}{11} \times \dfrac{2}{9} = \dfrac{12}{99} = \dfrac{4}{33}$

In the second example we could first cancel by 3, like this:

$$\dfrac{^2\cancel{6}}{11} \times \dfrac{2}{\cancel{9}_3} = \dfrac{2 \times 2}{11 \times 3} = \dfrac{4}{33}$$

Change any mixed numbers into improper fractions, before multiplying.

Example $3\frac{1}{3} \times 1\frac{1}{5} = \dfrac{^2\cancel{10}}{\cancel{3}_1} \times \dfrac{^2\cancel{6}}{\cancel{5}_1} = \dfrac{2 \times 2}{1 \times 1} = \dfrac{4}{1} = 4$

Write any integers as improper fractions with 1 as denominator, before multiplying.

Example $2\frac{1}{4} \times 6 = \dfrac{9}{_2\cancel{4}} \times \dfrac{\cancel{6}^3}{1} = \dfrac{27}{2} = 13\frac{1}{2}$

Division of fractions

To divide by a fraction, we invert the fraction (turn it upside down) and multiply by it.

Examples $\dfrac{3}{4} \div \dfrac{5}{7} = \dfrac{3}{4} \times \dfrac{7}{5} = \dfrac{21}{20} = 1\frac{1}{20}$

(Note that only the second fraction, $\dfrac{5}{7}$, is inverted.)

$$1\frac{1}{2} \div 2\frac{2}{5} = \dfrac{3}{2} \div \dfrac{12}{5} = \dfrac{3}{2} \times \dfrac{5}{12} = \dfrac{5}{8}$$

$$\dfrac{4}{5} \div 3 = \dfrac{4}{5} \div \dfrac{3}{1} = \dfrac{4}{5} \times \dfrac{1}{3} = \dfrac{4}{15}$$

Exercise 3.3

Simplify, giving the answers in their lowest terms:

1. $\dfrac{3}{7} \times \dfrac{2}{5}$ **2.** $\dfrac{1}{2} \times \dfrac{3}{4}$ **3.** $\dfrac{6}{7} \times \dfrac{1}{4}$ **4.** $\dfrac{3}{5} \times \dfrac{5}{6}$

5. $2\frac{1}{5} \times \frac{2}{3}$ **6.** $\frac{3}{4} \times 2\frac{1}{2}$ **7.** $3\frac{1}{2} \times 1\frac{2}{3}$ **8.** $1\frac{2}{3} \times 1\frac{2}{5}$

9. $3\frac{2}{5} \times 1\frac{1}{2}$ **10.** $3\frac{2}{3} \times 1\frac{3}{4}$ **11.** $6 \times \frac{2}{9}$ **12.** $\frac{3}{5} \times 10$

13. $1\frac{7}{12} \times 4$ **14.** $2\frac{4}{9} \times 6$ **15.** $\frac{1}{4} \div \frac{2}{3}$ **16.** $\frac{4}{7} \div \frac{3}{5}$

17. $\frac{2}{3} \div \frac{5}{6}$ **18.** $\frac{3}{5} \div \frac{9}{10}$ **19.** $1\frac{4}{5} \div 3$ **20.** $2\frac{2}{3} \div 4$

21. $\frac{3}{8} \div 1\frac{1}{2}$ **22.** $2\frac{1}{2} \div 1\frac{2}{3}$ **23.** $1\frac{1}{9} \div \frac{5}{6}$ **24.** $6\frac{1}{4} \div 2\frac{1}{7}$

$\frac{3}{4}$ of $100p = 75p$ and $\frac{3}{4} \times \frac{100}{1} = \frac{75}{1} = 75$

This illustrates that 'of' can be changed to '×'.

Example $\frac{5}{7}$ of $4\frac{2}{3} = \frac{5}{7} \times \frac{14}{3} = \frac{10}{3} = 3\frac{1}{3}$

Some questions on fractions contain brackets. The calculation inside the brackets should be worked first.

Example $1\frac{3}{5} \times \left(4\frac{1}{4} - 1\frac{1}{3}\right) = 1\frac{3}{5} \times \left(\frac{17}{4} - \frac{4}{3}\right) = 1\frac{3}{5} \times \left(\frac{51}{12} - \frac{16}{12}\right)$

$$= \frac{8}{5} \times \frac{35}{12} = \frac{14}{3} = 4\frac{2}{3}$$

Exercise 3.4

Simplify, giving your answers in their lowest terms:

1. $\frac{1}{4}$ of $\frac{8}{9}$ **2.** $\frac{1}{7}$ of $5\frac{1}{4}$ **3.** $\frac{2}{5}$ of $3\frac{1}{8}$ **4.** $\frac{4}{9}$ of $5\frac{5}{8}$

5. $\left(\frac{2}{3} - \frac{1}{2}\right) \div \frac{2}{3}$ **6.** $\left(\frac{3}{4} - \frac{2}{5}\right) \times \frac{4}{7}$ **7.** $3\frac{1}{3} \times \left(2\frac{1}{4} - \frac{3}{5}\right)$

8. $4\frac{1}{4} \div \left(\frac{5}{7} - \frac{1}{2}\right)$ **9.** $\left(\frac{4}{5} + \frac{1}{3}\right) + \left(\frac{4}{5} \times \frac{1}{3}\right)$ **10.** $\left(\frac{1}{2} - \frac{1}{3}\right) \div \left(\frac{1}{2} + \frac{1}{3}\right)$

11. $\frac{1}{3} \div \left(\frac{1}{4} + \frac{5}{6}\right)$ **12.** $\frac{7}{16} \div \left(\frac{4}{5} - \frac{3}{4}\right)$ **13.** $\left(2 - \frac{2}{3}\right) \div \frac{8}{9}$

14. $\dfrac{1\frac{1}{3} - \frac{1}{6}}{1\frac{1}{3} + \frac{1}{6}}$ **15.** $\dfrac{1}{\frac{2}{3} + \frac{3}{4}}$ **16.** $\dfrac{1\frac{1}{2} \times \frac{3}{5}}{1\frac{1}{2} + \frac{3}{5}}$

17. What fraction is halfway between:

(i) $\frac{1}{3}$ and $\frac{3}{5}$ (ii) $\frac{7}{12}$ and $\frac{3}{4}$ (iii) $\frac{1}{2}$ and $\frac{2}{3}$?

4 Decimals

$$\frac{3}{10} = 0.3, \quad \frac{7}{100} = 0.07, \quad \frac{9}{1000} = 0.009$$

In 54.608 each digit has a certain value according to its position. We have 5 tens, 4 units, 6 tenths, 0 hundredths and 8 thousandths.

$$\frac{83}{1000} = \frac{80}{1000} + \frac{3}{1000} = \frac{8}{100} + \frac{3}{1000} = 0.083$$

So to write $\frac{83}{1000}$ as a decimal, we place the 3 in the thousandths position and the 8 next to it.

Similarly, $\frac{907}{100} = 9.07$, the 7 being placed in the hundredths position.

Changing decimals to fractions in their lowest terms

$$0.48 = \frac{48}{100} = \frac{12}{25}, \quad 0.0035 = \frac{35}{10\,000} = \frac{7}{2000}, \quad 0.704 = \frac{704}{1000} = \frac{88}{125}$$

Changing fractions to decimals

Some fractions can be changed to decimals by making the denominators powers of 10.

$$\frac{13}{25} = \frac{13 \times 4}{25 \times 4} = \frac{52}{100} = 0.52, \quad \frac{9}{125} = \frac{9 \times 8}{125 \times 8} = \frac{72}{1000} = 0.072$$

Addition and subtraction of decimals

When setting out an addition or a subtraction sum, make sure that the decimal points are in line. Digits with the same place value will then lie under each other.

14.7 + 0.36 + 2.009
is written as:

```
  14.7
   0.36
   2.009
 ------
  17.069
```

23.6 − 9.83
is written as:

```
  23.60 ◄──── The nought in the top line
   9.83        is just to fill the space.
 ------
  13.77
```

Exercise 4.1

No calculating aids should be used.

1. Write as decimals: $\dfrac{9}{10}, \dfrac{3}{100}, \dfrac{7}{1000}, \dfrac{53}{100}, \dfrac{21}{1000}$.

2. Write as decimals: $\dfrac{33}{100}, \dfrac{209}{1000}, \dfrac{9}{100}, \dfrac{67}{10}, \dfrac{803}{100}$.

3. Write as fractions: 0.7, 0.09, 0.13, 0.003, 0.207.

4. Express as fractions in their lowest terms:

0.6, 0.06, 0.35, 0.045, 0.008.

5. Express as fractions in their lowest terms:

0.12, 0.05, 0.066, 0.555, 0.204.

6. Express as decimals: $\dfrac{1}{5}, \dfrac{1}{2}, \dfrac{3}{5}, \dfrac{1}{20}, \dfrac{7}{20}$.

7. Express as decimals: $\dfrac{1}{25}, \dfrac{13}{25}, \dfrac{37}{50}, \dfrac{11}{500}, \dfrac{213}{250}$.

8. State the decimal which has a value halfway between:

(i) 0.36 and 0.4   (ii) 0.8 and 0.9   (iii) 0.06 and 0.054.

Simplify the following:

9. $5.5 + 2.8$

10. $8.4 + 1.37$

11. $0.98 + 2.6$

12. $1.33 + 3.042$

13. $7.3 - 4.6$

14. $8.2 - 3.9$

15. $4.6 - 1.23$

16. $0.7 - 0.24$

17. $5 - 0.73$

18. $32 - 8.6$

19. $4.4 + 0.44 + 0.044$

20. $0.67 + 6.7 - 0.67$

Multiplying by powers of 10

When multiplying by 10, move the digits one place to the left in relation to the decimal point;
when multiplying by 100, move the digits two places to the left;
when multiplying by 1000, move the digits three places to the left;
and so on.

Examples

Number	0.7	0.03	0.009	0.34	0.0608
Number × 10	7	0.3	0.09	3.4	0.608
Number × 100	70	3	0.9	34	6.08

General multiplication

$$0.06 \times 0.7 = \frac{6}{100} \times \frac{7}{10} = \frac{42}{1000} = 0.042$$

Notice that 2 decimal places in the first number and 1 decimal place in the second number give 3 decimal places in the answer $(2 + 1 = 3)$. This illustrates the general method for multiplying two decimal numbers:

(i) multiply the two numbers, ignoring the decimal points $(6 \times 7 = 42)$

(ii) count the total number of decimal places in the given numbers $(2 + 1 = 3)$ and set the decimal point in the answer to give this number of places.

Examples $0.7 \times 0.008 = 0.0056$ $(1 + 3 = 4$ places$)$
$2.404 \times 0.03 = 0.07212 \, (3 + 2 = 5$ places$)$
$380 \times 0.15 = 57.00$ $(0 + 2 = 2$ places$)$
$= 57$

Division by powers of 10

When dividing by 10, move the digits 1 place to the right in relation to the decimal point;
when dividing by 100, move the digits 2 places to the right;
and so on.

Examples

Number	0.6	0.08	5.4	0.708
Number ÷ 10	0.06	0.008	0.54	0.0708
Number ÷ 100	0.006	0.0008	0.054	0.00708

$$9.28 \div 400 = (9.28 \div 4) \div 100 = 2.32 \div 100 = 0.0232$$

General division

Example 1 $0.141 \div 6 = 0.0235$ $6 \lfloor 0.1410$
$ 0.0235$

Example 2 $2.66 \div 0.7$

First write this as $\dfrac{2.66}{0.7}$. Then multiply both numbers by 10 so that the denominator becomes a whole number.

$$\frac{2.66}{0.7} = \frac{2.66 \times 10}{0.7 \times 10} = \frac{26.6}{7} = 3.8$$

Example 3 $37.8 \div 0.09$

$$\frac{37.8}{0.09} = \frac{37.8 \times 100}{0.09 \times 100} = \frac{3780}{9} = 420$$

Example 4 $0.0342 \div 0.6$

$$\frac{0.0342}{0.6} = \frac{0.0342 \times 10}{0.6 \times 10} = \frac{0.342}{6} = 0.057$$

Exercise 4.2

No calculating aids should be used.
Simplify the following:

1. 0.03×10	**2.** 0.7×100	**3.** 0.0056×100
4. 0.29×1000	**5.** 0.14×20	**6.** 3.6×200
7. 0.044×300	**8.** 3.5×40	**9.** $0.3 \div 10$
10. $0.8 \div 100$	**11.** $0.053 \div 10$	**12.** $7.6 \div 1000$
13. $4.5 \div 5$	**14.** $0.21 \div 3$	**15.** $0.088 \div 4$
16. $0.96 \div 6$	**17.** $0.3 \div 2$	**18.** $0.17 \div 2$
19. $1.8 \div 20$	**20.** $0.24 \div 30$	**21.** $0.14 \div 40$
22. $2.8 \div 500$	**23.** 0.3×0.2	**24.** 0.7×0.8
25. 0.06×0.4	**26.** 0.8×0.01	**27.** 4×0.03
28. 0.5×1.7	**29.** 0.4×0.05	**30.** 0.005×0.6
31. $0.06 \div 0.3$	**32.** $8 \div 0.02$	**33.** $1.4 \div 0.07$
34. $0.18 \div 0.3$	**35.** $0.45 \div 0.05$	**36.** $6.2 \div 0.01$
37. $0.008 \div 0.04$	**38.** $0.9 \div 0.05$	

Express the answers to the following as fractions in their simplest forms:

39. $\dfrac{1.2}{1.6}$ **40.** $\dfrac{0.2}{0.32}$ **41.** 0.8×1.5 **42.** 0.2×0.25

43. Multiply 23 by 17 and then write down the answers to
2.3×1.7, 0.23×1.7 and 0.23×0.17

44. Multiply 37 by 29 and then write down the answers to
37×2.9, 3.7×0.29 and 0.037×2.9.

Simplify the following:

45. $\dfrac{0.3 \times 0.08}{0.4}$ **46.** $\dfrac{0.3 \times 0.115}{0.05}$ **47.** $\dfrac{(0.4)^2 \times 0.3}{(0.2)^2}$

Approximations

All measurements are approximate. 'Canhill 7 km' on a signpost means
that the distance is really somewhere between 6.5 km and 7.5 km.

6 km   7 km   8 km

If the mass of a parcel is 240 g to the nearest 10 g, then it is between
235 g and 245 g. Notice that when the mass is given to the nearest 10 g
the two limits differ by 10 g. The limits are 5 g each side of 240 g.

230 g   240 g   250 g

Significant figures

The population of a town may be 46 283. We can give this to the nearest
thousand as 46 000. We say that we have two *significant figures*, the 4
and the 6. The noughts are just 'place fillers'. To give the population
to the nearest hundred, we note that 46 283 is nearer to 46 300 than to
46 200, so we give the answer as 46 300.

46200   46300

46283

The general rule is:

To give a number to 2 significant figures, look at the third figure. If
this is 5 or more, add 1 to the second figure. If it is less than 5, add
nothing to the second figure.

Similarly, for 3 significant figures, we look at the fourth figure, and
so on.

The decimal number 0.002 038 4 is 0.002 04 correct to 3 significant
figures.   1st 2nd 3rd

In 0.002 04, the 0 between the 2 and the 4 is significant but the other
0's are just place fillers.

The term 'significant figure' is often shortened to s.f.

Examples	Number	Correct to 3 s.f.	Correct to 2 s.f.	Correct to 1 s.f.
	4279	4280	4300	4000
	5.038	5.04	5.0	5
	0.085036	0.0850	0.085	0.09

Decimal places

Numbers are often rounded off, or given correct to a certain number
of *decimal places* (d.p.). A calculator might display a result as 3.2682544.
Correct to 2 decimal places this is 3.27. Because the third decimal place
is more than 5, the value is nearer to 3.27 than to 3.26.

Examples	Number	Correct to 3 d.p.	Correct to 2 d.p.	Correct to 1 d.p.
	17.8624	17.862	17.86	17.9
	0.5796	0.580	0.58	0.6
	0.0834	0.083	0.08	0.1

↑
Here 0 takes up
the first decimal place

Exercise 4.3

1. Give to the nearest metre: 2.7 m, 3.4 m, 8.38 m, 29.7 m.

2. Give to the nearest 10 km: 472 km, 638 km, 57 km, 2306 km.

3. Give to the nearest tenth of a second:
5.36 s, 8.22 s, 0.083 s, 6.077 s.

4. Give correct to 1 decimal place:
15.69, 26.52, 3.98, 47.464.

5. Give correct to 2 decimal places:
3.742, 7.438, 0.916, 0.0714.

6. Give correct to 3 decimal places:
0.6158, 0.8076, 0.02342, 0.0048.

7. Give correct to 2 s.f.: 8728, 542, 0.0687, 6.34.

8. Give correct to 1 s.f.: 372, 2148, 0.079, 0.0054.

9. Give correct to 3 s.f.: 7.428, 93841, 71.06, 0.03498.

10. Express 3.141593 correct to:
(i) 4 s.f. (ii) 2 s.f. (iii) 2 d.p. (iv) 4 d.p.

11. Express 0.008176 correct to:
(i) 3 d.p. (ii) 3 s.f. (iii) 2 d.p. (iv) 2 s.f.

12. State the number of significant figures in each of the following:
0.046 mm, 5.03 g, 4070 kg, 38000 km.

13. Between what limits do the following numbers lie?
(i) 4300 to the nearest hundred
(ii) 172000 to the nearest thousand

    (iii) 0.7 to the nearest tenth

    (iv) 36 to the nearest unit

14. Between what limits do the following lie?

    (i) 26 km to the nearest km   (ii) 820 g to the nearest 10 g

    (iii) £3200 to the nearest £100   (iv) 3 hours to the nearest $\frac{1}{2}$h.

15. Between what limits do the following lie?

    (i) 0.34 to 2 d.p.   (ii) 0.07 to 2 d.p.

    (iii) 6300 to 2 s.f.   (iv) 4.6 to 2 s.f.

16. Correct to the nearest centimetre, the length and width of a rectangle are 8 cm and 6 cm.

    (i) State the greatest possible length and the greatest possible width. Using these, calculate the greatest possible area.

    (ii) State the least possible length and the least possible width. Using these, calculate the least possible area.

17. The side of a square is 9 cm, correct to the nearest centimetre. Between what limits does the area lie? (Use the method of question 16.)

Changing fractions to decimals

On page 26 some fractions were converted to decimals by making the denominators powers of 10. Most fractions cannot be converted in this way.

    However, *any* fraction can be converted to a decimal by dividing the numerator by the denominator.

Examples $\frac{7}{8} = 7 \div 8 = 0.875$ and $\frac{57}{64} = 57 \div 64 = 0.890625$

For some fractions the division never ends.

Examples $\frac{1}{3} = 0.333\ldots\ldots,$ $\frac{7}{11} = 0.636363\ldots\ldots$

                $\frac{4}{7} = 0.571428571\ldots\ldots,$ $\frac{12}{55} = 0.2181818\ldots\ldots$

Such decimals have a repeating pattern. They are called *recurring decimals*. The repeating pattern can be shown by placing a dot over the first and last digits in the group of digits that recur. Thus

$$\frac{4}{7} = 0.5\dot{7}142\dot{8}, \quad \frac{12}{55} = 0.2\dot{1}\dot{8} \text{ and } \frac{1}{3} = 0.\dot{3}$$

If asked to give $\frac{12}{55}$ as a decimal, correct to 4 d.p., we give 0.2182 as the answer.

Approximate values

Example 1 By rounding each number to 1 s.f., find the value of
6.9 × 0.058 correct to 1 s.f.
6.9 is nearly 7 and 0.058 is nearly 0.06.
7 × 0.06 = 0.42 which is 0.4 correct to 1 s.f.
So 6.9 × 0.058 is 0.4 correct to 1 s.f.

Example 2 Find the value of 1.214 ÷ 0.0293 correct to 1 s.f.
1.214 can be rounded off to 1.2 and 0.0293 can be rounded
off to 0.03.
So the division becomes: $\dfrac{1.2}{0.03} = \dfrac{120}{3} = 40$

This method is useful for obtaining a rough estimate of the answer to
a calculation.

Exercise 4.4

1. Express as decimals, correct to 3 d.p.: $\dfrac{7}{9}, \dfrac{6}{7}, \dfrac{3}{11}, \dfrac{5}{13}, \dfrac{9}{17}$.

2. Express $\dfrac{2}{3}, \dfrac{4}{7}$ and $\dfrac{5}{9}$ as decimals, correct to 2 d.p., and then list the
fractions in order of size, with the smallest first.

3. Express as recurring decimals: $\dfrac{2}{3}, \dfrac{3}{7}, \dfrac{7}{9}, \dfrac{5}{11}, \dfrac{7}{12}$.

4. Express as decimals, correct to 4 d.p.: $\dfrac{13}{27}, \dfrac{6}{31}, \dfrac{2}{43}$.

5. Express as recurring decimals: $\dfrac{11}{101}, \dfrac{4}{27}, \dfrac{7}{33}, \dfrac{8}{55}$.

Round each number to 1 s.f. and then find an approximate value, correct
to 1 s.f., for the product or quotient:

6. 3.2 × 5.9	**7.** 1.9 × 3.1	**8.** 7.1 × 8.8	**9.** 98 × 5.1
10. 102 × 2.8	**11.** 0.39 × 0.22	**12.** 0.062 × 1.8	**13.** 6.1 ÷ 2.1
14. 8.2 ÷ 1.9	**15.** 39.3 ÷ 5.1	**16.** 19.6 ÷ 3.8	**17.** 0.63 ÷ 0.33

18. Which of the following is nearest to the value of 0.592 × 51.8?
0.03, 0.3, 3, 30, 300, 3000.

19. Which of the following is nearest to the value of 0.813 ÷ 39.7?
0.02, 0.2, 2, 20, 200, 2000.

20. Which of the following is nearest to the value of $\dfrac{0.0814 \times 3.23}{0.426}$?

0.006, 0.06, 0.6, 6, 60, 600.

5 Powers, roots, reciprocals and standard form

Squares and square roots

$$3^2 = 9, \qquad 7^2 = 49, \qquad 15^2 = 225, \qquad 34^2 = 1156$$

9, 49, 225 and 1156 are examples of *perfect squares*.

$(-3)^2$ is also 9, therefore 9 has two *square roots*, 3 and -3.

Similarly, 225 has two square roots, 15 and -15.

$\sqrt{}$ is the sign for the positive square root, so we can write:

$$\sqrt{225} = 15.$$

Using factors to find square roots of perfect squares

For $\sqrt{n}$ we need to find a number r such that $n = r \times r$.

We express n as the product of prime factors, and then regroup these primes to find r.

Examples $441 = 3 \times 3 \times 7 \times 7 = (3 \times 7) \times (3 \times 7) = 21 \times 21$
   so $\sqrt{441} = 21$.
   $2025 = 3 \times 3 \times 3 \times 3 \times 5 \times 5$
   $\qquad = (3 \times 3 \times 5) \times (3 \times 3 \times 5) = 45 \times 45$
   so $\sqrt{2025} = 45$.

The square root of a perfect square that ends in an even number of noughts can be found as follows:

$$490000 = 49 \times 10000 = 7 \times 7 \times 100 \times 100$$
$$= (7 \times 100) \times (7 \times 100) = 700 \times 700$$

so $\sqrt{490000} = 700$.

Cubes and cube roots

$4^3 = 64$ and $7^3 = 343$. 64 and 343 are examples of *perfect cubes*. 4 and 7 are their *cube roots*.

$\sqrt[3]{}$ is the symbol for cube root, so $\sqrt[3]{64} = 4$ and $\sqrt[3]{343} = 7$.

For $\sqrt[3]{n}$ we need a number r such that $n = r \times r \times r$.

Example $2744 = 2 \times 2 \times 2 \times 7 \times 7 \times 7 =$
   $= (2 \times 7) \times (2 \times 7) \times (2 \times 7) = 14 \times 14 \times 14$
   so $\sqrt[3]{2744} = 14$.

Powers and roots of fractions and decimals

$$\left(1\frac{3}{4}\right)^2 = \left(\frac{7}{4}\right)^2 = \frac{7^2}{4^2} = \frac{49}{16} = 3\frac{1}{16}; \qquad \sqrt{6\frac{1}{4}} = \sqrt{\frac{25}{4}} = \frac{\sqrt{25}}{\sqrt{4}} = \frac{5}{2} = 2\frac{1}{2}$$

$$0.3^2 = 0.3 \times 0.3 = 0.09; \qquad 0.8^2 = 0.8 \times 0.8 = 0.64$$

$$0.05^2 = 0.05 \times 0.05 = 0.0025;$$

$$\sqrt{0.0049} = \sqrt{\frac{49}{10000}} = \frac{\sqrt{49}}{\sqrt{10000}} = \frac{7}{100} = 0.07$$

Exercise 5.1

No calculating aids should be used.

1. Write down the squares of 4, 5, 8 and 9.

2. Write down the squares of 1, 10, 100 and 1000.

3. Write down the values of 2^2, 6^2, 7^2 and 10^2.

4. Find the values of 30^2, 40^2, 200^2 and 500^2.

5. State the square roots of 16, 36, 64 and 81.

6. Write down the values of $\sqrt{25}$, $\sqrt{49}$, $\sqrt{100}$ and $\sqrt{1}$.

7. State the value of 12^2. Hence write down the value of 120^2 and 1200^2.

8. Calculate 31^2. Hence write down the values of 310^2 and 3100^2.

Express each number as the product of prime factors, and then find its square roots:

9. 196   **10.** 1225   **11.** 324   **12.** 625
13. 484   **14.** 729   **15.** 784   **16.** 1089

17. Find the square roots of 1600, 160000 and 250000.

18. Find the square roots of 8100, 40000 and 9000000.

19. Find x if: (i) $x^2 = 64$ (ii) $x^2 = 3^2 + 4^2$ (iii) $x^2 = 8^2 + 6^2$.

20. Find y if: (i) $y^2 = 81$ (ii) $y^2 = 13^2 - 12^2$ (iii) $y^2 = 17^2 - 15^2$.

21. Find the cubes of 1, 2, 3, 4 and 5.

22. Find the values of 6^3, 7^3, 8^3 and 10^3.

23. State the cube roots of 1, 8, 27 and 1000.

24. State the values of $\sqrt[3]{64}$, $\sqrt[3]{125}$, $\sqrt[3]{216}$ and $\sqrt[3]{343}$.

Express each number as the product of prime factors and then find its cube root:

25. 729   **26.** 3375   **27.** 1728

Exercise 5.2

Simplify:

1. $\left(1\frac{1}{2}\right)^2$ 2. $\left(2\frac{2}{3}\right)^2$ 3. $\left(1\frac{4}{5}\right)^2$ 4. $\left(3\frac{1}{4}\right)^2$

5. $\sqrt{2\frac{1}{4}}$ 6. $\sqrt{5\frac{4}{9}}$ 7. $\sqrt{1\frac{9}{16}}$ 8. $\sqrt{12\frac{1}{4}}$

Evaluate:

9. 0.4^2 10. 0.2^2 11. 0.09^2 12. 0.012^2
13. $\sqrt{0.36}$ 14. $\sqrt{0.0009}$ 15. $\sqrt{0.01}$ 16. $\sqrt{0.0081}$

17. Calculate $\sqrt{196}$. Hence state the values of $\sqrt{0.0196}$ and $\sqrt{0.000196}$.

18. Calculate $\sqrt{225}$. Hence state the values of $\sqrt{2.25}$ and $\sqrt{0.0225}$.

19. Calculate the cubes of 0.2, 0.3, and 0.4.

20. Calculate the cubes of 0.1, 0.5 and 0.9.

Reciprocals

The *reciprocal* of n is $\frac{1}{n}$.

Thus the reciprocal of 5 is $\frac{1}{5}$ or 0.2 and that of 50 is $\frac{1}{50}$ or 0.02.

The reciprocal of 0.4 is $\frac{1}{0.4} = \frac{1}{\frac{4}{10}} = 1 \times \frac{10}{4} = \frac{10}{4} = 2.5$.

The reciprocal of $\frac{3}{4}$ is $\frac{1}{\frac{3}{4}} = 1 \times \frac{4}{3} = \frac{4}{3} = 1\frac{1}{3}$.

Exercise 5.3

State, as exact decimals, the reciprocals of:

1. 2, 20 and 200   2. 2, 0.2 and 0.02   3. 10, 0.1 and 1
4. 4, 40 and 0.4   5. $2\frac{1}{2}$, $3\frac{1}{3}$ and $1\frac{1}{4}$

State as fractions the reciprocals of:

6. 3, 7 and 9   7. $\frac{3}{7}$, $\frac{5}{3}$ and $3\frac{1}{2}$

Approximate values

$\sqrt{5}$ is an irrational number. It is 2.236068 correct to 6 d.p.

The reciprocal of 7 is $\frac{1}{7}$ which is 0.142857 to 6 d.p.

The squares of numbers are always exact but we often state them to a smaller number of significant figures.

For example, $8.267^2 = 68.343289$, which we might give as 68.3 correct to 3 s.f.

Exercise 5.4

Use a calculator (or tables) to find the values of the following, correct to 3 s.f.:

1. 7.36^2 2. 33.9^2 3. 0.253^2 4. 528^2 5. 0.904^2
6. 4270^2 7. 0.823^2 8. 0.0927^2 9. 0.183^2 10. 42.3^2
11. $\sqrt{6}$ 12. $\sqrt{29}$ 13. $\sqrt{250}$ 14. $\sqrt{3.5}$ 15. $\sqrt{35}$
16. $\sqrt{0.17}$ 17. $\sqrt{0.284}$ 18. $\sqrt{0.0068}$ 19. $\sqrt{4000}$ 20. $\sqrt{635.6}$

Find, correct to 3 s.f., the reciprocals of the following numbers:

21. 13 22. 57 23. 9.4 24. 0.78 25. 760
26. 4.85 27. 60.7 28. 0.187 29. 522 30. 0.0683

Find, correct to 3 s.f., the value of:

31. $\dfrac{1}{21.6} + \dfrac{1}{32.8}$ 32. $\dfrac{1}{0.638} + \dfrac{1}{0.547}$ 33. $\dfrac{1}{3.47} - \dfrac{1}{8.23}$
34. $8.6^2 + 7.4^2$ 35. $0.93^2 + 0.78^2$ 36. $4.28^2 - 2.31^2$

Negative indices

10^{-3} means $\dfrac{1}{10^3}$ which is $\dfrac{1}{1000}$ or 0.001. Also $10^0 = 1$.

Negative indices and the zero index are explained on page 97.

Powers of 10

Below is a table showing powers of 10. It can be extended indefinitely to left and right:

Power of 10	10^{-4}	10^{-3}	10^{-2}	10^{-1}	10^0	10^1	10^2	10^3	10^4
Value	0.0001	0.001	0.01	0.1	1	10	100	1000	10000

Standard form

Any positive number can be expressed in the form $a \times 10^n$ where n is a positive or negative integer and $1 \le a < 10$. This means that a can be equal to or greater than 1, but must be less than 10. So it could be 1, 1.43, 2.3, 6, 7.567 and so on.

A number expressed in this way is said to be in *standard form*.

Example $76300 = 7.63 \times 10000 = 7.63 \times 10^4$.

For numbers less than 1, n is a negative number.

Examples $0.054 = \dfrac{5.4}{100} = \dfrac{5.4}{10^2} = 5.4 \times \dfrac{1}{10^2} = 5.4 \times 10^{-2}$

$0.0008 = \dfrac{8}{10000} = \dfrac{8}{10^4} = 8 \times \dfrac{1}{10^4} = 8 \times 10^{-4}$

Exercise 5.5

Write the following in standard form:

1. 740   2. 86000   3. 4000   4. 309000   5. 42.76
6. 0.0053   7. 0.238   8. 0.0006   9. 0.00904   10. 0.7
11. 8 million   12. 6 thousandths

Write without powers of 10:

13. 4.2×10^2 14. 5.8×10^4 15. 6×10^3 16. 3.246×10^2
17. 3.8×10^{-2} 18. 1.3×10^{-4} 19. 5×10^{-3} 20. 1.256×10^{-1}

21. Explain why these are not examples of numbers in standard form:

$$35 \times 10^4, \quad 0.47 \times 10^5, \quad 8.3 \div 10^2, \quad 4.9 \times 10^{0.3}.$$

Calculations with numbers in standard form

In this work, powers of 10 have to be multiplied and divided.
When multiplying, add the indices. For example $10^2 \times 10^3 = 10^5$.
When dividing, subtract the indices. For example $10^7 \div 10^3 = 10^4$.
(Multiplication and division using indices is explained on page 84).

Examples $(7.3 \times 10^4) + (2.5 \times 10^4) = 9.8 \times 10^4$

$(2.4 \times 10^2) \times (7 \times 10^3) = 2.4 \times 7 \times 10^2 \times 10^3 = 16.8 \times 10^5$

$= 1.68 \times 10 \times 10^5 = 1.68 \times 10^6$

$\dfrac{2.2 \times 10^3}{5 \times 10^8} = \dfrac{2.2}{5} \times \dfrac{10^3}{10^8} = 0.44 \times 10^{-5}$

$= 4.4 \times 10^{-1} \times 10^{-5} = 4.4 \times 10^{-6}$

The square roots of numbers in standard form can also be found.

Examples Find the square root of: (i) 9×10^{-8} (ii) 4.9×10^7

$$\text{(i)} \ 9 \times 10^{-8} = 3 \times 3 \times 10^{-4} \times 10^{-4}$$
$$= (3 \times 10^{-4}) \times (3 \times 10^{-4})$$
so $\sqrt{(9 \times 10^{-8})} = 3 \times 10^{-4}$

$$\text{(ii)} \ 4.9 \times 10^7 = 4.9 \times 10 \times 10^6$$
$$= 49 \times 10^6 = 7 \times 7 \times 10^3 \times 10^3$$
$$= (7 \times 10^3) \times (7 \times 10^3)$$
so $\sqrt{(4.9 \times 10^7)} = 7 \times 10^3$

Exercise 5.6

Put into standard form:

1. 84×10^2 **2.** 670×10^3 **3.** 1730×10 **4.** 374×10^4

Evaluate the following, giving your answers in standard form:

5. $(2 \times 10^3) \times (3 \times 10^2)$ **6.** $(6 \times 10^5) \times (7 \times 10^3)$
7. $(5 \times 10^2) \times (4 \times 10^4)$ **8.** 30^3

Put into standard form:

9. 0.64×10^4 **10.** 0.3×10^5 **11.** 0.053×10^7
12. 0.16×10^6 **13.** 87×10^{-5} **14.** 0.022×10^{-2}

Evaluate the following, giving your answers in standard form:

15. $(8.8 \times 10^6) \div (4 \times 10^2)$ **16.** $(9.6 \times 10^5) \div (3 \times 10^3)$
17. $(4.5 \times 10^7) \div (5 \times 10^4)$ **18.** $(3.6 \times 10^{10}) \div (9 \times 10^4)$
19. $(9 \times 10^{-2}) \times (3 \times 10^{-4})$ **20.** $(8 \times 10^{-1}) \times (5 \times 10^{-3})$
21. $(0.02)^2$ **22.** $(0.7)^2$ **23.** $0.03 \div 5$ **24.** $30 \div 0.01$

25. Multiply 3.5×10^{-2} by 4.0×10^{-3}, expressing your answer:
   (i) in standard form   (ii) as a decimal.

26. If $a = 9 \times 10^{-6}$ and $b = 2 \times 10^{-6}$, calculate $a + b$, $a - b$, ab and $a \div b$, giving your answers in standard form.

27. If $p = 4 \times 10^2$ and $q = 5 \times 10^4$, calculate pq, $p \div q$ and $p + q$, giving your answers in standard form.

28. Find the square root of: (i) 16×10^6 (ii) 3.6×10^9.

29. Find the square root of: (i) 25×10^{-6} (ii) 6.4×10^{-9}.

30. If $c = 6.3 \times 10^8$ and $d = 7.0 \times 10^3$, find the value of $\sqrt{\dfrac{c}{d}}$, giving your answer in standard form.

Surds

A *surd* is an irrational number of the form $\sqrt{n}$ where n is a positive integer which is not a perfect square.

$\sqrt{2}$, $\sqrt{5}$ and $\sqrt{13}$ are surds but $\sqrt{9}$ and $\sqrt{16}$ are not.

$$\sqrt{a} \times \sqrt{b} = \sqrt{ab} \qquad (\sqrt{a} \text{ and } \sqrt{b} \text{ need not be surds}).$$

Examples $\sqrt{7} \times \sqrt{8} = \sqrt{7 \times 8} = \sqrt{56}$

$\sqrt{3} \times \sqrt{27} = \sqrt{3 \times 27} = \sqrt{81} = 9$

$3\sqrt{5} = \sqrt{9} \times \sqrt{5} = \sqrt{45}$

$\sqrt{12} = \sqrt{4 \times 3} = \sqrt{4} \times \sqrt{3} = 2 \times \sqrt{3} = 2\sqrt{3}$

$\dfrac{1}{\sqrt{5}} = \dfrac{\sqrt{5}}{\sqrt{5} \times \sqrt{5}} = \dfrac{\sqrt{5}}{5}$

$\dfrac{3}{\sqrt{2}} = \dfrac{3\sqrt{2}}{\sqrt{2} \times \sqrt{2}} = \dfrac{3\sqrt{2}}{2} \text{ or } \dfrac{\sqrt{18}}{2}$

Exercise 5.7

1. Simplify: $\sqrt{5} \times \sqrt{2}$, $\sqrt{7} \times \sqrt{3}$, $\sqrt{2} \times \sqrt{8}$, $\sqrt{12} \times \sqrt{3}$.

2. Express in the form $\sqrt{c}$:

   $2\sqrt{5}$, $3\sqrt{7}$, $5\sqrt{3}$, $4\sqrt{2}$, $2\sqrt{6}$, $6\sqrt{2}$.

3. Express in the form $a\sqrt{b}$ where a and b are integers and b is as small as possible:

   $\sqrt{18}$, $\sqrt{27}$, $\sqrt{50}$, $\sqrt{45}$, $\sqrt{98}$, $\sqrt{125}$.

4. Express in the form $\dfrac{\sqrt{c}}{d}$ where c and d are integers:

   $\dfrac{1}{\sqrt{3}}$, $\dfrac{1}{\sqrt{7}}$, $\dfrac{1}{2\sqrt{3}}$, $\dfrac{1}{3\sqrt{2}}$, $\dfrac{2}{\sqrt{6}}$, $\dfrac{5}{\sqrt{10}}$.

5. Express as $\dfrac{e}{f}$ where f is an integer:

   $\dfrac{\sqrt{3}}{\sqrt{2}}$, $\dfrac{\sqrt{5}}{\sqrt{7}}$, $\dfrac{\sqrt{2}}{\sqrt{8}}$, $\dfrac{\sqrt{3}}{\sqrt{27}}$, $\dfrac{\sqrt{3}}{\sqrt{6}}$, $\dfrac{\sqrt{2}}{\sqrt{10}}$.

6. Express as $k\sqrt{2}$ where k is an integer:

   $\sqrt{18}$, $\dfrac{10}{\sqrt{2}}$, $\sqrt{98} - \sqrt{32}$, $\sqrt{8} + \sqrt{50}$.

7. Given that $\sqrt{3} \approx 1.73205$, find $\sqrt{12}$, $\sqrt{27}$ and $\sqrt{300}$ to 4 d.p.

6 Percentages

Per cent means per hundred. A *percentage* is a fraction with a denominator of 100.

13 per cent means $\dfrac{13}{100}$ and can be written as 13% or 13 pc.

Changing percentages to fractions or decimals

To convert a percentage into a fraction or a decimal, divide it by 100.

Examples $36\% = \dfrac{36}{100} = \dfrac{9}{25}$ as a fraction in its lowest terms

$$= 0.36 \text{ as a decimal}$$

$$83\tfrac{1}{3}\% = \dfrac{83\tfrac{1}{3}}{100} = \dfrac{83\tfrac{1}{3} \times 3}{100 \times 3} \text{ (multiply top and bottom by 3 to obtain a whole number numerator)}$$

$$= \dfrac{250}{300} = \dfrac{5}{6} = 0.8\dot{3}$$

$$9.6\% = \dfrac{9.6}{100} = \dfrac{96}{1000} = \dfrac{12}{125} \text{ or } 0.096$$

Changing fractions or decimals to percentages

To convert a fraction or a decimal into a percentage, multiply it by 100.

Examples $\dfrac{3}{5} = \dfrac{3}{5} \times \dfrac{100}{1}\% = \dfrac{300}{5}\% = 60\%$

$$\dfrac{7}{8} = \dfrac{7}{8} \times \dfrac{100}{1}\% = \dfrac{700}{8}\% = 87\tfrac{1}{2}\%$$

$$0.062 = 0.062 \times 100\% = 6.2\%$$

Finding a percentage of a given quantity

Examples 28% of £150 $= \dfrac{28}{100} \times \dfrac{150}{1} = £42$

$\left(\text{Notice that '}28\% \text{ of' is replaced by '}\dfrac{28}{100} \times \text{'}\right)$

$17\dfrac{1}{2}\%$ of £60 $= \dfrac{17\frac{1}{2}}{100} \times \dfrac{60}{1} = \dfrac{35}{200} \times \dfrac{60}{1} = \dfrac{21}{2} = £10.50$

When using a calculator, you may find it easier to change the percentage into a decimal as in the next example.

39% of £482 $= 0.39 \times £482 = £187.98$

Expressing one quantity as a percentage of another

First express one as a fraction of the other. Then change the fraction to a percentage.

Example   Express £54 as a percentage of £225.

$$\dfrac{£54}{£225} = \dfrac{54}{225} = \dfrac{54}{225} \times \dfrac{100}{1}\% = 24\%$$

Some percentages to remember

$50\% = \dfrac{1}{2}$ $10\% = \dfrac{1}{10}$ $33\dfrac{1}{3}\% = \dfrac{1}{3}$

$25\% = \dfrac{1}{4}$ $5\% = \dfrac{1}{20}$ $66\dfrac{2}{3}\% = \dfrac{2}{3}$

$75\% = \dfrac{3}{4}$ $20\% = \dfrac{1}{5}$ $100\% = 1$

1% of £1 $= 1p$ so $x\%$ of £1 $= xp$

Exercise 6.1

No calculating aids should be used.
Express as fractions in their lowest terms:

1. 60% **2.** 55% **3.** 12% **4.** 15% **5.** 90%
6. 37% **7.** 70% **8.** 99% **9.** 9% **10.** 85%

Express as percentages:

11. $\dfrac{3}{10}$ **12.** $\dfrac{4}{5}$ **13.** $\dfrac{7}{20}$ **14.** $\dfrac{27}{50}$ **15.** $\dfrac{19}{25}$

Express as decimals:

16. 31%   **17.** 68%   **18.** 80%   **19.** 6%   **20.** 2%

Express as percentages:

21. 0.73   **22.** 0.55   **23.** 0.3   **24.** 0.07   **25.** 0.9

Express as fractions in their lowest terms:

26. $7\frac{1}{2}$%   **27.** $12\frac{1}{2}$%   **28.** $13\frac{1}{3}$%   **29.** $31\frac{1}{4}$%   **30.** $9\frac{3}{5}$%

Express as decimals:

31. 4.6%   **32.** 53.2%   **33.** 8.5%   **34.** 61.3%   **35.** 0.44%

Express as percentages:

36. 0.825   **37.** 0.035   **38.** 0.05   **39.** 0.015   **40.** 0.623

41. Express as fractions in their lowest terms:

50%,   25%,   10%, $33\frac{1}{3}$%.

42. Express as fractions in their lowest terms:

5%,   20%,   75%, $66\frac{2}{3}$%.

Exercise 6.2

No calculating aids should be used, for questions 1 to 28.

State the value of:

1. 50% of £12   **2.** 25% of £20   **3.** 10% of £70
4. $33\frac{1}{3}$% of £18   **5.** 20% of £15   **6.** 100% of £92
7. 3% of £1   **8.** 3% of £4   **9.** 11% of £1
10. 11% of £5

Find the value of:

11. 7% of £200   **12.** 8% of £350   **13.** 15% of £40
14. 23% of 200 g   **15.** 65% of 1 kg   **16.** 4% of 450 g
17. $2\frac{1}{2}$% of £360   **18.** $17\frac{1}{2}$% of £80

Express the first quantity as a percentage of the second:

19. £9, £100   **20.** £26, £200   **21.** £35, £500
22. 55p, £1   **23.** £3, £25   **24.** 90 g, 120 g
25. £7, £28   **26.** £1.40, £2   **27.** £4.20, £3.50
28. 54 minutes, 2 hours

Use a calculator for the remaining questions.
Find the value of the following:

29. 45% of £160   **30.** 34% of £45   **31.** 7% of £92

32. 13% of £88   **33.** 65% of £4.80   **34.** 8% of £6.25

35. $6\frac{1}{2}$% of £24   **36.** $37\frac{1}{2}$% of £6.16

Find the value of the following, correct to the nearest penny:

37. 9% of £8.60   **38.** 16% of £7.30   **39.** 83% of £14.40

Express the first quantity as a percentage of the second, correct to the nearest whole number:

40. £17, £23   **41.** £36, £92   **42.** 46p, £3

Exercise 6.3

No calculating aids should be used.

1. 25% of a sum of money is £40. Find the sum of money.

2. $33\frac{1}{3}$% of a sum of money is £12. Find the sum of money.

3. The pass mark in an examination is 45%. If the maximum possible number of marks is 120, how many are needed for a pass?

4. A salesman is allowed 6% commission. What is his commission on sales of £940?

5. A bank pays 8% interest on deposits. How much interest does it pay on £550?

6. At a sale the price of an article is reduced from £2 to £1.40. Express the reduction as a percentage of the original price.

7. A man bought a car for £4500. After one year it was worth £3690. Express the fall in value as a percentage of the price he paid.

8. The population of a town one year was 55000. Ten years later it was 77000. Express the increase as a percentage of the original figure.

9. During a sale, a shop allows a discount of 10% on the marked prices. Find the sale price of a coat marked at £45.

10. A manufacturer allows a retailer a trade discount of 30% of the catalogue prices. What does the retailer pay for a TV set listed at £240?

11. A pupil's marks in five examinations were: 42 out of 80, 59 out of 110, 18 out of 40, 137 out of 240, 39 out of 90. Express each mark as a percentage and then place the marks in order, with the highest mark first.

Percentage increases and decreases

Suppose that the price of a motor cycle is increased by 10%.

The original price is 100%. The increase is 10% of the original price, so the new price is 110% of the original.

In a sale, the price of a dress is reduced by 10%.

The original price is 100%. The reduction is 10% of the original price, so the sale price is 90% of the original.

Example 1　(i) Increase a price of £40 by 15%.

(ii) Decrease a price of £40 by 15%.

(i) The new price is 115% of the original price.

115% of £40 = 1.15 × £40 = £46

(ii) The new price is 85% of the original.

85% of £40 = 0.85 × £40 = £34

The next example shows how to obtain the original price from the new price.

Example 2　After an increase of 8% a cycle costs £81. Find the original price.

The new cost is 108% of the original.

108% of the original cost is £81

1% of the original cost is $\dfrac{£81}{108}$

100% of the original cost is $£\dfrac{81}{108} \times \dfrac{100}{1} = £75$

(*Warning*: The increase is NOT 8% of £81 (£6.48). It is 8% of the original price, that is 8% of £75, which is £6.)

Exercise 6.4

1. Increase £50 by 8%.　　**2.** Increase £60 by 5%.

3. Decrease £25 by 4%.　　**4.** Decrease £95 by 20%.

5. Increase £38 by 6%.　　**6.** Decrease £14 by 12%.

7. The price is £396 after an increase of 10%. Find the original price.

8. The price is £100 after an increase of 25%. Find the original price.

9. The price is £32 after a decrease of 20%. Find the original price.

10. The price is £54 after a decrease of 10%. Find the original price.

11. During a sale at a furniture shop, a discount of 20% is allowed on all marked prices.

   (i) What is the sale price of a chair marked at £40?

   (ii) A customer paid £120 for a sofa. What was the marked price?

12. A customer paid £357 for a television set, after its price had been reduced by 15%. Find the original price.

13. A man sold a car for £2840, which was 20% less than the price he paid for it. How much did he pay for it?

14. Value Added Tax at the rate of 15% is added to the marked prices of all articles in a shop.

   (i) An article is marked at £60. How much does the customer pay, after VAT is added?

   (ii) For a certain article, a customer paid £138, which included VAT. What was the marked price of the article?

15. A party of four people went to a restaurant for dinner. The cost of the meal was £5.50 per person. VAT at the rate of 15% was added. Then a service charge of 10% of the total was added. Find the final amount of the bill.

16. At a discount warehouse, a radio is labelled '£60 + VAT less 10% discount'. Anne says: 'To work out how much is paid, add 15% and then take 10% from this total'. Betty says: 'First take off 10%, and then add 15% of the answer'. Carry out each of these calculations. What do you notice about your results?
   Cathie says: 'Just add 5% to £60'. Is she correct?

Profit and loss

The price paid for an article is called the *cost price*.
The price at which it is sold is the *selling price*.
The difference between these prices is the *profit* or the *loss*.
The profit or loss is often expressed as a percentage of the cost price.

$$\text{Percentage profit (or loss)} = \frac{\text{profit (or loss)}}{\text{cost price}} \times \frac{100}{1}\%$$

Example 1   A dealer buys a cycle for £75 and sells it for £93. Find his percentage profit.

   Profit = £93 − £75 = £18

   $$\text{Percentage profit} = \frac{18}{75} \times \frac{100}{1}\% = 24\%$$

Example 2   A shopkeeper buys chairs for £120 each. At what price should he sell them, in order to make a profit of 25%?

$$\text{Profit} = 25\% \text{ of } £120 = \frac{1}{4} \text{ of } £120 = £30$$

Selling price = cost price + profit = £120 + £30 = £150
He should sell the chairs for £150 each.

The next example is similar to the second example on page 45.

Example 3 A second-hand car dealer wishes to buy a certain used car. He thinks he will be able to sell it again for £1500, and he wants to make a profit of at least 20%. What is the maximum price he should offer for the car?
The cost price or C.P. is 100%. The profit is 20% of the C.P. The selling price or S.P. is therefore 120% of the C.P.

120% of C.P. = £1500

$$1\% \text{ of C.P.} = £\frac{1500}{120}$$

$$100\% \text{ of C.P.} = £\frac{1500}{120} \times \frac{100}{1} = £1250$$

He should not offer more than £1250 for the car.
(Note: The profit is NOT 20% of £1500 (£300). This would give a wrong answer of £1200. The profit is 20% of the C.P., that is 20% of £1250, which is £250.)

Exercise 6.5

No calculating aids should be used.
Calculate the percentage profit or loss:

1. C.P. £12, profit £3 **2.** C.P. £600, profit £180
3. C.P. £10, profit £1.50 **4.** C.P. £400, loss £24
5. C.P. £20, S.P. £26 **6.** C.P. £75, S.P. £81
7. C.P. £70, S.P. £42 **8.** C.P. £20, S.P. £24.50

Calculate the selling price:

9. C.P. £36, profit 25% **10.** C.P. £8, profit 20%
11. C.P. £3500, loss 10% **12.** C.P. £450, profit 8%

Calculate the cost price:

13. S.P. £220, profit 10% **14.** S.P. £600, profit 20%
15. S.P. £600, loss 20% **16.** S.P. £91, loss 30%
17. S.P. £91, profit 30% **18.** S.P. £4.60, profit 15%

19. A greengrocer buys a box of 200 oranges for £15 and sells them at 9p each. Find his percentage profit.

20. Articles costing £16 per 100 are sold at 20p each. Find the percentage profit.

21. By selling a bicycle for £78, a shopkeeper makes a profit of 30% of his cost price. At what price should he sell it to make a profit of 25% of his cost price?

22. If he sells a radio for £60, a shopkeeper makes a profit of 25% of his cost price. At what price should he sell it if he is prepared to accept a profit of only 15%?

Simple interest

If you place money in a building society account, or a bank deposit account, you will be paid *interest*. Similarly, if you borrow money, you will be charged interest. The amount deposited or borrowed is called the *principal*. The interest is a percentage of this per year (per annum).

If you deposit £100, and the bank pays interest at a rate of 8% per annum, then you will received £8 interest at the end of the year. If you borrow £100, and the bank charges interest at the rate of 8% p.a., then you will have to pay the bank £8 interest at the end of the year.

If the interest is withdrawn (or paid) each year, the principal remains the same and we have an example of *simple interest*.

If £950 is deposited with a building society that pays 8% p.a., then each year the interest received is £950 × 0.08 = £76.

Over a period of 3 years, the total interest received is £76 × 3 = £228.

If a principal of £P is deposited for T years at a rate of R% p.a., the simple interest, £I, is given by the formula:

$$I = \frac{P \times R \times T}{100}$$

Example 1   Find the simple interest on £560 at 9% p.a. for 5 years. This can be worked either: (i) from basic ideas, or (ii) by using the formula.

(i) On £100 for 1 year the interest is £9
    On £1 for 1 year the interest is £0.09
    On £560 for 5 years it is £0.09 × 560 × 5 = £252.

(ii) $P = 560$, $R = 9$, $T = 5$

$$I = \frac{P \times R \times T}{100} = \frac{560 \times 9 \times 5}{100} = 252$$

The interest is £252.

Sometimes you are asked to find the principal, or the rate, or the time. For such questions, the formula should be used.

Example 2   What sum of money, invested for 4 years at 6% p.a., yields simple interest of £204?

Here $I = 204$, $R = 6$ and $T = 4$.

Since $\dfrac{PRT}{100} = I$, then $\dfrac{P \times 6 \times 4}{100} = 204$

$P = \dfrac{204 \times 100}{6 \times 4} = 850$

The principal is £850.

Exercise 6.6

Calculate the simple interest on:

1. £100 for 3 years at 8% p.a.
2. £300 for 2 years at 7% p.a.
3. £200 for 4 years at 9% p.a.
4. £400 for 5 years at 10% p.a.
5. £350 for 3 years at 6% p.a.
6. £225 for 8 years at 5% p.a.
7. £850 for 4 years at 11% p.a.
8. £55 for 5 years at 8% p.a.
9. £48 for 3 years at 9% p.a.
10. £1053 for 2 years at 6% p.a.

Calculate, correct to the nearest penny, the simple interest on:

11. £83.40 for 4 years at 12% p.a. 12. £264.60 for 3 years at $9\frac{1}{2}$% p.a.

13. What principal invested for 4 years at 11% p.a. yields simple interest of £77?

14. What principal invested at 7% p.a. for 2 years yields simple interest of £42?

15. How long does it take to obtain simple interest of £240 on £800 invested at 6% p.a.?

16. How long does it take to obtain simple interest of £144 on £240 invested at 10% p.a.?

17. Find the rate, if the simple interest obtained by investing £720 for 6 years is £216.

18. Find the rate, if the simple interest obtained by investing £225 for 4 years is £54.

19. A man deposited £450 in a bank. At the end of the first year £36 was added. What was the interest rate p.a.?

    He left the whole sum (£450 + £36) in the bank for a further year at the same rate of interest. What did the sum amount to at the end of that second year?

20. Sheila has £600 to invest for 2 years. She can invest it at $7\frac{1}{2}$% p.a. and withdraw the interest at the end of each year. Alternatively, she can invest it at 7% p.a. with the interest added to the capital at the end of each year. Which is more profitable and by how much?

Compound interest

In the case of *compound interest*, the interest is added to the principal at the end of each year (or other agreed period of time) and the interest for the next year is calculated on this new larger principal.

Example   £650 is invested at 8% p.a. compound interest. Find the amount at the end of 3 years.

First year:	principal £650	
	interest £ 52	(8% of £650)
Second year:	principal £702	(by adding the above two amounts)
	interest £ 56.16	(8% of £702)
Third year:	principal £758.16	(by adding the last two amounts)
	interest £ 60.65	(8% of £758.16, to nearest penny)

Amount after 3 years   £818.81

Using a calculator

Suppose that £400 is invested at 9% p.a.

Each year the interest on £1 is 9p, so at the end of a year each £1 becomes £1.09.

At the end of the first year, £400 becomes £400 $\times$ 1.09 = £436.

At the end of the second year, £436 becomes £436 $\times$ 1.09 = £475.24.

At the end of the third year, £475.24 becomes £475.24 $\times$ 1.09 = £518.01, to nearest penny.

Notice that, at the end of each year, the sum of money is multiplied by 1.09. After 3 years it becomes £400 $\times$ 1.09 $\times$ 1.09 $\times$ 1.09, that is £400 $\times$ 1.09^3. After n years it becomes £400 $\times$ 1.09^n.

Depreciation

Cars and other machines decrease in value each year, as they become more worn. This decrease in value is called *depreciation*.

Example   The value of a certain machine in a factory decreases each year, by 20% of its value at the start of the year. If the machine cost £15000 when new, calculate its value at the end of the second year.

Value at the start of year 1 is	£15000	
Loss in value in year 1 is	£ 3000	(20% of £15000)
Value at the start of year 2 is	£12000	(by subtraction)
Loss in value in year 2 is	£ 2400	(20% of £12000)
Value at the end of year 2 is	£ 9600	(by subtraction)

Alternatively, at the end of each year the value is 0.8 of its value at the beginning of that year.

$(100\% - 20\% = 80\% = 0.8)$

Therefore, after 2 years its value is £15000 × 0.8 × 0.8. After n years its value is £15000 × 0.8^n.

Exercise 6.7

Calculate the amount obtained by investing at compound interest:

1. £400 for 2 years at 8% p.a.   2. £600 for 2 years at 10% p.a.
3. £850 for 2 years at 6% p.a.   4. £250 for 2 years at 12% p.a.
5. £100 for 3 years at 10% p.a.   6. £400 for 3 years at 5% p.a.
7. £670 for 3 years at 9% p.a.   8. £214 for 3 years at 7% p.a.

9. A car was bought for £6000. Each year its value fell by 20% of the value at the start of the year. What was it worth: (i) at the end of the first year (ii) at the end of the second year?

10. Each year the value of the machinery in a certain factory decreases by 10% of its value at the beginning of the year. Calculate, correct to the nearest £100, the value at the end of:

(i) the first year (ii) the second year (iii) the third year

if the original value was £100000.

11. A company borrowed £40000, at an interest rate of 12% p.a. At the end of each year it repaid £15000, made up of interest on the loan plus partial repayment of the loan. How much of the loan was still owing, after 2 years?

12. The mass of a crystal growing in a 'mother' liquid is found to increase by 40% each day. Calculate the masses on days 1, 2, 3, 4 and 5 for a crystal with an initial mass of 100 g. Draw a graph of these values. From your graph, estimate: (i) the mass after $2\frac{1}{2}$ days (ii) when the mass has trebled.

7 Ratio and proportion

Ratio

A *ratio* enables us to compare two or more quantities of the same kind. Packets of a certain make of cornflakes are available in two sizes: large, containing 375 g of cornflakes, and extra large, containing 500 g.

$$\frac{\text{Mass of large}}{\text{Mass of extra large}} = \frac{375 \text{ g}}{500 \text{ g}} = \frac{3 \times 125}{4 \times 125} = \frac{3}{4}$$

We can say that the ratio of the large mass to the extra large mass is 3 to 4 and we write this ratio as $3:4$.

This means that the large mass is $\frac{3}{4}$ of the extra large mass and that the extra large mass is $\frac{4}{3}$ times the large mass.

We can also say that the ratio of the extra large mass to the large mass is $4:3$.

Quantities to be compared must always be in the same units. To find the ratio of the distances 3 km and 900 m, the 3 km must first be changed to metres. It is 3000 m.

So $3 \text{ km}:900 \text{ m} = 3000 \text{ m}:900 \text{ m} = 30:9 = 10:3$.

Notice that a ratio properly consists of two numbers; there are no units.

Division in a given ratio

To divide £60 among Jane, Jill and John in the ratio $3:7:2$, we must first divide it into 12 parts $(3 + 7 + 2 = 12)$ and then give Jane 3 parts, Jill 7 parts and John 2 parts.

Each part is $\frac{1}{12}$ of £60 = £5. Jane receives $3 \times £5 = £15$, Jill receives $7 \times £5 = £35$ and John receives $2 \times £5 = £10$.
(As a check: $£15 + £35 + £10 = £60$.)

Increasing and decreasing in a given ratio

Example Increase 30 cm in the ratio $5:4$.

$$\frac{\text{New length}}{\text{Old length}} = \frac{5}{4}$$

so new length $= \frac{5}{4}$ of old length $= \frac{5}{4}$ of 30 cm = 37.5 cm.

Scales

A tourist map has a *scale* of 1:50000.

This means that 1 cm on the map represents 50000 cm (or 500 m) on the ground. In the same way, 1 km on the ground is represented by $\frac{1}{50000}$ km on the map, which is $\frac{1}{50000}$ of 100000 cm, or 2 cm.

The scale could also be expressed as 2 cm to 1 km.

Exercise 7.1

Express each of the following ratios in its simplest form:

1. 9:15
2. 14:10
3. 20:36
4. 5:35
5. 44:11
6. 45:120
7. 40p:£1
8. 1 kg:650 g
9. 2500 m:3 km
10. 80p:£4
11. 1.8 km:800 m
12. 2.4 kg:900 g

13. A rectangle has a length of 16 cm and a width of 12 cm. State the ratio of: (i) its length to width   (ii) its width to length.

14. Two squares have sides of length 9 cm and 6 cm. State the ratio of: (i) their sides   (ii) their perimeters   (iii) their areas.

Express the following ratios in their simplest forms:

15. 1.5:3.5
16. $2\frac{1}{4}:1\frac{1}{2}$
17. $\frac{1}{3}:\frac{1}{5}$
18. 5.4:0.45

19. 0.96 g:1.2 g
20. 0.72 m:3.6 m
21. 1.2 g:0.25 g
22. 1.6 m:880 cm

23. A plan of a room has a scale of 1:50.
   What length on the plan represents:
   (i) 300 cm   (ii) 2 m   (iii) 2.8 m in the room?
   What distance in the room is represented by:
   (iv) 2 cm   (v) 3.2 cm   (vi) 9 mm on the plan?

24. A map has a scale of 1:50000.
   What distance in kilometres is represented on the map by:
   (i) 4 cm   (ii) 6.3 cm?
   What length in centimetres on the map represents:
   (iii) 1 km   (iv) 6 km   (v) 700 m?

Divide the given number or quantity in the given ratio:

25. 45, 3:2
26. 40, 5:3
27. £20, 3:7
28. 3 m, 1:4
29. £9, 5:1
30. £6, 7:13
31. £3.60, 2:3:4
32. £2.40, 1:4:5

33. Bronze is formed by mixing copper, tin and zinc in the ratio 95:4:1 by mass. Find the mass of each metal in 2 kg of bronze.

34. A certain pastry mixture consists of sugar, margarine and flour in the ratio 1:3:6. How much of each is needed to make 500 g of the mixture?

35. An alloy consists of 6 parts of metal A to 5 parts of metal B. How much of metal A should be mixed with 30 kg of metal B?

36. A sum of money is divided in the ratio 2:3. The smaller part is £12. What is the larger part?

37. Find the value of x if: (i) $x:12 = 5:3$ (ii) $21:x = 7:4$.

38. Find the value of x if: (i) $x:2 = 2.7:3.6$ (ii) $7:x = 0.56:0.72$.

39. (i) Increase £30 in the ratio 6:5.

   (ii) Decrease £30 in the ratio 5:6.

40. (i) Decrease 24 cm in the ratio 3:4.

   (ii) Increase 24 cm in the ratio 4:3.

41. A photograph is enlarged in the ratio 5:2. What are the new sizes of the following: a tree of height 2 cm, a bridge of length 3.4 cm and a house of height 4.6 cm?

42. The sides of a triangle measure 6 cm, 8 cm and 9 cm. It is enlarged in the ratio 7:2. What are the new side lengths?

43. Three partners share the profits of their business as follows: A has a fixed salary of £8000 p.a., B has £6000 p.a., and the remainder is shared among A, B and C in the ratio 2:3:5. One year the profits were £48000.

   (i) How much remained that year, after the two salaries had been paid?

   (ii) How much did each partner receive from this remainder?

Proportion

Example 1   4 metres of cloth cost £11.20. Find the cost of 5.5 metres.

4 m cost 1120p

$$1 \text{ m costs} \frac{1120}{4}\text{p} \qquad 5.5 \text{ m cost} \frac{1120}{4} \times \frac{11}{2}\text{p} = £15.40$$

This example involves *direct proportion*. An increase in length causes an increase in cost; a decrease in length causes a decrease in cost.

Example 2   It takes 6 days for 12 men to pick a crop of plums. How long would it take 9 men?

12 men take 6 days

1 man would take 12×6 days

$$9 \text{ men would take} \frac{12 \times 6}{9}\text{ days} = 8 \text{ days}$$

This example involves *inverse proportion*. A decrease in the number of men causes an increase in the time; an increase in the number of men causes a decrease in the time.

Exercise 7.2

1. 4 machines produce a batch of castings in 30 hours. How long would it take 6 machines to produce a batch of the same size?

2. 4 kg of grass seed are needed for 60 m² of ground. How much seed is needed for 105 m² of ground?

3. 300 metres of wire netting cost £34.80. Find the cost of 700 metres of the netting.

4. When 45 students shared a prize equally, each got £24. How much would each get if a prize of the same value was shared equally by 20 students?

5. A car travels 92 km on 12 litres of petrol. How far is it likely to go on 30 litres of petrol?

6. A bookshelf can hold 18 books of width 4 cm. How many books of width 4.5 cm can it hold?

7. A farm has sufficient grain to feed 30 hens for 4 days. How long would the grain last for 40 hens?

8. A bar of metal is 14 cm long and has a mass of 210 g. Another bar of the same metal, with the same cross-section, is 18 cm long. What is its mass?

9. A novel has 180 pages with an average of 365 words per page. If the size of the type is decreased so that there is an average of 450 words per page, how many pages will then be needed?

10. A man takes 115 steps to cover 100 metres. How many steps will he take to cover 360 metres? How far, to the nearest metre, will he go in 1000 steps?

Foreign exchange

The rates below show the *foreign currency* obtainable for £1 sterling on 3rd January 1986.

France	10.89 francs	Japan	292 yen
Switzerland	2.98 francs	United States	1.44 dollars
West Germany	3.55 marks		

Example (i) How many French francs could be bought for £14.80, that day?

£1   = 10.89 francs so

£14.80 = 14.80 × 10.89 francs

= 161.17 francs to the nearest centime

(ii) How much did it cost to buy 297 French francs, that day?

10.89 francs = £1 so

$$1 \text{ franc} = £\frac{1}{10.89}$$

$$297 \text{ francs} = £\frac{1}{10.89} \times 297$$

$$= £27.27 \text{ to the nearest penny}$$

Exercise 7.3

Using the above rates, find how much foreign currency will be obtained, if the following amounts of British currency are changed:

1. £34 into French francs
2. £216 into US dollars
3. £1258 into Japanese yen
4. £924 into Swiss francs
5. £16.50 into West German marks
6. £7.20 into French francs

Using the above rates, find the value in British money of:

7. 2376 West German marks
8. 453 Swiss francs
9. 79 US dollars
10. 3428 Japanese yen
11. 208 French francs
12. 8902 US dollars

13. In a Swiss shop a watch was marked at 106 francs. Change the price into British money.

14. A traveller in France changed £120 into francs at a rate of £1 = 10.52 fr. He spent 938 francs. Then he changed the rest back into British money at the rate of £1 = 11.07 fr. How much British money did he receive?

Speeds

If distance is given in kilometres, and time in hours, then speed is in km/h. If distance is given in metres, and time in seconds, then speed is in m/s.

At sea, speed is measured in *knots*. A knot is 1 nautical mile per hour.

$$\text{Average speed} = \frac{\text{distance}}{\text{time}}; \quad \text{distance} = \text{speed} \times \text{time}; \quad \text{time} = \frac{\text{distance}}{\text{speed}}$$

Example 1   An aircraft travels 350 km in 25 minutes. Find its average speed.

$$25 \text{ min} = \frac{25}{60}\text{h} = \frac{5}{12}\text{h}$$

$$\text{Average speed} = \frac{350}{\frac{5}{12}} = \frac{350}{1} \times \frac{12}{5} = 840 \text{ km/h}$$

Example 2   A boat has a speed of 18 km/h. How far does it travel in 35 minutes?

$$35 \text{ min} = \frac{35}{60} \text{h} = \frac{7}{12} \text{h}$$

$$\text{Distance} = \frac{18}{1} \times \frac{7}{12} \text{km} = \frac{21}{2} \text{km} = 10.5 \text{ km}$$

Exercise 7.4

1. An aircraft flies at 600 km/h. How far does it go in:

(i) $1\frac{1}{2}$h   (ii) 1 min   (iii) 7 min?

2. How long does it take an aircraft flying at 600 km/h to travel:

(i) 300 km   (ii) 60 km   (iii) 740 km?

3. Find the average speed in each of the following cases:

(i) a lorry travels 81 km in $1\frac{1}{2}$h (speed in km/h)

(ii) a cyclist travels 1 km in 3 min (speed in km/h)

(iii) a girl runs 90 m in 12 s (speed in m/s).

4. Find the average speed in km/h in the following cases:

(i) a train takes 8 min for 18 km

(ii) a racing car takes 8 s for 400 m

(iii) a spacecraft takes 4 s for 42 km.

5. How long does it take a cyclist to travel 66 km at a speed of 18 km/h?

6. An express train is travelling at 120 km/h. How far does it go between 09.55 h and 10.05 h?

7. A journey took 1 h 20 min at an average speed of 87 km/h. Find the distance travelled.

8. How long does it take a car travelling at 45 km/h to pass over a bridge of length 250 m?

9. Telegraph poles are 60 m apart on a certain road. A car took 28 s to pass from the first to the eighth pole. Find its speed in km/h.

10. (i) Express 27 km/h in m/s   (ii) Express 35 m/s in km/h.

11. A girl cycled 14 km in 55 min. Find her average speed in m/s, correct to 1 d.p.

12. Calculate, correct to the nearest minute, the time taken to travel 9 km at 38 km/h.

13. A motorist has to make a journey of 30 km. He estimates that his speed will be somewhere between 40 km/h and 60 km/h. Calculate the difference between his estimated longest and shortest times for the journey.

8 Length, area and volume

Units of length

1 centimetre (cm) = 10 millimetres (mm)

1 metre (m)   = 100 centimetres or 1000 millimetres

1 kilometre (km) = 1000 metres

Conversion of units

Lengths can be converted from one unit to another by multiplication or division by a power of 10. (See pages 27 and 28.)

Examples $0.0682 \text{ m} = 0.0682 \times 1000 \text{ mm} = 68.2 \text{ mm}$

$$740 \text{ cm} = 740 \div 100 \text{ m} = 7.4 \text{ m}$$

$$95\,300 \text{ m} = 95\,300 \div 1000 \text{ km} = 95.3 \text{ km}$$

Units of mass

1 kilogram (kg) = 1000 grams (g)

1 tonne (t)   = 1000 kilograms

Exercise 8.1

1. Express in millimetres: (i) 8 cm   (ii) 4.3 cm   (iii) 0.27 cm.

2. Express in centimetres: (i) 6 m   (ii) 9.4 m   (iii) 0.028 m.

3. Express in metres: (i) 5 km   (ii) 6.3 km   (iii) 0.047 km.

4. Express in metres: (i) 400 cm   (ii) 230 cm   (iii) 7600 cm.

5. Express in metres: (i) 7000 mm   (ii) 823 mm   (iii) 64 mm.

6. Express in kilometres: (i) 2000 m   (ii) 58 m   (iii) 3460 m.

7. Express in grams: (i) 4.8 kg   (ii) 0.075 kg   (iii) 0.0086 kg.

8. Express in kilograms: (i) 3200 g   (ii) 740 g   (iii) 92 g.

9. Express in tonnes: (i) 800 kg   (ii) 34 kg   (iii) 7.2×10^5 kg.

10. Express in kilograms: (i) 5.6 t   (ii) 0.37 t   (iii) 4.8×10^{-5} t.

Rectangles

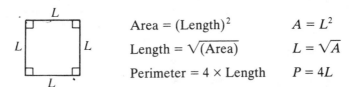

Area = Length × Breadth $A = L \times B$

Length = Area ÷ Breadth $L = A \div B$

Breadth = Area ÷ Length $B = A \div L$

Fig. 8.1

Perimeter (or distance round) = 2 × Length + 2 × Breadth
$P = 2L + 2B$

Squares

Area = (Length)² $A = L^2$

Length = √(Area) $L = \sqrt{A}$

Perimeter = 4 × Length $P = 4L$

Fig. 8.2

Example 1   Find the perimeter of a square of area 324 cm².
Length = $\sqrt{324}$ = 18 cm
Perimeter = 4 × 18 = 72 cm

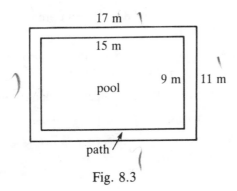

Fig. 8.3

Example 2   A rectangular swimming pool measuring 15 m by 9 m is surrounded by a path of width 1 m. Calculate the area of the path.
Area of pool + path (large rectangle) = 17 × 11 = 187 m²
Area of pool (small rectangle)   = 15 × 9 = 135 m²
Area of path = 187 − 135 = 52 m²

Units of area

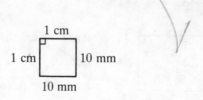

Fig. 8.4

The area of the square is $1 \, \text{cm}^2$ or $10 \times 10 \, \text{mm}^2$

Hence $1 \, \text{cm}^2 = 100 \, \text{mm}^2$

As $1 \, \text{m} = 100 \, \text{cm}$, $1 \, \text{m}^2 = 100 \times 100 \, \text{cm}^2 = 10\,000 \, \text{cm}^2$

For larger areas we use ares, hectares and square kilometres.

$1 \, \text{are} = 100 \, \text{m}^2$ $1 \, \text{hectare} = 100 \, \text{ares}$ $1 \, \text{km}^2 = 100 \, \text{hectares}$

Exercise 8.2

No calculating aids should be used.

1. Calculate the perimeter and area of each of the following rectangles:
   (i) length 7 cm, breadth 5 cm   (ii) length 18 cm, breadth 12 cm
   (iii) length 35 m, breadth 25 m   (iv) length 2.6 m, breadth 1.8 m.

2. Calculate the perimeters and areas of squares with sides:
   (i) 6 cm   (ii) 9 m   (iii) 5.4 cm.

3. A rectangle has a length of 8 cm and an area of 96 cm². Calculate its breadth and its perimeter.

4. A rectangle has a breadth of 3.5 m and an area of 16.8 m². Calculate its length and its perimeter.

5. A square has an area of 49 cm². Calculate its length and its perimeter.

6. A square has a perimeter of 36 cm. Calculate its area.

7. A rectangular piece of card, 35 cm by 20 cm, is cut into squares of side 5 cm. How many squares are obtained?

8. A room measures 6 m by 5 m. How many square tiles of side $\frac{1}{2}$ m are needed to cover the floor?

9. A rectangular lawn measuring 9 m by 7 m is surrounded by a path of width 1 m. Calculate the area of the path.

10. A picture 18 cm by 15 cm is mounted on a piece of card so that there is a border of 3 cm width on one long side and a border of 2 cm width on the other three sides. Calculate the area of the border.

11. A square has the same area as a rectangle that measures 50 cm by 18 cm. Find the side length of the square.

12. Express in mm²: (i) 1 cm² (ii) 3 cm² (iii) 5.7 cm².

13. Express in cm²: (i) 600 mm² (ii) 7000 mm² (iii) 85 mm².

14. Express in cm²: (i) 1 m² (ii) 4 m² (iii) 8.3 m².

15. Express in hectares: (i) 70 000 m² (ii) 46 000 m².

16. A rectangle measures 75 mm by 64 mm. Calculate its area in cm².

17. A field measures 450 m by 300 m. Calculate its perimeter in km and its area in hectares.

Parallelograms

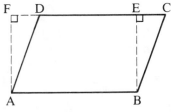

Fig. 8.5

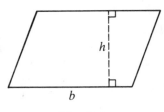

Fig. 8.6

Area of $\triangle$AFD = area of $\triangle$BEC

so area of quadrilateral ABCF − area of $\triangle$AFD

$\qquad$ = area of quadrilateral ABCF − area of $\triangle$BEC

Therefore area of parallelogram ABCD

$\qquad$ = area of rectangle ABEF

$\qquad$ = base(AB) × height(BE)

Thus area of parallelogram ABCD = base(AB) × height(BE).

So, in Fig. 8.6, $\qquad A = bh$

Triangles

A triangle is half a parallelogram.

Area of $\triangle = \dfrac{1}{2}$(base × height)

$\qquad A = \dfrac{1}{2}bh$

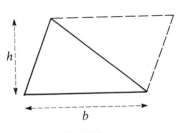

Fig. 8.7

Trapezia

The diagonal divides the trapezium into two triangles.

$$\text{Area of trapezium} = \frac{1}{2}ah + \frac{1}{2}bh$$

$$= \frac{1}{2}(a + b)h$$

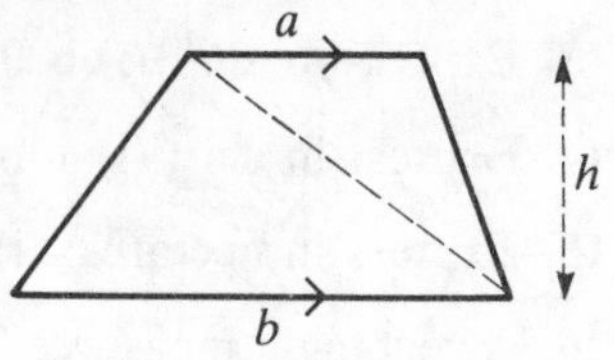

Fig. 8.8

Example 1   Find the area of a triangle with a base of 14 cm and a height of 10 cm.

$$\text{Area} = \frac{1}{2} \times 14 \times 10 = 7 \times 10 = 70 \text{ cm}^2$$

Example 2   Find the area of a trapezium with parallel sides of lengths 8.6 cm and 5.4 cm that are 3.5 cm apart.

$$\text{Area} = \frac{1}{2}(8.6 + 5.4) \times 3.5 = \frac{1}{2} \times 14.0 \times 3.5$$

$$= 7.0 \times 3.5 = 24.5 \text{ cm}^2$$

Exercise 8.3

No calculating aids should be used.

1. Calculate the area of a parallelogram with:
   (i) base 12 cm, vertical height 8 cm
   (ii) base 7.5 cm, vertical height 6 cm.

2. Calculate the area of a triangle with:
   (i) base 10 cm, vertical height 8 cm
   (ii) base 14 cm, vertical height 9 cm.

3. Calculate the areas of the triangles in Fig. 8.9:

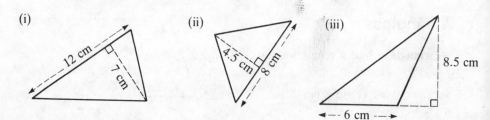

(i)   (ii)   (iii)

12 cm   4.5 cm   8 cm   8.5 cm

7 cm   6 cm

Fig. 8.9

4. Find the area of a trapezium with:

(i) parallel sides 6 cm and 8 cm, distance apart 4 cm

(ii) parallel sides 11 cm and 5 cm, distance apart 6.5 cm.

5. Find the area of each of the following figures by dividing it into a rectangle and a triangle:

(i)   (ii)   (iii)

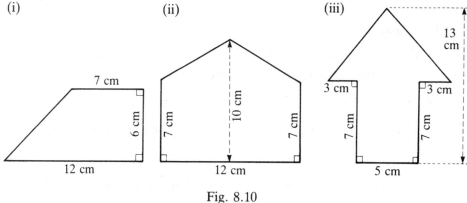

Fig. 8.10

6. The area of a triangle is 36 cm² and the base is 9 cm. Calculate the height.

7. The area of a triangle is 60 cm² and the height is 10 cm. Calculate the base.

8. The area of a trapezium is 42 cm². Its parallel sides are of length 9 cm and 5 cm. Calculate the height.

Circles

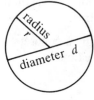

Fig. 8.11

Circumference = $\pi \times$ diameter

$$c = \pi d$$

or $\quad c = 2\pi r \qquad$ since $d = 2r$

The area, A, of a circle is given by: $\qquad A = \pi r^2$

The value of π cannot be given exactly. It is an irrational number. Correct to 3 d.p. it is 3.142. Your calculator will probably give you several more decimal places.

Sometimes it is convenient to use $\dfrac{22}{7}$ as an approximate value of π.

Example   Find the circumference and area of a circle of radius 9.3 cm.

$$c = 2\pi r = 2 \times \pi \times 9.3 = 58.4 \text{ cm to 3 s.f.}$$

$$A = \pi r^2 = \pi \times 9.3^2 = \pi \times 86.49 = 272 \text{ cm}^2 \text{ to 3 s.f.}$$

Notice that $\pi \times 9.3^2$ is not the same as $(\pi \times 9.3)^2$. You must square 9.3 first and then multiply the answer by π.

Exercise 8.4

1. Find, correct to 3 s.f., the circumference of a circle with:
   (i) diameter 5.8 cm   (ii) radius 7.6 cm   (iii) radius 48 cm.

2. Find, correct to 3 s.f., the area of each circle in question 1.

3. Without using a calculator, and taking π as $\dfrac{22}{7}$, find the circumference of a circle with:

   (i) diameter 14 cm   (ii) radius 21 cm   (iii) radius $3\dfrac{1}{2}$ m.

4. Without using a calculator, and taking π as $\dfrac{22}{7}$, find the area of a circle with:

   (i) radius 7 cm   (ii) radius 35 cm   (iii) radius $3\dfrac{1}{2}$ m.

5. Calculate the circumference of a car tyre of radius 28 cm, using $\dfrac{22}{7}$ for π. How far does the car move when the wheel makes 100 revolutions?

6. The diameter of a bicycle wheel is 64 cm. Calculate the circumference, using a decimal value for π. How many complete revolutions does the wheel make, in a distance of 2 kilometres?

7. A circular pond of diameter 6 m is surrounded by a concrete path of width 1 m. Calculate, correct to 3 s.f., the area of:
   (i) the pond
   (ii) the pond and path together
   (iii) the path alone.

8. A running track has two semi-circular ends of radius 28 m, and two straight sides. The perimeter of the track is 400 m. Using $\dfrac{22}{7}$ for π, calculate:
   (i) the length of each curve
   (ii) the length of each straight side.

9. The formula $c = \pi d$ can be rearranged as $d = \dfrac{c}{\pi}$. Use this to calculate, correct to 3 s.f., the diameter of a circle of circumference 62 cm.

10. Calculate the radius of a circle with a circumference of 85 cm.

11. The formula $A = \pi r^2$ can be rearranged as $r = \sqrt{\dfrac{A}{\pi}}$. Use this to calculate the radius of a circle of area 326 cm².

12. Calculate the diameter of a circle of area 2000 cm².

13. The circumference of a circle is 48 cm. Calculate:
    (i) its diameter   (ii) its area.

Length of arc

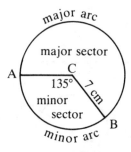

Fig. 8.12

In Fig. 8.12, the points A and B divide the circumference into two arcs, the *minor arc* and the *major arc*. We say that arc AB *subtends* an angle of 135° at the centre C.

Since 135° is $\dfrac{135}{360}$, or $\dfrac{3}{8}$, of a complete revolution,

the minor arc AB $= \dfrac{3}{8}$ of the circumference

$$= \dfrac{3}{8} \text{ of } \pi d \approx \dfrac{3}{8} \times \dfrac{22}{7} \times \dfrac{14}{1}$$

$$= \dfrac{33}{2} = 16\dfrac{1}{2} \text{ cm}$$

Area of sector

In Fig. 8.12, the inside of the circle is divided into two parts by the radii AC and BC. The smaller part is the *minor sector* and the larger part is the *major sector*.

Area of minor sector $= \dfrac{3}{8}$ of area of circle $= \dfrac{3}{8}$ of πr^2

$$\approx \frac{3}{8} \times \frac{22}{7} \times \frac{7}{1} \times \frac{7}{1} = \frac{231}{4} = 57\frac{3}{4}\ \text{cm}^2$$

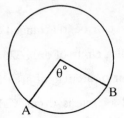

Fig. 8.13

In general, if arc AB subtends an angle $\theta°$ at the centre:

$$\text{length of arc} = \frac{\theta}{360} \times \pi d$$

$$\text{area of sector} = \frac{\theta}{360} \times \pi r^2$$

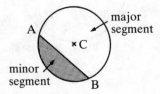

Fig. 8.14

In Fig. 8.14, the chord AB divides the inside of the circle into two parts called *segments*. The smaller one (shaded) is the *minor segment* and the larger is the *major segment*. From Fig. 8.15 it is clear that:

area of minor segment = area of minor sector − area of $\triangle$ABC

Fig. 8.15

Exercise 8.5

1. Using $\dfrac{22}{7}$ for π, find the length of each arc in Fig. 8.16.

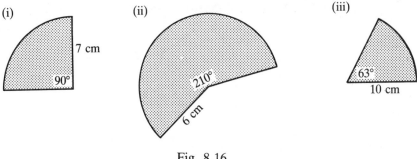

(i)

7 cm

90°

(ii)

210°

6 cm

(iii)

63°

10 cm

Fig. 8.16

2. Find the area of each of the sectors in Fig. 8.16.

For questions 3–6 sketch the sector, and then calculate, correct to 3 s.f.:
(i) the length of the arc   (ii) the area of the sector.

3. Radius 6 cm, angle at centre 40°

4. Radius 10 cm, angle at centre 70°

5. Radius 5.4 cm, angle at centre 45°

6. Radius 12.6 cm, angle at centre 136°

7. The minute hand of a watch is 1.5 cm long. Find the circular distance moved by the tip of the hand in 10 minutes.

8. Taking the earth as a sphere of diameter 12750 km, find the length of an arc of the equator which subtends an angle of 1° at the centre of the earth.

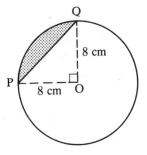

Q

8 cm

P

8 cm   O

Fig. 8.17

9. For Fig. 8.17, calculate the area of:
   (i) the sector bounded by arc PQ, radius PO and radius QO
   (ii) △POQ
   (iii) the shaded segment.

Square PQRS has its corners on a circle of radius 8 cm. Calculate the area inside the circle but outside the square.

Volumes and surface areas of prisms

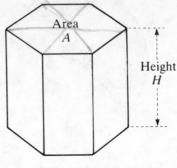

Fig. 8.18

A *prism* is a solid which has a uniform cross-section.

Its volume = (area of cross-section) × height

= (area of top or base) × height

$V = AH$

Hence $H = V \div A$ and $A = V \div H$

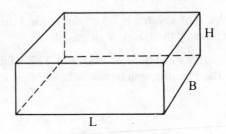

Fig. 8.19

A *cuboid* is a prism with a rectangular base.

Since base area = length × breadth,

volume = length × breadth × height

$V = LBH$

Hence $L = \dfrac{V}{BH}$, $B = \dfrac{V}{LH}$ and $H = \dfrac{V}{LB}$

A *cube* is a cuboid with all edges the same length, so $L = B = H$.

volume = (length of edge)³

$V = L^3$

Hence $L = \sqrt[3]{V}$

The volumes of liquids are usually measured in litres:

$$1 \text{ litre} = 1000 \text{ cm}^3$$

$$1 \text{ millilitre} = \frac{1}{1000} \text{ litre} = 1 \text{ cm}^3$$

The *capacity* of a bottle or other container is the volume of liquid it can hold.

Example 1 Find the volume and surface area of the cuboid in Fig. 8.19 if $L = 7.5 \text{ cm}$, $B = 6 \text{ cm}$ and $H = 5 \text{ cm}$.

$$V = 7.5 \times 6 \times 5 = 45.0 \times 5 = 225 \text{ cm}^3$$

Area of back and front $= 2 \times \text{front area}$
                        $= 2 \times 7.5 \times 5$
                        $= 75 \text{ cm}^2$

Area of top and bottom $= 2 \times \text{top area}$
                        $= 2 \times 7.5 \times 6$
                        $= 90 \text{ cm}^2$

Area of the two sides $= 2 \times \text{right side}$
                        $= 2 \times 6 \times 5$
                        $= 60 \text{ cm}^2$

Total surface area $= 75 + 90 + 60$
                        $= 225 \text{ cm}^2$

Example 2 A glass fish tank is a cuboid of length 50 cm, breadth 32 cm and height 28 cm. 24 litres of water are poured into the tank. Find the depth of the water.

Volume of water $= 24 \times 1000 = 24\,000 \text{ cm}^3$

For a cuboid $V = LBH$ so $H = \dfrac{V}{LB}$

$$H = \frac{24\,000}{50 \times 32} = \frac{24\,000}{1600} = 15$$

The depth of water is 15 cm.

Example 3 Calculate the edge length of a cube of volume 600 cm³.
For a cube $V = L^3$ so $L = \sqrt[3]{V} = \sqrt[3]{600} = 8.4$ to 2 s.f.
The edge length is 8.4 cm to 2 s.f.

Exercise 8.6

Without using any calculating aids, find the volumes of the following solids:

1. A cuboid measuring 8 cm by 6 cm by 5 cm.

2. A cuboid measuring 10 cm by 7 cm by 4 cm.

3. A cube of edge 4 cm.

4. A cuboid with a square base of side 9 cm and a height of 10 cm.

Calculate the total surface area of each of the following:

5. A cube of edge 5 cm.

6. A cuboid measuring 5 cm by 4 cm by 3 cm.

7. A cuboid measuring 8 cm by 6 cm by 4.5 cm.

8. A cube of edge 4 cm is painted green and then cut into cubes of edge 1 cm. How many of them have:
   (i) 3 green faces   (ii) 2 green faces   (iii) 1 green face
   (iv) no green faces?

9. A rectangular tank measures 60 cm by 40 cm by 30 cm. How many litres of water can it hold?

10. A room measures 500 cm by 480 cm by 350 cm. Find its volume in cubic metres.

11. When some cubes of edge 2 cm were lowered into a tank full of water, 136 ml of water overflowed. How many cubes were used?

12. (i) Find the total surface area of a cube of edge 3 cm.
    (ii) The total surface area of a cube is 96 cm². Find the length of an edge.

13. How many cubes of edge 4 cm can be made from 2000 cm³ of metal, and how much metal will be left over?

Calculate the volume of each of the following, giving your answers correct to 3 s.f.:

14. A cube of edge 7.3 cm.

15. A cuboid measuring 12.8 cm by 9.6 cm by 8.2 cm.

16. A cuboid measuring 23 cm by 14 cm by 8 cm.

17. A cuboid with a square base of side 5.5 cm, and height 16 cm.

18. Calculate, correct to 3 s.f., the total surface area of a cube of edge 9.4 cm.

19. Calculate, correct to 3 s.f., the total surface area of the cuboid in question 17.

Exercise 8.7

1. A prism has a triangular base of area 56 cm² and a volume of 364 cm³. Calculate its height.

2. A cuboid has a length of 8 cm, a breadth of 6 cm and a volume of 216 cm³. Calculate its height.

3. 33 litres of water are poured into a rectangular tank 55 cm long and 40 cm wide. Calculate the depth of the water.

4. A metal cuboid measures 55 cm by 40 cm by 30 cm. Calculate its mass to the nearest 10 kg, if each cubic centimetre has a mass of 8.4 g.

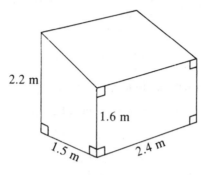

Fig. 8.20

5. Figure 8.20 shows a garden shed. Calculate:
   (i) the area of an end wall
   (ii) the volume.

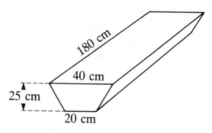

Fig. 8.21

6. Figure 8.21 shows a water trough. Each end is a trapezium with parallel sides of 40 cm and 20 cm which are 25 cm apart. The length is 180 cm. Calculate the capacity of the trough in litres.

7. (i) A building has a rectangular flat roof measuring 10 m by 8 m. State its area.

   (ii) During a storm, 1.5 cm of rain fell on the roof. Calculate the volume of rain in m^3.

   (iii) The rain from the roof ran into a rectangular water tank of base 1.6 m by 1.5 m. Calculate the rise in the level of water in the tank.

8. Water flows into a rectangular tank of length 90 cm and width 50 cm, at the rate of 240 ml per second. Calculate the rate at which the water level in the tank rises, in cm/min.

Cylinders

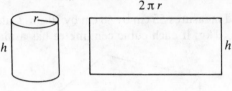

Fig. 8.22

A cylinder is a prism with a circle as its base.

$$\text{Volume} = \text{base area} \times \text{height}$$
$$= \pi r^2 \times h = \pi r^2 h$$

A piece of paper covering the curved surface can be opened out to give a rectangle of height h and length $2\pi r$. ($2\pi r$ is the circumference of the circular top.)

Thus area of curved surface $= 2\pi r \times h = 2\pi r h$.

Example   Find the volume and total surface area of a cylinder of radius 8.3 cm and height 15.6 cm.

Volume $= \pi r^2 h = \pi \times 8.3^2 \times 15.6 = 3380 \text{ cm}^2$ to 3 s.f.

Area of curved surface $= 2\pi r h$
$$= 2 \times \pi \times 8.3 \times 15.6 \approx 813.55 \text{ cm}^2$$

Area of two flat ends $= 2\pi r^2$
$$= 2 \times \pi \times 8.3^2 \approx 432.85 \text{ cm}^2$$

Total surface area $\approx 813.55 + 432.85 = 1246.40 \text{ cm}^2$
$$= 1250 \text{ cm}^2 \text{ to 3 s.f.}$$

Exercise 8.8

1. Using $\frac{22}{7}$ for π, calculate the volume of a cylinder with:

(i) radius 7 cm, height 10 cm   (ii) radius 4 cm, height $3\frac{1}{2}$ cm.

2. Using a decimal value for π, calculate, correct to 3 s.f., the volume of a cylinder with:

(i) radius 9 cm, height 25 cm   (ii) radius 6.7 cm, height 17.3 cm.

3. Using $\frac{22}{7}$ for π, calculate the curved surface area of each cylinder in question 1.

4. Using a decimal value for π, calculate the total surface area of the cylinder in question 2(ii).

5. A soup tin has a diameter of 7 cm and a height of 12 cm. A label is wrapped round the curved surface, with an overlap of 1 cm. State the dimensions of the label.

6. An open cylindrical can (no top) has a radius of 21 cm and a height of 50 cm. Calculate the total surface area (base and curved surface) of the outside of the can.

7. A cylindrical container of internal diameter 10 cm and height 14 cm is full of salt. Calculate the volume of the salt using $\dfrac{22}{7}$ for π.

If 10 cm^3 of salt has a mass of 12 g, find the mass of salt in the container.

8. A cylindrical jug has an internal diameter of 30 cm and an internal height of 36 cm. Show that its capacity is 8100 π cm^3. Find a similar expression for the capacity of a cylindrical glass of internal diameter 6 cm and internal height 10 cm. How many such glasses can be filled from the jug? (You do not need to use a value for π.)

9. A closed cylindrical water tank has a height of 1.4 m and a diameter of 40 cm. Calculate:

(i) the total surface area in square metres

(ii) the capacity in litres.

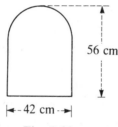

56 cm

|←- 42 cm -→|

Fig. 8.23

10. Figure 8.23 shows the upper surface of a metal plate of thickness 4 mm. Its shape can be divided into a rectangle and a semi-circle. Calculate:

(i) the area of this surface   (ii) the volume of the plate.

Use $\dfrac{22}{7}$ for π.

11. Water is flowing at 3 m/s in a pipe of inside diameter 4 cm. How many litres pass a certain point in 1 minute?

12. For a cylinder, volume = base area × height.
Hence height = volume ÷ base area.
A cylinder has a radius of 7 cm and a volume of 1386 cm^3. Using $\dfrac{22}{7}$ for π, find:

(i) the base area   (ii) the height.

13. 2.31 litres of water are poured into a cylinder of radius 7 cm. Calculate the depth of water in the cylinder.

14. The formula $V = \pi r^2 h$ can be rearranged as $r^2 = \dfrac{V}{\pi h}$. Use this to calculate the radius of a cylinder with a volume of 792 cm³ and a height of 7 cm. (Use $\dfrac{22}{7}$ for π.)

15. An upright cylindrical tank, with an open top, has a radius of 70 cm and a height of 120 cm, and is half full of water. Calculate:

   (i) the internal surface area of the tank

   (ii) the volume of water in the tank.

   112 heavy bricks, each measuring 15 cm by 11 cm by 5 cm, are carefully placed inside the tank so that all the bricks are completely immersed in the water. Calculate the rise in the level of the water.

   (Use $\dfrac{22}{7}$ for π.)   (W)

16. During a storm, the depth of the rainfall was 15.4 mm. The rain which fell on a horizontal roof measuring 7.5 m by 3.6 m was collected in a cylindrical tank of radius 35 cm, which was empty before the storm began. Calculate:

   (a) the area, in cm², of the roof

   (b) the volume, in cm³, of the rain which fell on the roof.

   Taking π as $\dfrac{22}{7}$, find:

   (c) the area, in cm², of the cross-section of the tank

   (d) the height, in cm, of the rain water in the tank.

   Given that a watering can holds five litres, how many times could it be filled completely from the rain water?   (L)

Pyramids

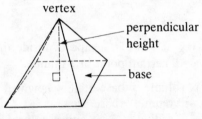

vertex

perpendicular height

base

Fig. 8.24

For a pyramid on a base of any shape:

$$\text{volume} = \frac{1}{3} \times \text{base area} \times \text{perpendicular height}$$

$$V = \frac{1}{3} Ah$$

The perpendicular height is the perpendicular distance of the vertex from the base.

Example Find the volume of a pyramid with a rectangular base measuring 6 cm by 8 cm, and a height of 10 cm.

$$\text{Volume} = \frac{1}{3} \times 6 \times 8 \times 10 = 2 \times 80 = 160 \text{ cm}^3$$

Cones

A cone is a pyramid on a circular base.

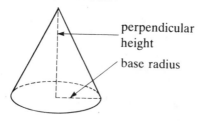

perpendicular height

base radius

Fig. 8.25

$$\text{Volume} = \frac{1}{3} \times \text{base area} \times \text{perpendicular height} = \frac{1}{3} \times \pi r^2 \times h$$

$$V = \frac{1}{3} \pi r^2 h$$

Example Find the volume of a cone of base radius 6 cm and height 14 cm.

$$\text{Volume} = \frac{1}{3} \pi r^2 h \approx \frac{1}{3} \times \frac{22}{7} \times 36 \times 14 = 528 \text{ cm}^3$$

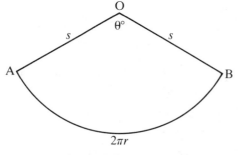

Fig. 8.26

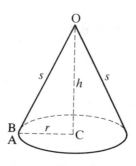

Fig. 8.27

If a piece of paper having the shape of the sector AOB in Fig. 8.26 is bent so that the edge OA joins the edge OB and arc AB becomes a circle, then the hollow cone of Fig. 8.27 is formed. The radius, s, of the sector becomes the slant height of the cone.

Example   A sector of radius 24 cm and 150° is bent to form a hollow cone. Find the radius, vertical height and curved surface area of the cone.

As circumference of base of cone = length of arc of sector,

$$2\pi r = \frac{150}{360} \times 2\pi \times 24$$

from which $r = 10$ cm.

Applying Pythagoras' theorem (page 258) to triangle OAC in Fig. 8.27,

$$r^2 + h^2 = s^2, \qquad 10^2 + h^2 = 24^2$$

from which $h = 21.8$ cm to 3 s.f.

Curved surface area of cone = area of sector

$$= \frac{150}{360} \times \pi \times 24^2 = 754 \text{ cm}^2 \text{ to 3 s.f.}$$

In general, $r = \dfrac{\theta}{360} \times s$

$$r^2 + h^2 = s^2$$

and   curved surface area of a cone $= \pi r s$.

Spheres

Volume of a sphere $= \dfrac{4}{3} \pi r^3$

Surface area of a sphere $= 4\pi r^2$

Example   Find the volume and surface area of a sphere of radius 2.4 cm.

$$\text{Volume} = \frac{4}{3} \times \pi \times 2.4^3 = 57.9 \text{ cm}^3 \text{ to 3 s.f.}$$

$$\text{Surface area} = 4 \times \pi \times 2.4^2 = 72.4 \text{ cm}^2 \text{ to 3 s.f.}$$

Exercise 8.9

Pyramids

1. Calculate the volume of a pyramid with a rectangular base measuring 9 cm by 7 cm, and a height of 10 cm.

2. Calculate the volume of a pyramid with a rectangular base measuring 7.5 cm by 5.5 cm, and a height of 8 cm.

3. Calculate the volume of a pyramid with a square base of side 15 cm and a height of 24 cm.

4. The Great Pyramid of Egypt has a square base of side 230 m and a height of 104 m. Calculate its volume, giving your answer in standard form correct to 3 s.f.

Cones

5. Using $\frac{22}{7}$ for π, calculate the volume of a cone of base radius 10 cm and height 21 cm.

6. Using $\frac{22}{7}$ for π, calculate the volume of a cone of base radius 14 cm and height 6 cm.

7. Using a decimal value for π, calculate, correct to 3 s.f., the volume of a cone with a base radius of 15 cm and a height of 40 cm.

8. Using a decimal value for π, calculate, correct to 3 s.f., the volume of a cone with a base radius of 6.2 cm and a height of 19.3 cm.

9. A wine glass is conical in shape. Its largest diameter is 6 cm and its depth is 8 cm. How many millilitres of wine can it hold?

10. A sector of radius 30 cm and angle 216° is cut from a sheet of tin and bent to form a hollow cone. Calculate the radius, vertical height and curved surface area of the cone.

11. A clown's hat of radius 10 cm and vertical height 30 cm is to be made from a sector of thin cardboard. Find the radius and angle of the sector needed.

12. A cone of height 8 cm is cut from a wooden cone of height 12 cm and base radius 9 cm (Fig. 8.28). The shape remaining is called a frustum. Calculate the area of the curved surface of: (i) the original cone (ii) the small cone (iii) the frustum.

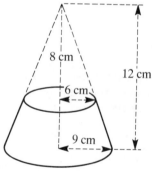

Fig. 8.28

13. A lampshade has the shape of a frustum of a cone. The diameter of the top is 10 cm and of the bottom is 20 cm. The height is 12 cm. Use the method of question 12 to calculate, correct to 2 s.f., the area of material used.

14. The diameter of the top of a bucket is 24 cm and the diameter of the base is 16 cm. The height is 20 cm. Calculate the capacity in litres, correct to 2 s.f.

Spheres

15. Calculate the volume of a sphere of radius:

    (i) 5 cm   (ii) 7.4 cm   (iii) 13.5 cm.

16. Calculate the surface area of each of the spheres in question 15.

17. Calculate the volume of metal needed to make 1000 ball bearings of diameter 4 mm.

18. The formula $V = \frac{4}{3}\pi r^3$ can be rearranged as $r = \sqrt[3]{\frac{3V}{4\pi}}$. Use this to calculate, correct to 2 s.f., the radius of a sphere of volume 660 cm³.

Two shapes

19. A building has the shape of a cylinder with a conical roof, as shown in Fig. 8.29. Calculate the volume of the building.

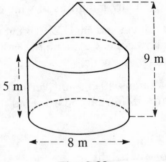

9 m

5 m

8 m

Fig. 8.29

20. A boiler consists of a cylinder with a hemisphere at each end. The diameter of the boiler is 140 cm and its total length is 320 cm. Calculate, correct to the nearest litre, the capacity of the boiler.

ice
cream
cornet

Fig. 8.30

21. Figure 8.30 shows an ice cream cornet. The cone has a diameter of 6 cm and a height of 11 cm. Calculate the volume of the ice cream, assuming that the cone is completely filled and the part above it is a hemisphere.

22. A metal sphere of radius 6 cm is melted down and cast into a cylinder of radius 6 cm. Show that the volume of metal is $288\,\pi\,\text{cm}^3$. Hence calculate the height of the cylinder. You do not need to use a value for π.

9 Household and personal finance

Exercise 9.1

1. A shop advertises "10 months interest free credit and a deposit of only 20% of the price".

   The price of a computer is £185. What is the deposit? The balance is paid in ten monthly instalments. What is each instalment?

2. A camera is advertised at £129.99 cash or a deposit of £12.99 and 18 monthly payments of £8. Calculate the difference between the two methods of payment.

   Express this difference as a percentage of the cash price.

3. The cash price of a bicycle is £75. If you make an initial payment of 20%, how much should you pay each month to clear the balance in twelve months, if no interest is charged?

   If you were asked to pay £5.60 per month, how much extra would you pay over the twelve months? Express this extra as a percentage of the balance.

4. A man estimates the value of his house at £32 000 and its contents at £13 000. Find his annual insurance premium if the rate for the building is £1.40 per £1000 and the rate for the contents is £2.90 per £1000.

5. A man is paid at the basic rate of £4.20 per hour. If he works more than 38 hours in a week, each extra hour is paid for at $1\frac{1}{2}$ times the basic rate. One week he worked for 43 hours. What was his wage for that week?

   Another week his wage was £178.50. How many hours overtime did he do?

6. A job is advertised at a starting salary of £5280 p.a. rising by annual increments of £320 to a maximum of £9120 p.a.

   (i) Find the difference between the starting and maximum salaries.

   (ii) How many increments are there?

   (iii) What is the salary for the fifth year?

7. For electricity in a certain area in 1985 there was a quarterly charge of £7.02 and a charge of 5.07p for each unit used.

   The readings on the Lumens' meter at the beginning and end of a quarter were 32 928 and 33 885. Calculate the bill for that quarter.

   Their bill for the next quarter was £34.40. How many units did they use?

8. For a domestic telephone in 1985, there was a quarterly charge of £15.15 and a charge of 4.7p for each dialled unit. VAT at 15% was added to the total. How much was paid by the Chatter family who used 563 dialled units?

9. Mr Green's wage was £120 per week.

(i) Find his total income for a year.

(ii) The first £3700 of his income was free of tax. Find his taxable income for the year.

(iii) He paid tax at the rate of 30%. Find his tax for the year.

(iv) Find his tax per week, correct to the nearest penny.

(v) Find his pay per week after deduction of tax.

10. A householder takes out a mortgage for £16 000 on 1 January 1985. Interest at the rate of 12% p.a. on the amount owing at the beginning of a year is added to the account on 31 December of that year. The householder pays £240 on the 28th of each month. How much is owing on 1 January 1986 and on 1 January 1987?

11. A gas meter registers the volume of gas, in hundreds of cubic feet, which has passed through it. In September 1985 the reading was 6398 and in December 1985 it was 6534. The charge for gas was 35p per therm. A therm is a unit of heat and for that gas the number of therms = hundreds of cubic feet × 1.042. There was also a standing charge of £9.90. Calculate the total bill for the quarter.

The bill for the next quarter was £43.15. How many therms were used and what was the reading on the meter at the end of that quarter.

12. A finance company calculates its hire purchase terms as follows: The deposit is 10% of the price. Interest at the rate of 16% p.a. is added to the balance (price − deposit). The resulting sum (balance + interest) is divided into an equal number of monthly payments.

Show that for a washing machine, price £250, the HP terms for payment over 3 years are a deposit of £25 and monthly payments of £9.25.

Find the HP terms for payments over 2 years for furniture priced at £440.

13. A store offers a television set by three methods:

A: basic price of £350
B: a deposit of £62 and 12 monthly payments of £27.50
C: hire of the set for £12.50 per month rental.

(i) Find the total paid by method B.

(ii) How much more is paid by method B than by method A? Express this as a percentage of the basic price.

(iii) By method C, how long is it before the rental payments are equal to the basic price? (*contd over page*)

Another store offers the same set for rental as follows:

Deposit £33; nothing to pay for 3 months; then £12.75 per month. For one year, is this cheaper than method C? Is it cheaper for two years?

14. The 29th Issue National Savings certificates were on sale from 15 October 1984. They cost £25 each and grow in value as follows:

At the end of the 1st year by £1.50 to £26.50.

During the 2nd year by 46p for each completed 3 months to £28.34.

During the 3rd year by 56p for each completed 3 months to £30.58.

During the 4th year by 69p for each completed 3 months to £33.34.

During the 5th year by 85p for each completed 3 months to £36.74.

(i) Olive bought 4 certificates on 19 November 1984. What will they be worth on 20 November 1988?

(ii) Derek bought 1 certificate on 16 January 1985. What will it be worth on 21 April 1987?

(iii) Ann bought 12 certificates on 27 February 1985. What will they be worth on 8 October 1989?

(iv) If a certificate is held for 5 complete years, what is the percentage increase in value?

10 Basic processes of algebra

Addition and subtraction

$3 + 3 + 3 + 3 + 3 = 5 \times 3 = 15$

Similarly: $a + a + a + a + a = 5 \times a$ which is written as $5a$

$2a + 4a = (a + a) + (a + a + a + a)$

$\qquad = a + a + a + a + a + a = 6a \qquad (2 + 4 = 6)$

$9b + 7b = 16b$

$7c - 3c = (c + c + c + c + c + c + c) - (c + c + c)$

$\qquad = c + c + c + c + c + c + c - c - c - c$

$\qquad = c + c + c + c = 4c \qquad (7 - 3 = 4)$

$10d - 4d = 6d$

$6k - 6k = 0$

$2f + 6f + 5g + g = (2f + 6f) + (5g + g) = 8f + 6g$

$8h + 3n - 2h = (8h - 2h) + 3n = 6h + 3n$

Only like terms can be added or subtracted. We cannot obtain a single term by adding $3a$ and $2b$.

Multiplication and division

$2a \times 3 = 2 \times a \times 3 = 2 \times 3 \times a = 6 \times a = 6a$

$4b \times 7c = 4 \times b \times 7 \times c = 4 \times 7 \times b \times c = 28 \times b \times c = 28bc$

$5d \times 2d = 5 \times d \times 2 \times d = 5 \times 2 \times d \times d = 10 \times d^2 = 10d^2$

$12f \div 3 = \dfrac{12f}{3} = \dfrac{4 \times 3 \times f}{3} = 4f$

$15g \div 5g = \dfrac{15g}{5g} = \dfrac{3 \times 5 \times g}{5 \times g} = 3$

$7h \div 7h = \dfrac{7 \times h}{7 \times h} = 1$

$\dfrac{6y}{12} = \dfrac{6 \times y}{6 \times 2} = \dfrac{y}{2}$ or $\dfrac{1}{2}y$

$$\frac{4}{8t} = \frac{4}{4 \times 2 \times t} = \frac{1}{2t}$$

$$20np^2 \div 4np = \frac{4 \times 5 \times n \times p \times p}{4 \times n \times p} = 5p$$

Multiplication and division using indices

a^5 means $a \times a \times a \times a \times a$

$a^5 \times a^2 = (a \times a \times a \times a \times a) \times (a \times a)$

$\qquad = a \times a \times a \times a \times a \times a \times a = a^7 \qquad (5 + 2 = 7)$

In general: $a^m \times a^n = a^{m+n}$

That is, when multiplying powers of the same letter, add the indices.

We cannot multiply powers of different letters in this way. All we can do with $a^5 \times b^2$ is to write it as $a^5 b^2$.

$$a^5 \div a^2 = \frac{a \times a \times a \times a \times a}{a \times a} = a \times a \times a = a^3 \qquad (5 - 2 = 3)$$

In general: $a^m \div a^n = a^{m-n}$

That is, when dividing powers of the same letter, subtract the indices.

$$(a^5)^2 = a^5 \times a^5 = a^{5+5} = a^{10} \qquad (5 \times 2 = 10)$$

In general: $(a^m)^n = a^{mn}$

That is, when raising a power of a letter to another power, multiply the indices.

$3b^3$ means $3 \times b^3$, or $3 \times b \times b \times b$

and not $3b \times 3b \times 3b$ which is $27b^3$.

$7b^7 \times 3b^3 = 7 \times b^7 \times 3 \times b^3 = 7 \times 3 \times b^7 \times b^3 = 21b^{10}$

$$10c^{10} \div 2c^2 = \frac{10 \times c^{10}}{2 \times c^2} = 5 \times c^8 = 5c^8$$

$3a^5 b \times 2a^3 b^2 = 3 \times 2 \times a^5 \times a^3 \times b \times b^2 = 6a^8 b^3$

$$12c^4 d^3 \div 4cd^2 = \frac{12 \times c^4 \times d^3}{4 \times c \times d^2} = 3 \times c^3 \times d = 3c^3 d$$

Exercise 10.1

Simplify where possible:

1. $6a + 2a$	**2.** $6b - 2b$	**3.** $6c \times 2c$	**4.** $6d \div 2d$
5. $e^6 \times e^2$	**6.** $f^6 \div f^2$	**7.** $7f \times 3g$	**8.** $7h + 3k$
9. $n^7 \times k^3$	**10.** $p^7 \times p^3$	**11.** $3a - 3a$	**12.** $3b \div 3b$

13. $c^6 \div c$ **14.** $d^4 \div d^4$ **15.** $(e^4)^2$ **16.** $(f^3)^5$

17. $12g + 4g$ **18.** $12h \div 4h$ **19.** $12k \div 4$ **20.** $12n + 4$

21. $(3p^2)^3$ **22.** $18q^6 \div 9q^2$ **23.** $5t \times 2v$ **24.** $w^2x \div wx$

25. $3a^2 \times 5b^4$ **26.** $14c^5 \div 2c$ **27.** $4de^3 \times 2e^3$ **28.** $15f^5g^2 \div 3f^3g^2$

29. $3a + 4a - a$ **30.** $5b - b - 4b$ **31.** $3c + 5d - 2c$

32. $4f - g + 3g$ **33.** $5h^2 - 2h^2$ **34.** $k^2 + 3k^2 - 2k^2$

35. $5m^2 \div 5m^2$ **36.** $7n^3 - 7n^3$ **37.** $3a + 3 + 2a$

38. $5b + 6 - 4b - 5$ **39.** $3 \times 7c$ **40.** $2d \times 3 \times 4e$

41. $5f \div 15$ **42.** $3g \div 6h$ **43.** $9k \div 15k$

44. $3m \div mp$

45. Find the product of $3y$, y and $2y^2$.

46. Find the sum of $2n$, n and $5n$.

Directed numbers

On the Celsius temperature scale, the freezing point of water is 0°C. A temperature of 3° above zero can be written as +3°C and a temperature of 3° below zero can be written as −3°C.

Numbers which have a sign attached to them are called *directed numbers*. +3 is a *positive number* and −3 is a *negative number*.

There are many other situations where directed numbers are useful. A height of 100 m above sea level can be written as +100 m and a depth of 50 m below sea level as −50 m. If +20 km means 20 km north of a certain town, then −20 km means 20 km south of that town.

Addition using directed numbers

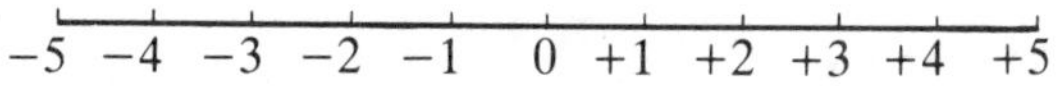

$$-5 \quad -4 \quad -3 \quad -2 \quad -1 \quad 0 \quad +1 \quad +2 \quad +3 \quad +4 \quad +5$$

Fig. 10.1

Consider Fig. 10.1. Start at zero, move 7 units to the right and then 2 more to the right. The result is 9 units to the right.

We can write this as :

$$(+7) + (+2) = (+9). \tag{A}$$

Now start at zero, move 7 units to the right and then 2 units to the left. The result is 5 units to the right.

We can write this as :

$$(+7) + (-2) = (+5) \tag{B}$$

Using Fig. 10.1, check the following:

$$(-7) + (+2) = (-5) \qquad\qquad (C)$$

and

$$(-7) + (-2) = (-9) \qquad\qquad (D)$$

Subtraction using directed numbers

From $14 = 8 + 6$ we have $14 - 8 = 6$. Similarly:

from $(+9) = (+7) + (+2)$ we have $(+9) - (+7) = (+2)$ \qquad (E)

from $(+5) = (+7) + (-2)$ we have $(+5) - (+7) = (-2)$ \qquad (F)

from $(-5) = (-7) + (+2)$ we have $(-5) - (-7) = (+2)$ \qquad (G)

from $(-9) = (-7) + (-2)$ we have $(-9) - (-7) = (-2)$ \qquad (H)

Look again at Fig. 10.1. To go from -5 to $+2$, you must move 7 units to the right. In other words you must add 7. That is, $(-5) + (+7) = (+2)$. Now compare this with line G above. Both answers are the same. This shows that $-(-7)$ must mean the same as $+(+7)$. So $-(-7)$ can be replaced by $+(+7)$ in both lines G and H.

The brackets are often left out, and for positive numbers we usually leave out the $+$ sign too. The lines E, F, G and H become:

$$9 - 7 = 2$$

$$5 - 7 = -2$$

$$-5 + 7 = 2$$

$$-9 + 7 = -2$$

and lines A and D become:

$$7 + 2 = 9$$

$$-7 - 2 = -9$$

We have the following working rules for such statements as those above:

1. When the numbers have opposite signs, take the smaller from the larger and give the answer the sign of the larger.
2. When both numbers have the same sign, add them and give this sign to the answer.

Multiplication using directed numbers

A man deposits £5 each week in a savings bank account. Three weeks from now his account will have £15 more than it has now. We can write this as:

$$(+5) \times (+3) = (+15)$$

Three weeks ago his account had £15 less than it has now. We used
+3 for three weeks from now, so we use −3 for three weeks ago. We
can write:
$$(+5) \times (-3) = (-15)$$
Another man withdraws £5 each week from a savings account. We write
this as (−5). Three weeks from now his account will have £15 less than
it has now. We write this as:
$$(-5) \times (+3) = (-15)$$
Three weeks ago his account had £15 more than it has now. We can
write:
$$(-5) \times (-3) = (+15).$$
The above four statements illustrate the rules for multiplying two
directed numbers:

1. If the signs are the same, the product is positive.

2. If the signs are opposite, the product is negative.

Examples $(+7) \times (-2) = (-14)$
$(-7) \times (-2) = (+14)$
$(-7) \times (+2) = (-14)$
or, leaving out the + signs and some of the brackets:
$7 \times (-2) = -14$
$(-7) \times (-2) = 14$
$(-7) \times 2 = -14$

Division using directed numbers

From the four multiplication statements above, we obtain the following
division statements:

$$\frac{(+15)}{(+3)} = (+5), \quad \frac{(-15)}{(-3)} = (+5), \quad \frac{(-15)}{(+3)} = (-5), \quad \frac{(+15)}{(-3)} = (-5)$$

As for multiplication, like signs give a positive answer and unlike signs
give a negative answer.

Examples $12 \div (-3) = -4, \quad (-12) \div (-3) = 4, \quad (-12) \div 3 = -4$

Exercise 10.2

State the value of:

1. $5 + (-3), \ (-5) + (-3), \ (-5) + 3, \ 5 + 3$

2. $(-1) + (-6), \ 1 + 6, \ (-1) + 6, \ 1 + (-6)$

3. $(-8) - 3, \ (-8) - (-3), \ 8 - (-3), \ 8 - 3$

4. $4 - (-9), \ (-4) - 9, \ (-4) - (-9), \ 4 - 9$

5. $(-2) + 3 + (-4)$, $2 + (-3) + (-4)$, $(-2) + (-3) + (-4)$

6. $(-1) + (-5) + (-6)$, $1 + (-5) + 6$, $1 + 5 + (-6)$

7. $3 - (-2) + (-1)$, $(-3) - 2 - (-1)$, $(-3) - (-2) - (-1)$

8. $-4 + 7$, $-4 - 7$, $4 - 7$

9. $5 - 2$, $-5 + 2$, $-5 - 2$

10. $-3 - 1 + 6$, $3 - 1 - 6$, $3 + 1 - 6$

11. $(-4) \times 2$, $4 \times (-2)$, $(-4) \times (-2)$

12. $(-15) \div 3$, $15 \div (-3)$, $(-15) \div (-3)$

13. $(-2)^2$, $(-2)^3$, $(-2)^4$

14. $(-2) \times (-4) \times (-1)$, $2 \times (-4) \times (-1)$, $(-2) \times 4 \times (-1)$

15. $3 \times (-5) \times 2$, $(-3) \times (-5) \times 2$, $(-3) \times (-5) \times (-2)$

16. $(-2) \times (-3) \div (-6)$, $2 \times (-3) \div 6$

17. $(-3)^2 \times (-1)^3$, $(-3)^3 \times (-1)^2$

18. $(-a) + (-a)$, $a + (-a)$, $(-a) - (-a)$, $(-a) - a$

19. $(-b) \times (-b)$, $b \times (-b)$, $(-b) \div (-b)$, $b \div (-b)$

20. $(-c) \times (-c) \times (-c)$, $(-2) \times (-3) \times (-d)$, $4 \times (-3) + f$

21. $(-3g)^2$, $(-2h)^3$, $4k \times (-k)$, $(-5m) \times (-3m)$

Evaluating expressions

If $a = -3$, $b = -5$ and $c = 4$, then

$ab = (-3) \times (-5) = 15$, $a + b = (-3) + (-5) = -8$

$bc = (-5) \times 4 = -20$, $a - b = (-3) - (-5) = -3 + 5 = 2$

$b^2 = (-5)^2 = 25$, $abc = (-3) \times (-5) \times 4 = 60$

$(2a)^2 = (-6)^2 = 36$ but $2a^2 = 2 \times (-3)^2 = 2 \times 9 = 18$

$\frac{1}{2}(bc)^2 = \frac{1}{2}(-20)^2 = \frac{1}{2}(400) = 200$ but $(\frac{1}{2}bc)^2 = (-10)^2 = 100$

$a^2 - 6a - 7 = (-3)^2 - 6(-3) - 7 = 9 + 18 - 7 = 20$

$9 - a^3 = 9 - (-3)^3 = 9 - (-27) = 9 + 27 = 36$

$(-a)^2 = (-(-3))^2 = (3)^2 = 9$ but $-a^2 = -(-3)^2 = -(9) = -9$

If $f(x) = 5x - x^2$, then $f(3)$ means the value of $f(x)$ obtained by replacing x by 3.

Hence $f(3) = 5 \times 3 - 3^2 = 15 - 9 = 6$.

Also $f(-3) = 5(-3) - (-3)^2 = -15 - 9 = -24$

Exercise 10.3

1. If $x = 3$, $y = 2$ and $z = 1$, find the values of:

$2x + y$, $3y - 4z$, $5yz$, $x^2 - y^2$, $\dfrac{x}{3}$,

xyz, $x - y - z$, $2x - 3y$, $\dfrac{x}{y} - z$, $\dfrac{x^2}{y^2}$.

2. If $p = 2$, $q = 3$ and $r = 4$, find the values of:

pr, q^2, $2q^2$, $(2q)^2$, $5pq$, $5p - 2q$, $p + q - r$, $\dfrac{p^2}{r}$, $(1 + q)^2$, $(r - q)^2$.

3. If $k = -3$, $n = 6$, $p = -1$ and $t = 0$, find the values of:
$n + p$, np, $n - p$, $p - n$, kp, kt, $k + t$,
$n \div k$, $k \div p$, $t \div n$, $2k^2$, $(2k)^2$, $3p$, $6 - 3p$,
$(t + p)^2$, p^3, $(kp)^3$, $(k + p)^3$, $2n + 3k$, $(n + 5p)^2$.

4. If $x = 6$, $y = -2$ and $z = \dfrac{1}{3}$, find the values of:

xy, xz, $x - y$, $y - x$, $x^2 - y^2$, $(x - y)^2$,

$x \div y$, $x^2 \div y^2$, xyz, $(3y)^2$, $3y^2$, $(2y)^3$, $2y^3$, $\dfrac{x^2}{z}$,

$\left(\dfrac{x}{z}\right)^2$, $xz + y$, $x + 3y$, $z + \dfrac{y}{x}$, $\dfrac{x + y}{z}$, $\sqrt{x + y}$.

5. (i) State the values of x^2 for $x = 5, 1, 0, -1, -5$.
(ii) State the values of $-y^2$ for $y = 5, 1, 0, -1, -5$.
(iii) State the values of n^3 for $n = 2, 1, 0, -1, -2$.
(iv) State the values of $-n^3$ for $n = 2, 1, 0, -1, -2$.

6. State the values of $\dfrac{6}{x}$ for $x = 6, 4, 2, -2, -4, -6$.

7. State the values of $2x - 3$ for $x = 2, 1, 0, -1$.

8. State the values of $5 - 3y$ for $y = 0, 1, 2, 3$.

9. State the values of $10 - x^2$ for $x = 4, 2, 0, -2, -4$.

10. State the values of $(2 - y)^2$ for $y = 5, 1, 0, -1, -5$.

11. State the values of $5 - n^3$ for $n = 2, 1, 0, -1, -2$.

12. State the values of $\dfrac{6}{x - 2}$ for $x = 5, 4, 3, 1, 0, -1$.

13. Find the values of $x^2 - x$ for $x = 0, 1, 2, 3, 4, -1$.

14. Find the values of $y^2 - 2y - 3$ for $y = 4, 2, 0, -2, -4$.

15. If $y = 3 - x - x^2$, find the values of y when $x = 2, 1, 0, -1, -2$.

16. If $y = 5 + 2x - 3x^2$, find the values of y when $x = 2, 1, 0, -1, -2$.

17. If $f(x) = x^3 + 2x$, find the values of $f(3), f(1), f(-1), f(-3)$.

18. If $g(x) = 3x^2 - x^3$, find the values of $g(2), g(1), g(-1), g(-2)$.

19. If $h(x) = 6 - x^2 + x^3$, find the values of $h(2), h(1), h(0),$ $h(-1), h(-2)$.

20. If $p(x) = \dfrac{6}{2-x}$, find the values of $p(0), p(1), p(3), p(4),$ $p(-1), p(-2)$.

21. If $x = \dfrac{1}{2}$, find the values of $\dfrac{1}{x}, \dfrac{1}{x^2}, \dfrac{1}{x^3}$.

22. If $x = -\dfrac{2}{3}$, find the values of $\dfrac{1}{x}, \dfrac{1}{x^2}, \dfrac{1}{x^3}$.

23. Find the values of $\dfrac{6}{x}$ when $x = \dfrac{1}{2}, \dfrac{1}{3}, \dfrac{1}{4}, -\dfrac{1}{2}$.

24. Find the values of $\dfrac{4}{x^2}$ when $x = 1, \dfrac{1}{3}, \dfrac{2}{3}, -1, -\dfrac{1}{3}$.

25. If $q(x) = 5 - \dfrac{2}{x}$, find the values of $q(1),\ q(-1),\ q\left(\dfrac{1}{2}\right),\ q\left(-\dfrac{1}{2}\right),$ $q\left(\dfrac{1}{3}\right), q\left(-\dfrac{1}{3}\right)$.

26. If $r(x) = 6 - \dfrac{1}{x^2}$, find the values of $r(1),\ r(-1),\ r\left(\dfrac{1}{2}\right),\ r\left(-\dfrac{1}{2}\right),$ $r\left(\dfrac{1}{3}\right), r\left(-\dfrac{1}{3}\right)$.

Brackets

When removing brackets, multiply each term inside the brackets by the term outside the brackets.

If there is a minus sign in front of the brackets, all signs must be changed when the brackets are removed.

Examples $5(a + 7) = 5a + 35$

$\qquad 3(2b - 4c) = 3 \times 2b - 3 \times 4c = 6b - 12c$

$\qquad (6 - d)f = 6f - df$

$\qquad -g^2(3g^3 - g + 4) = -3g^5 + g^3 - 4g^2$

$\qquad -2h(h + 3n - 5p) = -2h^2 - 6hn + 10hp$

$\qquad 5(w - 4) - 3(w + 7) = 5w - 20 - 3w - 21 = 2w - 41$

Exercise 10.4

Simplify:

1. $2(a + 3)$
2. $3(b - 4)$
3. $c(c + 1)$
4. $d(3 - f)$
5. $(g + 5)h$
6. $(k - 6)k$
7. $3m(m + 5)$
8. $4n(2n - p)$
9. $r^2(r^2 + t)$
10. $wx(w^2 - x^2)$
11. $-2(3 - a)$
12. $-5b(b - 2)$
13. $\frac{1}{2}(6c - 8)$
14. $\frac{1}{3}(12 - 9d)$
15. $f(g - h + 2)$
16. $3k(k + 2m - n)$
17. $-2(p - 2q + 3)$
18. $-rt(r + t + w)$
19. $3(a + 2) + 2(a + 1)$
20. $4(b - 3) + 5(b + 2)$
21. $3(c - 7) - 2(c + 3)$
22. $5(d + 1) - (d - 1)$
23. $x - (3 - x)$
24. $y - (y + 5)$
25. $7 - (n + 3)$
26. $1 - (k - 1)$
27. $f(f - 3) - 2f(1 - f)$
28. $3g(g + 4) - g(2g - 1)$
29. $h(3h + k) + k(2k - h)$
30. $m(5 - 2m) - (m - 7)$

Binomial products

$$(a + b)p = ap + bp$$

Replacing p with $(c + d)$, this statement becomes:

$$(a + b)(c + d) = a(c + d) + b(c + d) = ac + ad + bc + bd$$

Fig. 10.2 illustrates this expansion. The three large rectangles each have the same area, but the area is calculated in a different way each time:

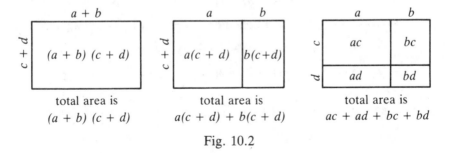

Fig. 10.2

Note that when $(a + b)(c + d)$ is expanded to give $ac + ad + bc + bd$, each term in the second bracket, $(c + d)$, is multiplied by a. Then each term in $(c + d)$ is multiplied by b.

For the expansion of $(x + 3)(y + 5)$, each term in $(y + 5)$ is multiplied by x. Then each term in $(y + 5)$ is multiplied by 3. This gives:

$$(x + 3)(y + 5) = x(y + 5) + 3(y + 5) = xy + 5x + 3y + 15$$

Similarly:

$$(n + 2)(n + 7) = n(n + 7) + 2(n + 7)$$
$$= n^2 + 7n + 2n + 14 = n^2 + 9n + 14$$

$$(3p + 5)(2p + 9) = 3p(2p + 9) + 5(2p + 9)$$
$$= 6p^2 + 27p + 10p + 45 = 6p^2 + 37p + 45$$

Exercise 10.5

Expand:

1. $(a + 3)(b + 2)$
2. $(c + 5)(d + 1)$
3. $(e + 6)(e + 2)$
4. $(f + 3)(f + 4)$
5. $(2 + g)(7 + g)$
6. $(3 + h)(6 + m)$
7. $(n + 5)(n + 5)$
8. $(2 + p)(2 + p)$
9. $(3a + 2)(b + 5)$
10. $(2c + 5)(c + 3)$
11. $(3d + 4)(2f + 1)$
12. $(5g + 2)(3g + 4)$
13. $(h + 2k)(h + 3k)$
14. $(2m + 3p)(m + 4p)$
15. $(x + 2y)^2$
16. $(3q + t)^2$
17. $(m + 1)(m + 2) - m^2 - 2$
18. $(t + 3)(t + 4) + (t + 1)(t + 5)$

Binomial products involving negative signs

$$(d - f)q = dq - fq$$

Replacing q with $(g + h)$ this becomes:

$$(d - f)(g + h) = d(g + h) - f(g + h) = dg + dh - fg - fh$$

Each term in $(g + h)$ is multiplied by d and then by $-f$.
Replacing q in $(d - f)q = dq - fq$ with $(k - m)$ we have:

$$(d - f)(k - m) = d(k - m) - f(k - m) = dk - dm - fk + fm$$

Each term in $(k - m)$ is multiplied by d and then by $-f$. Similarly:

$$(x - 5)(y + 4) = x(y + 4) - 5(y + 4) = xy + 4x - 5y - 20$$

$$(n - 3)(n - 7) = n(n - 7) - 3(n - 7) = n^2 - 7n - 3n + 21$$
$$= n^2 - 10n + 21$$

$$(4 + p)(8 - p) = 4(8 - p) + p(8 - p) = 32 - 4p + 8p - p^2$$
$$= 32 + 4p - p^2$$

$$(2f - 3g)(7f + 4g) = 2f(7f + 4g) - 3g(7f + 4g)$$
$$= 14f^2 + 8fg - 21fg - 12g^2 = 14f^2 - 13fg - 12g^2$$

Exercise 10.6

Expand:

1. $(a - 3)(b + 4)$ 2. $(c - 2)(c + 9)$ 3. $(d - 5)(f - 2)$

4. $(g - 7)(g - 1)$ **5.** $(h + 2)(h - 8)$ **6.** $(k + 1)(k - 7)$
7. $(3 - m)(2 + m)$ **8.** $(5 - n)(6 - n)$ **9.** $(p - 7)(p + 7)$
10. $(q - 4)^2$ **11.** $(3a - 1)(a + 4)$ **12.** $(2b - 5)(b - 3)$
13. $(5c - 3)(2c - 1)$ **14.** $(2d - 5)(3d + 7)$ **15.** $(3f - 2)(3f + 2)$
16. $(2g - 5)^2$ **17.** $(8 - 3h)(2 - 3h)$ **18.** $(7 - 5k)(7 + 5k)$
19. $(5m - p)(2m - p)$ **20.** $(3x - 2y)(4x + 3y)$
21. $(n - 3)(n - 4) - n(n - 7)$ **22.** $p(p - 5) + (p - 2)(p - 3)$

Important identities

$$(a + b)^2 = (a + b)(a + b) = a(a + b) + b(a + b)$$
$$= a^2 + ab + ab + b^2 = a^2 + 2ab + b^2$$
$$(a - b)^2 = (a - b)(a - b) = a(a - b) - b(a - b)$$
$$= a^2 - ab - ab + b^2 = a^2 - 2ab + b^2$$
$$(a + b)(a - b) = a(a - b) + b(a - b)$$
$$= a^2 - ab + ab - b^2 = a^2 - b^2$$

We thus have the three identities:

$$(a + b)^2 = a^2 + 2ab + b^2 \qquad (1)$$
$$(a - b)^2 = a^2 - 2ab + b^2 \qquad (2)$$
$$(a + b)(a - b) = a^2 - b^2 \qquad (3)$$

Examples   These examples show the use of the above identities:

$$(y + 5)^2 = y^2 + 2 \times 5 \times y + 5^2 = y^2 + 10y + 25$$
$$(4n - 7)^2 = (4n)^2 - 2 \times (4n) \times 7 + 7^2 = 16n^2 - 56n + 49$$
$$\left(3 - \frac{2}{w}\right)^2 = 3^2 - 2 \times 3 \times \frac{2}{w} + \left(\frac{2}{w}\right)^2 = 9 - \frac{12}{w} + \frac{4}{w^2}$$
$$(5 + 3p)(5 - 3p) = 5^2 - (3p)^2 = 25 - 9p^2$$

Example   Using identities (1) and (2), find the values of:
(i) 53^2 (ii) 9.8^2
(i) Putting $a = 50$ and $b = 3$ in identity (1),
$$53^2 = (50 + 3)^2 = 50^2 + 2 \times 50 \times 3 + 3^2$$
$$= 2500 + 300 + 9 = 2809$$
(ii) Putting $a = 10$ and $b = 0.2$ in identity (2),
$$9.8^2 = (10 - 0.2)^2 = 10^2 - 2 \times 10 \times 0.2 + 0.2^2$$
$$= 100 - 4 + 0.04 = 96.04$$

Exercise 10.7

Expand:

1. $(a + 3)^2$ **2.** $(b - 4)^2$ **3.** $(6 + c)^2$
4. $(5 - d)^2$ **5.** $(3f + 1)^2$ **6.** $(2g - 1)^2$

7. $\left(h + \dfrac{1}{2}\right)^2$ **8.** $\left(k - \dfrac{2}{3}\right)^2$ **9.** $(m + 3)(m - 3)$

10. $(n - 5)(n + 5)$ **11.** $(7 - p)(7 + p)$ **12.** $(4 + q)(4 - q)$
13. $(3r + 1)(3r - 1)$ **14.** $(5t - 1)(5t + 1)$ **15.** $(w + 3x)(w - 3x)$
16. $(2y - z)(2y + z)$ **17.** $(2c - 5d)(2c + 5d)$ **18.** $(3f + 4g)(3f - 4g)$

19. $(h^3 - 1)(h^3 + 1)$ **20.** $\left(k + \dfrac{2}{3}\right)\left(k - \dfrac{2}{3}\right)$ **21.** $(n + 3p)^2$

22. $(q - 5r)^2$ **23.** $(3t + 5)^2$ **24.** $(4v + 3)^2$

25. $(2w - 5x)^2$ **26.** $\left(3y - \dfrac{1}{3}z\right)^2$ **27.** $(p^2 - 4)^2$

28. $(8 + q^3)^2$ **29.** $\left(r - \dfrac{1}{r}\right)^2$ **30.** $\left(t + \dfrac{3}{5}\right)^2$

Use identities (1) and (2), and the method in the second example above, to calculate:

31. 31^2 **32.** 29^2 **33.** 53^2 **34.** 48^2 **35.** 202^2
36. 198^2 **37.** 9.7^2 **38.** 19.8^2

Use identities (1) and (2) to calculate, correct to 2 d.p.:

39. 4.03^2 **40.** 1.04^2 **41.** 4.98^2 **42.** 3.96^2

Expand and simplify:

43. $(x + 5)^2 - (x - 5)^2$ **44.** $(7 - y)^2 - (7 + y)^2$
45. $(n + 3)^2 + (n - 3)^2$

Completing the square

The expressions $x^2 + 12x + 36$, $y^2 - 18y + 81$ and $9n^2 + 30n + 25$ are *perfect squares*, since, by using the identities (1) and (2), we find that they are equal to $(x + 6)^2$, $(y - 9)^2$ and $(3n + 5)^2$.

Example   Find the values of n, p and q so that (i) $x^2 + 22x + n$,
(ii) $y^2 - 7y + p$ and (iii) $25t^2 - 40t + q$ are perfect squares.

(i) We have $a^2 + 2ab + b^2 = (a + b)^2$.
Comparing $x^2 + 22x + n$ with $a^2 + 2ab + b^2$, we find
$a = x$ and $2ab = 22x$.
Hence $b = 11$ and $n = b^2 = 11^2 = 121$.

(ii) Also $a^2 - 2ab + b^2 = (a - b)^2$.
Comparing $y^2 - 7y + p$ with $a^2 - 2ab + b^2$, we find
$a = y$ and $2ab = 7y$.

$$\text{Hence } b = \frac{7}{2} \text{ and } p = b^2 = \left(\frac{7}{2}\right)^2 = \frac{49}{4} = 12\frac{1}{4}.$$

(iii) Comparing $25t^2 - 40t + q$ with $a^2 - 2ab + b^2$, we find
$a = 5t$ and $2ab = 40t$.
Hence $b = 4$ and $q = b^2 = 4^2 = 16$.

Exercise 10.8

Copy and complete the following identities.

1. $x^2 + 6x + 9 = (x + \quad)^2$ 2. $y^2 - 8y + \quad = (y - \quad)^2$
3. $n^2 - 14n \quad = (n \quad)^2$ 4. $p^2 + 20p \quad = (\quad)^2$
5. $x^2 - 5x + \quad = \left(x - \frac{5}{2}\right)^2$ 6. $y^2 + 3y + \quad = (y \quad)^2$
7. $n^2 + 11n + \quad = (\quad)^2$ 8. $q^2 - q \quad = (\quad)^2$

State the number which must be added to each of the following, to form a perfect square.

9. $x^2 + 10x$ 10. $h^2 + 16h$ 11. $m^2 - 4m$ 12. $n^2 - 10n$
13. $k^2 + 5k$ 14. $p^2 - 7p$

Expressing functions of the type $ax^2 + bx + c$ in the form $a(x + p)^2 + q$.

We use identities (1) and (2) above as follows:

$$x^2 + 10x + 32 = x^2 + 10x + 25 + 7 = (x + 5)^2 + 7$$

$$y^2 - 6y + 2 = y^2 - 6y + 9 - 7 = (y - 3)^2 - 7$$

$$3n^2 + 12n + 7 = 3(n^2 + 4n) + 7 = 3(n^2 + 4n + 4) - 12 + 7$$

$$= 3(n + 2)^2 - 5$$

$$13 + 8x - x^2 = 13 - (x^2 - 8x) = 13 - (x^2 - 8x + 16) + 16$$

$$= 29 - (x - 4)^2$$

Maximum and minimum values

The function $x^2 + 10x + 32$ has different values, for different values of x.

If $x = 3$, $x^2 + 10x + 32 = 9 + 30 + 32 = 71$
If $x = 0$, $x^2 + 10x + 32 = 32$
If $x = -8$, $x^2 + 10x + 32 = 64 - 80 + 32 = 16$

By choosing a suitable value of x, can $x^2 + 10x + 32$ have any value, however large or small?

$x^2 + 10x + 32$ can be expressed as $(x + 5)^2 + 7$.

Now $(x + 5)^2$ cannot be negative. Its smallest value is 0, when $x = -5$. So the smallest value of $(x + 5)^2 + 7$ is $0 + 7 = 7$. Thus $x^2 + 10x + 32$ cannot have a value less than 7. 7 is its *minimum value*.

Now consider $13 + 8x - x^2$. This can be expressed as $29 - (x - 4)^2$. $(x - 4)^2$ cannot be negative. Its smallest value is 0, when $x = 4$. So the largest value of $29 - (x - 4)^2$ is 29. This is the *maximum value* of $13 + 8x - x^2$.

Exercise 10.9

Write each of the following in the form $(x + p)^2 + q$ where p and q are positive or negative integers.

1. $x^2 + 10x + 28$ **2.** $x^2 + 8x + 10$ **3.** $x^2 - 4x + 7$
4. $x^2 - 12x + 20$ **5.** $x^2 - 2x - 5$ **6.** $x^2 + 6x - 3$

Write each of the following in the form $(x + p)^2 + q$ where p and q are positive or negative rational numbers.

 7. $x^2 + 5x + 2$ **8.** $x^2 + 3x - 7$ **9.** $x^2 - 7x + 4$
10. $x^2 - 9x - 5$

Express each of the following in the form $a(x + p)^2 + q$ where a, p and q are positive or negative integers.

11. $3x^2 + 30x + 17$ **12.** $2x^2 + 12x + 5$ **13.** $5x^2 + 10x - 3$
14. $3x^2 - 6x - 7$ **15.** $4 - x^2 - 6x$ **16.** $5 - 2x^2 + 16x$

Find the maximum or minimum value of each of the following functions, together with the corresponding value of x.

17. $x^2 + 6x + 17$ **18.** $x^2 - 4x + 9$ **19.** $x^2 - 2x - 6$
20. $x^2 + 8x + 11$ **21.** $7 - 4x - x^2$ **22.** $3 - x^2 + 10x$
23. $3x^2 + 6x + 14$ **24.** $10 + 20x - 2x^2$

11 Indices: negative, fractional and zero

On page 84, the following rules for indices are stated:

Rule 1	$a^m \times a^n = a^{m+n}$
Rule 2	$a^m \div a^n = a^{m-n}$
Rule 3	$(a^m)^n = a^{mn}$

Negative indices

Using rule 2, $a^3 \div a^5 = a^{3-5} = a^{-2}$

But $a^3 \div a^5 = \dfrac{a \times a \times a}{a \times a \times a \times a \times a} = \dfrac{1}{a \times a} = \dfrac{1}{a^2}$

so $a^{-2} = \dfrac{1}{a^2}$

In general: $\qquad\qquad a^{-n} = \dfrac{1}{a^n}$ **Rule 4**

Example $\quad 3^{-4} = \dfrac{1}{3^4} = \dfrac{1}{81}, \quad 5^{-1} = \dfrac{1}{5^1} = \dfrac{1}{5}$

$2b^{-3} = 2 \times b^{-3} = 2 \times \dfrac{1}{b^3} = \dfrac{2}{b^3}$

but $(2b)^{-3} = \dfrac{1}{(2b)^3} = \dfrac{1}{2^3 b^3} = \dfrac{1}{8b^3}$

The zero index

$a^5 \div a^5 = a^{5-5} = a^0$. But $a^5 \div a^5 = 1$.

Hence: $\qquad\qquad a^0 = 1$ **Rule 5**

Examples $\quad 6^0 = 1, \; y^0 = 1, \; \left(\dfrac{3}{5}\right)^0 = 1, \; (7x)^0 = 1$

Fractional indices

Using rule 3, $(a^{\frac{1}{2}})^2 = a^{\frac{1}{2} \times 2} = a^1 = a$

Now, if $k^2 = p$, then $k = \sqrt{p}$

so, since $(a^{\frac{1}{2}})^2 = a$, $a^{\frac{1}{2}} = \sqrt{a}$

Similarly $(a^{\frac{1}{3}})^3 = a^{\frac{1}{3}\times 3} = a^1 = a$ so $a^{\frac{1}{3}} = \sqrt[3]{a}$

In general: $a^{\frac{1}{q}} = \sqrt[q]{a}$ **Rule 6**

Examples $49^{\frac{1}{2}} = \sqrt{49} = 7$, $8^{\frac{1}{3}} = \sqrt[3]{8} = 2$

         $a^{\frac{2}{5}} = (a^{\frac{1}{5}})^2 = (\sqrt[5]{a})^2$ or $a^{\frac{2}{5}} = (a^2)^{\frac{1}{5}} = \sqrt[5]{(a^2)}$

In general: $a^{\frac{p}{q}} = (\sqrt[q]{a})^p$ or $\sqrt[q]{(a^p)}$ **Rule 7**

Example $(32)^{\frac{2}{5}} = (\sqrt[5]{32})^2 = (2)^2 = 4$

Exercise 11.1

1. Express as fractions without indices (for example $3^{-4} = \dfrac{1}{3^4} = \dfrac{1}{81}$):

   5^{-2}, 3^{-1}, 2^{-3}, 4^{-2}.

2. Express in the form x^n where n is a fraction:

   $\sqrt[3]{x}$, $\sqrt[5]{x}$, $\sqrt[4]{x}$, $\sqrt{x}$.

3. Find the values of: $9^{\frac{1}{2}}$, $25^{\frac{1}{2}}$, $8^{\frac{1}{3}}$, $27^{\frac{1}{3}}$, $16^{\frac{1}{4}}$.

4. Express in the form y^n where n is a fraction:

   $(\sqrt[3]{y})^2$, $(\sqrt[4]{y})^3$, $(\sqrt[3]{y})^4$, $\sqrt[3]{(y^2)}$, $\sqrt[4]{(y^5)}$.

5. Find the value of: $4^{\frac{3}{2}}$, $8^{\frac{2}{3}}$, $16^{\frac{3}{4}}$, $8^{\frac{5}{3}}$, $27^{\frac{4}{3}}$.

6. State the meaning of: b^{-3}, b^0, $b^{\frac{1}{3}}$, $b^{\frac{2}{3}}$.

7. State the meaning of: $c^{\frac{1}{4}}$, c^0, c^{-4}, $c^{\frac{3}{4}}$.

8. Express in the form x^n: $\dfrac{1}{x^3}$, $\sqrt[3]{x}$, $\dfrac{1}{x^5}$, $\sqrt[4]{x}$, $(\sqrt[5]{x})^2$.

9. Find the value of: $16^{\frac{1}{2}}$, 9^0, 7^{-1}, 2^{-3}, $64^{\frac{1}{3}}$.

10. Find the value of: 2^{-2}, 1^{-3}, 6^0, 4^{-1}, $4^{\frac{5}{2}}$.

11. Find the value of: $9^{-\frac{1}{2}}$, $8^{-\frac{1}{3}}$, $\left(\dfrac{5}{4}\right)^{-1}$, $\left(\dfrac{6}{7}\right)^0$, $\left(\dfrac{2}{3}\right)^{-2}$.

12. Find the value of: $\left(\dfrac{4}{9}\right)^{\frac{1}{2}}$, $\left(\dfrac{8}{27}\right)^{\frac{1}{3}}$, $\left(\dfrac{4}{9}\right)^{-\frac{1}{2}}$, $\left(\dfrac{4}{9}\right)^{-2}$.

13. Remove the brackets from:

   $(2x)^{-3}$, $(3y)^{-1}$, $(4k)^0$, $(p^{-4})^2$, $(2r^2)^{-1}$.

14. Find the value of:

(i) $9^{\frac{1}{2}} \times 8^{\frac{1}{3}}$ (ii) $4^{\frac{1}{2}} \times 5^0$ (iii) $2^{-1} \times 16^{\frac{1}{2}}$.

15. Simplify:

(i) $a^{\frac{1}{4}} \times a^{\frac{1}{2}}$ (ii) $b^{\frac{3}{5}} \div b^{\frac{1}{5}}$ (iii) $c^2 \times c^{\frac{1}{2}}$ (iv) $d^2 \div d^{1\frac{1}{2}}$.

16. Simplify:

(i) $3f^{\frac{1}{2}} \times 2f^{1\frac{1}{2}}$ (ii) $5g^{\frac{1}{3}} \times 2g^{\frac{1}{2}}$ (iii) $6h^{\frac{1}{2}} \div 2h^{\frac{1}{4}}$ (iv) $10k \div 2k^{\frac{1}{2}}$.

12 Factors

Type 1, common factors: form $ax \pm ay$

If a is multiplied by $(x + y)$, the product is $ax + ay$.

$$a(x + y) = ax + ay$$

Each term inside the brackets is multiplied by a. If we write:

$$ax + ay = a(x + y)$$

we are expressing $ax + ay$ as the product of two factors, a and $(x + y)$. In other words, we are *factorising* the expression $ax + ay$. Notice that a is a factor of both ax and ay. When the terms of an expression such as $ax + ay$ have a common factor, the factor is placed in front of the brackets.

Example 1 $3n + 12$

3 is a factor of both $3n$ and 12.

$3n + 12 = 3 \times n + 3 \times 4 = 3(n + 4)$

Example 2 $c^2 - cd$

c is a factor of both c^2 and cd.

$c^2 - cd = c \times c - c \times d = c(c - d)$

Example 3 $pt + t^2 - tx$

t is a factor of pt, t^2 and tx.

$pt + t^2 - tx = t(p + t - x)$

Sometimes two or more factors can be placed in front of the brackets.

Example 4 $6kx - 15nx$

3 and x are factors of both $6kx$ and $15nx$.

$6kx - 15nx = 3x \times 2k - 3x \times 5n = 3x(2k - 5n)$

Note 1: After factorising an expression, always check that no other factor can be taken out of the brackets in your answer. If you write $6kx - 15nx = 3(2kx - 5nx)$, you have not finished the factorisation, because x is a factor of $2kx - 5nx$.

Note 2: Check your answers mentally (or on paper) by multiplying the factors out.

Exercise 12.1

Factorise the following:

1. $5a + 5b$	**2.** $cd - cf$	**3.** $gh + kh$	**4.** $7m - 7p$
5. $2r - 6$	**6.** $3t + 9$	**7.** $10 + 2x$	**8.** $15 - 5y$
9. $a^2 + 3a$	**10.** $5b - b^2$	**11.** $3c + 3$	**12.** $5 - 5d$
13. $ef + f$	**14.** $gh - g$	**15.** $6k + 8$	**16.** $15 + 6m$
17. $2ab + 2ac$	**18.** $3d^2 - 3de$	**19.** $5fg - 10f$	**20.** $2h^2 - 4h$
21. $ab + ac + ad$	**22.** $ef - eg - eh$	**23.** $k^2 - km + 3k$	
24. $5n - 5p - 5q$	**25.** $3r + 3t + 6v$	**26.** $2x + 4y + 8$	

Type 2, difference of two squares: form $a^2 - b^2$

On page 93 we had the identity $(a + b)(a - b) = a^2 - b^2$.
Using this identity, we can factorise any expression of the type $a^2 - b^2$.

Example 1 $x^2 - 25$. Here $a = x$ and $b = 5$.
$$x^2 - 25 = x^2 - 5^2 = (x + 5)(x - 5)$$

Example 2 $49 - y^2$. Here $a = 7$ and $b = y$.
$$49 - y^2 = 7^2 - y^2 = (7 + y)(7 - y)$$

Example 3 $9n^2 - 100$. Here $a = 3n$ and $b = 10$.
$$9n^2 - 100 = (3n)^2 - 10^2 = (3n + 10)(3n - 10)$$

Exercise 12.2

Factorise:

1. $a^2 - 9$	**2.** $b^2 - 16$	**3.** $9 - c^2$	**4.** $16 - d^2$
5. $e^2 - 1$	**6.** $1 - f^2$	**7.** $9g^2 - 16$	**8.** $4h^2 - 25$
9. $16k^2 - 1$	**10.** $1 - 36n^2$	**11.** $25p^2 - 81$	**12.** $100 - 49t^2$

Further examples of the type $a^2 - b^2$

Example 4 $7y^2 - 63$. Here 7 is a factor of both $7y^2$ and 63 so the first
step is to place 7 in front of the brackets.
$$7y^2 - 63 = 7(y^2 - 9) = 7(y^2 - 3^2) = 7(y + 3)(y - 3)$$

Example 5 $n^2 - 2\frac{7}{9} = n^2 - \frac{25}{9} = n^2 - \left(\frac{5}{3}\right)^2 = \left(n + \frac{5}{3}\right)\left(n - \frac{5}{3}\right)$

Example 6 $x^6 - 169 = (x^3)^2 - 13^2 = (x^3 + 13)(x^3 - 13)$

It is sometimes possible to make use of the factors of $a^2 - b^2$ in a
calculation. Here are two examples.

Example 7 $77^2 - 23^2 = (77 + 23)(77 - 23) = (100)(54) = 5400$

Example 8 $0.7^2 - 0.5^2 = (0.7 + 0.5)(0.7 - 0.5) = (1.2)(0.2) = 0.24$

Exercise 12.3

Factorise:

1. $3x^2 - 12$ 2. $5y^2 - 45$ 3. $2n^2 - 32$ 4. $4p^2 - 4$
5. $km^2 - 9k$ 6. $h^3 - g^2h$ 7. $r - rt^2$ 8. $64 - 4q^2$
9. $c^2 - \dfrac{4}{9}$ 10. $d^2 - \dfrac{1}{4}$ 11. $f^2 - 2\dfrac{1}{4}$ 12. $g^2 - 6\dfrac{1}{4}$
13. $h^2 - 0.09$ 14. $k^2 - 0.25$ 15. $n^6 - 49$ 16. $p^{10} - 4$

Use the method of Examples 7 and 8 to calculate the following:

17. $83^2 - 17^2$ 18. $55^2 - 45^2$ 19. $113^2 - 13^2$ 20. $97^2 - 87^2$
21. $0.7^2 - 0.3^2$ 22. $4.3^2 - 2.3^2$ 23. $17^2 - 15^2$ 24. $5.2^2 - 4.8^2$

Type 3, simple trinomials: form $x^2 + bx + c$ where b and c are numbers

The expression $x^2 + bx + c$ is called a *trinomial* because it has three terms.
  We obtained trinomials on page 92 by expanding expressions such as $(x + 3)(x + 7)$.

$$(x + 3)(x + 7) = x^2 + 3x + 7x + 21 = x^2 + 10x + 21$$

$$(x + 3) \text{ and } (x + 7) \text{ are the factors of } x^2 + 10x + 21.$$

In general, an expression of the type $x^2 + bx + c$ has factors $(x + n)$ and $(x + p)$ where n and p are numbers.

Example 1 $x^2 + 7x + 12$
  Let $x^2 + 7x + 12 = (x + n)(x + p) = x^2 + nx + px + np$
  $\qquad\qquad\qquad\qquad\qquad\qquad = x^2 + (n + p)x + np$

  Then $n + p = 7$ and $np = 12$.
  For $np = 12$ we can have:
  $\qquad n = 12$ and $p = 1$, giving $n + p = 13$
  or $n = 6$ and $p = 2$, giving $n + p = 8$
  or $n = 4$ and $p = 3$, giving $n + p = 7$

  $n = 4$ and $p = 3$ gives the correct value for $n + p$, so
  $x^2 + 7x + 12 = (x + 4)(x + 3)$

Example 2 $x^2 - 8x + 15$
  Let $x^2 - 8x + 15 = (x + n)(x + p) = x^2 + (n + p)x + np$
  Then $n + p = -8$ and $np = 15$.
  Since np is positive, n and p are either both positive or both negative.

$n + p$ is negative, so n and p must both be negative.
For $np = 15$ we can have:
$n = -15$ and $p = -1$,   giving $n + p = -16$
or $n = -5$ and $p = -3$, giving $n + p = -8$

Since the correct value of $n + p$ is -8, then $n = -5$ and $p = -3$, so
$x^2 - 8x + 15 = (x - 5)(x - 3)$

Example 3 $y^2 + 3y - 10$
Let $y^2 + 3y - 10 = (y + n)(y + p) = y^2 + (n + p)y + np$
Then $n + p = 3$ and $np = -10$.
Since np is negative, n and p must have opposite signs.
As $n + p$ is positive, the positive number must be larger than the negative number.
For $np = -10$, we can have:
$n = 10$ and $p = -1$, giving $n + p = 9$
or $n = 5$ and $p = -2$, giving $n + p = 3$

Since the correct value of $n + p$ is 3, then $n = 5$ and $p = -2$, so
$y^2 + 3y - 10 = (y + 5)(y - 2)$

Exercise 12.4

Factorise:

1. $a^2 + 5a + 6$
2. $b^2 + 6b + 8$
3. $c^2 + 4c + 3$
4. $d^2 + 6d + 5$
5. $e^2 - 9e + 14$
6. $f^2 - 11f + 18$
7. $g^2 - 8g + 12$
8. $h^2 - 7h + 12$
9. $k^2 + 2k - 8$
10. $m^2 + m - 6$
11. $n^2 - 5n - 14$
12. $p^2 - 2p - 15$
13. $q^2 - 4q - 12$
14. $r^2 + 4r - 12$
15. $t^2 + 11t - 12$
16. $u^2 - u - 12$
17. $15 + 8w + w^2$
18. $10 - 7x + x^2$
19. $6 - y - y^2$
20. $7 + 6z - z^2$
21. $1 - 4n + 3n^2$
22. $1 + 4p - 5p^2$
23. $k^2 + 4km + 3m^2$
24. $g^2 - 8gh + 7h^2$

Type 4, harder trinomials: form $ax^2 + bx + c$ where a, b and c are numbers

Trinomials with numbers before the x^2 were obtained on page 92, by expanding expressions such as $(3x + 5)(x + 2)$ and $(2x - 7)(5x + 4)$. For example $(3x + 5)(x + 2) = 3x^2 + 6x + 5x + 10 = 3x^2 + 11x + 10$. We now factorise some trinomials of this type.

Example 1 $5x^2 + 17x + 14$
To obtain $5x^2$ we must have $(5x\ \ )(x\ \ )$.
Let $5x^2 + 17x + 14 = (5x + n)(x + p) = 5x^2 + nx + 5px + np$
$ = 5x^2 + (n + 5p)x + np$

Then $np = 14$ and $n + 5p = 17$

For the pair of numbers (n, p) to satisfy $np = 14$, we can have $(2, 7)$, $(7, 2)$, $(14, 1)$ or $(1, 14)$.

For $n = 2$ and $p = 7$, $n + 5p = 2 + 35 = 37$, which is wrong.

For $n = 7$ and $p = 2$, $n + 5p = 7 + 10 = 17$, which is right.

Hence $5x^2 + 17x + 14 = (5x + 7)(x + 2)$

Example 2 $3x^2 - 2x - 5$

Let $3x^2 - 2x - 5 = (3x + n)(x + p) = 3x^2 + nx + 3px + np$
$$= 3x^2 + (n + 3p)x + np$$

Then $np = -5$ and $n + 3p = -2$.

For the pair of numbers (n, p) to satisfy $np = -5$, we can have $(5, -1)$, $(-5, 1)$, $(1, -5)$ or $(-1, 5)$.

$n = -5$ and $p = 1$ give the correct value for $n + 3p$.

Hence $3x^2 - 2x - 5 = (3x - 5)(x + 1)$

Exercise 12.5

1. Expand $(3x + 2)(x + 5)$, $(3x + 5)(x + 2)$, $(3x + 1)(x + 10)$ and $(3x + 10)(x + 1)$.
   Hence state the factors of: (i) $3x^2 + 17x + 10$ (ii) $3x^2 + 13x + 10$.

2. Expand $(2x - 7)(x - 3)$, $(2x - 3)(x - 7)$, $(2x - 21)(x - 1)$ and $(2x - 1)(x - 21)$.
   Hence state the factors of: (i) $2x^2 - 13x + 21$ (ii) $2x^2 - 43x + 21$.

3. Expand $(5x - 3)(x + 1)$, $(5x + 3)(x - 1)$, $(5x - 1)(x + 3)$ and $(5x + 1)(x - 3)$.
   Hence state the factors of: (i) $5x^2 + 14x - 3$ (ii) $5x^2 - 2x - 3$.

Factorise:

4. $3x^2 + 8x + 5$	5. $3x^2 - 8x + 5$	6. $3x^2 + 16x + 5$
7. $3x^2 - 16x + 5$	8. $2x^2 + 9x + 7$	9. $2x^2 - 15x + 7$
10. $2x^2 - 5x + 3$	11. $2x^2 - 7x + 3$	12. $2x^2 + 3x - 5$
13. $2x^2 - 3x - 5$	14. $2x^2 + 9x - 5$	15. $2x^2 - 9x - 5$
16. $5x^2 - 2x - 3$	17. $5x^2 + 2x - 3$	18. $5x^2 - 14x - 3$
19. $5x^2 + 14x - 3$	20. $3a^2 - a - 2$	21. $3b^2 - 5b + 2$
22. $3c^2 + 7c + 2$	23. $3d^2 - 5d - 2$	24. $5e^2 + 12e + 7$
25. $5f^2 - 2f - 7$	26. $7g^2 + 2g - 5$	27. $7h^2 - 36h + 5$
28. $2k^2 - 9k + 10$	29. $3m^2 - 7m - 6$	30. $5n^2 + 11n + 6$
31. $3p^2 + p - 10$	32. $6a^2 + 11a + 3$	33. $10b^2 - 21b + 9$

34. Factorise $2x^2 + 5x + 3$. Then, by putting $x = 10$, express 253 as the product of two prime numbers.

35. Factorise $3p^2 + 10p + 7$. Then, by putting $p = 100$, express 31007 as the product of two prime numbers.

36. Factorise $4q^2 + 8q + 3$ and use the result to express 40803 as the product of four prime numbers.

Type 5, factors by grouping: expressions with four terms

Consider the following working for the product of $(a + b)$ and $(c + d)$.

$$(a + b)(c + d) = a(c + d) + b(c + d) = ac + ad + bc + bd.$$

To factorise $ac + ad + bc + bd$, we reverse the above working thus:

$$ac + ad + bc + bd = a(c + d) + b(c + d) = (a + b)(c + d).$$

It may help you to understand the last step if we write a single letter for $(c + d)$:

$ac + ad + bc + bd = a(c + d) + b(c + d)$
$\qquad\qquad\qquad = an + bn \qquad$ where $n = (c + d)$
$\qquad\qquad\qquad = (a + b)n \qquad$ since n is the common factor
$\qquad\qquad\qquad = (a + b)(c + d) \qquad$ replacing n with $(c + d)$.

Example 1 $gh + 7g + 2h + 14$
g is the common factor of gh and $7g$ and 2 is the common factor of $2h$ and 14.
So $gh + 7g + 2h + 14 = g(h + 7) + 2(h + 7)$
Now $(h + 7)$ is the common factor of these two terms.
So $gh + 7g + 2h + 14 = (g + 2)(h + 7)$.

Example 2 $x^2 + 3xy - 5x - 15y$
x is the common factor of x^2 and $3xy$ and -5 is the common factor of $-5x$ and $-15y$.
So $x^2 + 3xy - 5x - 15y = x(x + 3y) - 5(x + 3y)$
$\qquad\qquad\qquad\qquad\qquad = (x - 5)(x + 3y)$

Example 3 $18 - 3n - 6p + np$
$= 3(6 - n) - p(6 - n)$
$= (3 - p)(6 - n)$

Example 4 $12cd - 8c + 3d - 2$
$= 4c(3d - 2) + 1(3d - 2)$
$= (4c + 1)(3d - 2)$
The 1 on the second line helps us to see that the first bracket on the next line is $(4c + 1)$.

Exercise 12.6

Factorise:

1. $ab + 3a + 2b + 6$
2. $cd + cf + 5d + 5f$
3. $g^2 - gh + 3g - 3h$
4. $km - 7k - 5m + 35$
5. $n^2 - n - np + p$
6. $5 - 5t + v - vt$
7. $12 + 4x + 3w + wx$
8. $7u + 7y - uy - y^2$
9. $15 - 5c + 3d - cd$
10. $2gh - 12g + 3h - 18$

11. $2km + 5k - 14m - 35$
12. $3p^2 - 3pr + 4pq - 4qr$
13. $2ab + 3ac - 10bc - 15c^2$
14. $3d^2 - 2df + 9de - 6ef$
15. $3 - 3h - m + mh$
16. $2 - k + 2n - kn$
17. $pq + 5q - p - 5$
18. $15wx + 20w - 6x - 8$
19. $7y^2 - 21by + 2ay - 6ab$
20. $8np + 12n - 2pt - 3t$
21. $ab - cd + ad - bc$ (First write this as $ab + ad - cd - bc$.)
22. $xy + 10 + 2x + 5y$ (First change the order so that your first two terms have a common factor.)
23. $4h + m^2 + 4hm + m$ **24.** $5xy + 10 - 2x^2 - x^3y$

Exercise 12.7   Miscellaneous questions

Factorise:

1. $3x + 3y$
2. $h^2 - 25$
3. $n^2 - np$
4. $1 - 9k^2$
5. $m^2 + 7m + 10$
6. $n^2 - n - 12$
7. $t^2 - 6t + 8$
8. $2a^2 - 6ab$
9. $3x^2 - 75$
10. $1 + 7a + 10a^2$
11. $cd^2 - 9c$
12. $15 - 2b - b^2$
13. $32 - 2y^2$
14. $3x^2 - 6x - 9$
15. $3y^2 - 6y + 3$
16. $6n^2 + n - 2$
17. $5h^2 - 8h - 4$
18. $8a^2 + 2a - 15$
19. $3b^3 - 3b$
20. $2c^2 + 4c - 30$
21. $3x^2 - 11x + 10$
22. $ab + 5a - 2b - 10$
23. $3 - 2c - 3d + 2cd$
24. $4xy + 8x + 3y + 6$
25. $km - 4m - 3k + 12$
26. $x^2y^2 + 3xy - 10$
27. $3p^2 - 2pq - 8q^2$
28. $ab - cd - bc + ad$
29. $e^2 - 6fg + 2eg - 3ef$

13 Simple algebraic fractions

The methods used for fractions in algebra are similar to those used in arithmetic.

Equivalent fractions

$$\text{Just as}\quad \frac{3}{7} = \frac{3 \times 5}{7 \times 5} = \frac{15}{35} \quad \text{so} \quad \frac{a}{b} = \frac{a \times c}{b \times c} = \frac{ac}{bc}$$

Some other fractions equivalent to $\dfrac{a}{b}$ are:

$$\frac{5a}{5b}, \quad \frac{\frac{1}{3}a}{\frac{1}{3}b}, \quad \frac{a(f+g)}{b(f+g)}, \quad \frac{a^2}{ab}, \quad \frac{ab}{b^2}$$

Reduction to lowest terms

$$\frac{6hm}{15mp} = \frac{2 \times 3 \times h \times m}{3 \times 5 \times m \times p} = \frac{2h}{5p}, \qquad \frac{3y}{y^3} = \frac{3 \times y}{y \times y \times y} = \frac{3}{y^2}$$

$$\frac{2x-6}{x^2-3x} = \frac{2(x-3)}{x(x-3)} = \frac{2}{x}, \qquad \frac{10u^2}{5u} = \frac{5 \times 2 \times u \times u}{5 \times u} = 2u$$

Exercise 13.1

Copy and complete:

1. $\dfrac{1}{a} = \dfrac{}{ab}$ **2.** $\dfrac{1}{c} = \dfrac{}{c^2}$ **3.** $\dfrac{1}{d} = \dfrac{}{3d}$ **4.** $\dfrac{2}{e} = \dfrac{}{3e}$

5. $\dfrac{5}{g} = \dfrac{}{gh}$ **6.** $\dfrac{7}{k} = \dfrac{}{k^2}$ **7.** $\dfrac{n}{p} = \dfrac{}{pq}$ **8.** $\dfrac{r}{t} = \dfrac{}{t^2}$

9. $\dfrac{w}{5} = \dfrac{}{15}$ **10.** $\dfrac{3x}{7} = \dfrac{}{14}$ **11.** $\dfrac{1}{a} = \dfrac{3}{}$ **12.** $\dfrac{1}{b} = \dfrac{c}{}$

13. $\dfrac{1}{d} = \dfrac{d}{}$ **14.** $\dfrac{5}{e} = \dfrac{15}{}$ **15.** $\dfrac{f}{g} = \dfrac{fh}{}$ **16.** $\dfrac{k}{m} = \dfrac{3k}{}$

17. $\dfrac{n}{p} = \dfrac{n^3}{}$ **18.** $\dfrac{q}{r} = \dfrac{5qt}{}$ **19.** $\dfrac{2v}{w} = \dfrac{}{2w^2}$ **20.** $\dfrac{2x}{5y} = \dfrac{}{15y^3}$

21. $3 = \dfrac{}{5}$ **22.** $a = \dfrac{}{b}$ **23.** $c = \dfrac{}{3}$ **24.** $5 = \dfrac{}{g}$

25. $1 = \dfrac{}{4}$ **26.** $1 = \dfrac{}{f}$ **27.** $g = \dfrac{}{g}$ **28.** $h = \dfrac{}{5k}$

29. $3 = \dfrac{}{4n}$ **30.** $p = \dfrac{}{p^2}$ **31.** $2r = \dfrac{}{3t}$ **32.** $5w = \dfrac{}{2w^2}$

Reduce to their lowest terms:

33. $\dfrac{ab}{bc}$ **34.** $\dfrac{5d}{7d}$ **35.** $\dfrac{3e}{3f}$ **36.** $\dfrac{g^2}{gh}$ **37.** $\dfrac{km}{m^2}$

38. $\dfrac{3p}{9}$ **39.** $\dfrac{4}{4q}$ **40.** $\dfrac{r}{rs}$ **41.** $\dfrac{t}{t^2}$ **42.** $\dfrac{v^2}{v^5}$

43. $\dfrac{10w}{15}$ **44.** $\dfrac{14}{21x}$ **45.** $\dfrac{ab^2}{a^2b}$ **46.** $\dfrac{12cd}{18d^2}$ **47.** $\dfrac{15ef^2}{20e^2f}$

Simplify:

48. $\dfrac{7a}{a}$ **49.** $\dfrac{5b}{5}$ **50.** $\dfrac{c^3}{c}$ **51.** $\dfrac{3de}{d}$

52. $\dfrac{6e^2}{2e}$ **53.** $\dfrac{15fg}{3f}$ **54.** $\dfrac{14h^3}{7h^2}$ **55.** $\dfrac{6k^2m^2}{6km}$

Reduce to their lowest terms:

56. $\dfrac{a(b-c)}{b(b-c)}$ **57.** $\dfrac{d(3+f)}{e(3+f)}$ **58.** $\dfrac{6(g-h)}{9(g-h)}$ **59.** $\dfrac{k(m+p)}{k(n-q)}$

60. $\dfrac{2a+2b}{3a+3b}$ **61.** $\dfrac{c^2-5c}{cd-5d}$ **62.** $\dfrac{3f-12}{f^2-4f}$ **63.** $\dfrac{6+3g}{14+7g}$

64. $\dfrac{3(h+2)}{(h+2)}$ **65.** $\dfrac{4k-8}{k-2}$ **66.** $\dfrac{m(x+y)}{4m}$ **67.** $\dfrac{3p-3q}{6}$

Copy and complete:

68. $\dfrac{a}{b} = \dfrac{}{b(c+3)}$ **69.** $\dfrac{5}{7} = \dfrac{}{7(d-4)}$ **70.** $\dfrac{e}{(f+g)} = \dfrac{}{h(f+g)}$

71. $\dfrac{k}{m+n} = \dfrac{}{5m+5n}$ **72.** $\dfrac{3}{4} = \dfrac{}{4p+4q}$ **73.** $\dfrac{r}{t} = \dfrac{}{tu+tw}$

Harder questions

Exercise 13.2

Reduce to their lowest terms:

1. $\dfrac{3x+12}{x^2+4x}$ **2.** $\dfrac{5n-5p}{n-p}$ **3.** $\dfrac{a+2}{(a+2)(a-2)}$ **4.** $\dfrac{b+3}{b^2-9}$

5. $\dfrac{c^2-25}{c-5}$ **6.** $\dfrac{e}{e^2-ef}$ **7.** $\dfrac{4}{4g+4h}$ **8.** $\dfrac{k^2-k}{3k}$

Copy and complete:

9. $\dfrac{1}{a+b} = \dfrac{}{(a+b)^2}$ **10.** $\dfrac{1}{c-3} = \dfrac{}{(c-3)(c+3)}$

11. $\dfrac{1}{d-5} = \dfrac{}{d^2-25}$ **12.** $\dfrac{3}{e+2} = \dfrac{}{(e+2)^2}$

13. $\dfrac{4}{f+3} = \dfrac{}{f^2-9}$ **14.** $\dfrac{g}{g-h} = \dfrac{}{g^2-h^2}$

Addition and subtraction

$$\frac{a}{b} + \frac{c}{d} = \frac{ad}{bd} + \frac{bc}{bd} = \frac{ad+bc}{bd}; \quad \frac{x}{6} + \frac{x}{9} = \frac{3x}{18} + \frac{2x}{18} = \frac{5x}{18}$$

$$\frac{3}{2f} - \frac{1}{5g} = \frac{15g}{10fg} - \frac{2f}{10fg} = \frac{15g-2f}{10fg}$$

$$\frac{a-3}{2} + \frac{a+4}{3} = \frac{3(a-3)}{6} + \frac{2(a+4)}{6} = \frac{3a-9}{6} + \frac{2a+8}{6} = \frac{5a-1}{6}$$

$$\frac{b+2c}{10} - \frac{2b-c}{15} = \frac{3(b+2c)}{30} - \frac{2(2b-c)}{30} = \frac{3b+6c}{30} - \frac{4b-2c}{30}$$

$$= \frac{(3b+6c)-(4b-2c)}{30} = \frac{3b+6c-4b+2c}{30} = \frac{8c-b}{30}$$

Exercise 13.3

Simplify:

1. $\dfrac{a}{2} + \dfrac{a}{5}$ **2.** $\dfrac{b}{3} - \dfrac{b}{7}$ **3.** $\dfrac{c}{d} - \dfrac{e}{f}$ **4.** $\dfrac{1}{g} + \dfrac{1}{h}$ **5.** $\dfrac{4}{k} + \dfrac{3}{k}$

6. $\dfrac{8}{m} - \dfrac{2}{m}$ **7.** $\dfrac{1}{3n} + \dfrac{1}{2n}$ **8.** $\dfrac{2}{p} + \dfrac{3}{q}$ **9.** $\dfrac{5}{rt} - \dfrac{2}{tv}$ **10.** $\dfrac{w}{5x} + \dfrac{3}{10x}$

11. $\dfrac{a}{bc} - \dfrac{e}{cd}$ **12.** $\dfrac{3}{f} - \dfrac{1}{f^2}$ **13.** $1 + \dfrac{a}{b}$ **14.** $3 - \dfrac{c}{d}$ **15.** $\dfrac{e}{f} + g$

16. $\dfrac{2}{h} - h$ **17.** $\dfrac{5}{6c} - \dfrac{2}{3c^2}$ **18.** $\dfrac{1}{a} + \dfrac{1}{2a} + \dfrac{1}{3a}$

19. $\dfrac{5}{y} - \dfrac{2}{y^2} + \dfrac{1}{y^3}$ **20.** $\dfrac{1}{3b} + \dfrac{1}{2b} - \dfrac{1}{b}$ **21.** $\dfrac{a+2}{3} + \dfrac{a+5}{2}$

22. $\dfrac{b-1}{2} - \dfrac{b+3}{4}$ **23.** $\dfrac{c+5}{5} + \dfrac{c+3}{3}$ **24.** $\dfrac{d-2}{7} - \dfrac{d-5}{2}$

25. $\dfrac{e+f}{9} + \dfrac{e-2f}{6}$ **26.** $\dfrac{7-g}{5} - \dfrac{3+g}{10}$ **27.** $\dfrac{2x}{3} + \dfrac{x-1}{2}$

28. $\dfrac{3x}{4} - \dfrac{x-2}{3}$ **29.** $\dfrac{2h-5}{8} - \dfrac{h+3}{12}$ **30.** $\dfrac{3p+2}{10} + \dfrac{p+4}{15}$

Express as two fractions added together:

31. $\dfrac{2a + 3b}{6}$ **32.** $\dfrac{cd + de}{ce}$ **33.** $\dfrac{3 + f}{g}$ **34.** $\dfrac{10 + h^2}{5h}$

35. $\dfrac{5k - 2m}{10}$ **36.** $\dfrac{3p - 4q}{pq}$ **37.** $\dfrac{t - 7}{t^2}$ **38.** $\dfrac{9w - 12x}{6wx}$

Harder questions

Example $\dfrac{3}{x + 1} - \dfrac{2}{x + 5} = \dfrac{3(x + 5)}{(x + 1)(x + 5)} - \dfrac{2(x + 1)}{(x + 1)(x + 5)}$

$$= \dfrac{3x + 15}{(x + 1)(x + 5)} - \dfrac{2x + 2}{(x + 1)(x + 5)}$$

$$= \dfrac{(3x + 15) - (2x + 2)}{(x + 1)(x + 5)} = \dfrac{3x + 15 - 2x - 2}{(x + 1)(x + 5)}$$

$$= \dfrac{x + 13}{(x + 1)(x + 5)}$$

Exercise 13.4

Simplify:

1. $\dfrac{1}{a + 2} + \dfrac{1}{a + 3}$ **2.** $\dfrac{1}{b - 3} + \dfrac{1}{b - 2}$ **3.** $\dfrac{1}{3 + c} + \dfrac{1}{4 - c}$

4. $\dfrac{3}{d + 2} + \dfrac{2}{d + 1}$ **5.** $\dfrac{2}{e - 1} + \dfrac{1}{e - 2}$ **6.** $\dfrac{1}{f + 3} - \dfrac{1}{f + 5}$

7. $\dfrac{1}{g - 2} - \dfrac{1}{g + 2}$ **8.** $\dfrac{3}{h - 2} - \dfrac{5}{h + 1}$ **9.** $\dfrac{4}{k - 3} - \dfrac{1}{k + 2}$

10. $\dfrac{2}{m + 1} + \dfrac{3}{m}$ **11.** $\dfrac{1}{n} - \dfrac{1}{n + 3}$ **12.** $\dfrac{1}{3 - y} + \dfrac{1}{3 + y}$

13. $\dfrac{a}{a - b} - \dfrac{b}{a + b}$ **14.** $\dfrac{c}{d - 5} + \dfrac{c}{d + 5}$ **15.** $\dfrac{1}{g(g - 2)} + \dfrac{1}{g(g + 1)}$

16. $\dfrac{3}{h + 3} - \dfrac{2}{h + 2}$ **17.** $\dfrac{1}{k^2 - k} - \dfrac{1}{k^2 + k}$ **18.** $\dfrac{2}{m} - \dfrac{6}{m(m + 3)}$

19. $\dfrac{1}{n^2 - n} + \dfrac{1}{n}$ **20.** $\dfrac{1}{x + 2} + \dfrac{2}{x(x + 2)}$ **21.** $\dfrac{y}{y^2 - 4} + \dfrac{2}{y - 2}$

Multiplication and division

$$\frac{ab}{c^2} \times \frac{c}{a} = \frac{abc}{ac^2} = \frac{b}{c}; \quad \frac{d^2}{6} \div \frac{d}{8} = \frac{d^2}{6} \times \frac{8}{d} = \frac{8d^2}{6d} = \frac{4d}{3}$$

$$\frac{f^2 - 9}{f^2} \div \frac{f + 3}{f} = \frac{(f + 3)(f - 3)}{f^2} \times \frac{f}{(f + 3)}$$

$$= \frac{(f + 3)(f - 3)f}{f^2(f + 3)} = \frac{f - 3}{f}$$

Exercise 13.5

Simplify:

1. $\dfrac{n}{p} \times \dfrac{t}{w}$ **2.** $\dfrac{k^2}{mr} \times \dfrac{m}{k}$ **3.** $\dfrac{3}{h} \times \dfrac{h}{6}$ **4.** $\dfrac{a}{b} \times \dfrac{c}{a}$ **5.** $\dfrac{5}{d} \times \dfrac{e}{f}$

6. $\dfrac{g}{4} \times \dfrac{6}{g}$ **7.** $\dfrac{h}{5} \times \dfrac{5}{h^2}$ **8.** $\dfrac{k^2}{3} \times \dfrac{6}{k}$ **9.** $\dfrac{m}{p} \div \dfrac{q}{r}$ **10.** $\dfrac{t}{5} \div \dfrac{v}{4}$

11. $\dfrac{w}{3} \div \dfrac{w}{2}$ **12.** $\dfrac{y}{7} \div \dfrac{z}{7}$ **13.** $\dfrac{a}{b} \div \dfrac{c}{b}$ **14.** $\dfrac{d}{e} \div \dfrac{f}{d}$ **15.** $\dfrac{g^2}{h^2} \div \dfrac{g}{h}$

16. $\dfrac{n}{(p + q)} \times \dfrac{(p + q)}{t}$ **17.** $\dfrac{3(w + x)}{2} \times \dfrac{1}{(w + x)}$

18. $\dfrac{a + b}{5} \div \dfrac{a + b}{3}$ **19.** $\dfrac{4}{c - d} \div \dfrac{6}{c - d}$

20. $\dfrac{5a + 5b}{3} \times \dfrac{2}{a + b}$ **21.** $\dfrac{c^2 - cd}{6} \times \dfrac{8}{c - d}$

22. $\dfrac{2e + 2f}{15} \div \dfrac{4e + 4f}{5}$ **23.** $\dfrac{2}{3g + 3h} \div \dfrac{3}{2g + 2h}$

24. $\dfrac{k}{k + 5} \times \dfrac{k + 5}{k^2}$ **25.** $\dfrac{(m - 2)^2}{8} \times \dfrac{4}{m - 2}$

26. $\dfrac{n - 3}{6} \times \dfrac{4}{n^2 - 9}$ **27.** $\dfrac{p^2 - 25}{p^2} \div \dfrac{p - 5}{p}$

28. $\dfrac{r}{r - 7} \div \dfrac{r^2}{r^2 - 49}$ **29.** $\dfrac{3h^2}{h^2 - 2h} \div \dfrac{h}{h^2 - 4}$

30. $\left(\dfrac{1}{x} + y\right) \div \left(x + \dfrac{1}{y}\right)$

14 Linear equations and inequalities

Linear equations

The equations $x - 7 = 2$ and $x + 9 = 5$ are examples of *linear equations*. They contain no term in x^2 or any higher power of x. The *solution* of each equation is the value of x which makes the equation true. To obtain this value you must *solve* the equation, as shown in the examples below.

Examples involving one step

1.
$$x - 7 = 2$$

Adding 7 to each side:
$$x - 7 + 7 = 2 + 7$$
$$x = 9$$

2.
$$x + 9 = 5$$

Subtracting 9 from each side:
$$x + 9 - 9 = 5 - 9$$
$$x = -4$$

3.
$$\frac{x}{4} = 6$$

Multiplying each side by 4:
$$\frac{x}{4} \times 4 = 6 \times 4$$
$$x = 24$$

4.
$$3x = 11$$

Dividing each side by 3:
$$\frac{3x}{3} = \frac{11}{3}$$
$$x = 3\frac{2}{3}$$

Examples with two or more steps

5.
$$2x - 9 = 5$$
$$2x - 9 + 9 = 5 + 9$$
$$2x = 14$$
$$\frac{2x}{2} = \frac{14}{2}$$
$$x = 7$$

Check: when $x = 7$,
$$2x - 9 = 14 - 9 = 5$$

6.
$$7x + 11 = 2x + 3$$
$$7x - 2x + 11 = 2x - 2x + 3$$
$$5x + 11 = 3$$
$$5x + 11 - 11 = 3 - 11$$
$$5x = -8$$
$$\frac{5x}{5} = -\frac{8}{5}$$
$$x = -1\frac{3}{5} \text{ or } -1.6$$

Check: when $x = -1.6$,
$$7x + 11 = -11.2 + 11 = -0.2$$
$$2x + 3 = -3.2 + 3 = -0.2$$

7. $3(x - 5) = 7(x - 3)$

Removing the brackets:

$$3x - 15 = 7x - 21$$

$$3x - 15 + 21 = 7x - 21 + 21$$

$$3x + 6 = 7x$$

$$3x + 6 - 3x = 7x - 3x$$

$$6 = 4x$$

$$\frac{6}{4} = \frac{4x}{4}$$

$$x = 1\frac{1}{2}$$

Check: when $x = 1\frac{1}{2}$,

$$3(x - 5) = 3\left(1\frac{1}{2} - 5\right)$$

$$= 3\left(-3\frac{1}{2}\right) = -10\frac{1}{2}$$

$$7(x - 3) = 7(1\frac{1}{2} - 3)$$

$$= 7\left(-1\frac{1}{2}\right) = -10\frac{1}{2}$$

8. $5(2 - 3x) = x + 14$

Removing the brackets:

$$10 - 15x = x + 14$$

$$10 - 15x + 15x = x + 14 + 15x$$

$$10 = 16x + 14$$

$$10 - 14 = 16x + 14 - 14$$

$$-4 = 16x$$

$$-\frac{4}{16} = \frac{16x}{16}$$

$$x = -\frac{1}{4}$$

Check: when $x = -\frac{1}{4}$,

$$5(2 - 3x) = 5\left(2 + \frac{3}{4}\right) = 5\left(2\frac{3}{4}\right) = 13\frac{3}{4}$$

$$x + 14 = -\frac{1}{4} + 14 = 13\frac{3}{4}$$

Exercise 14.1

Solve the following equations:

1. $x - 3 = 5$

2. $x + 4 = 9$

3. $\dfrac{x}{5} = 7$

4. $2x = 16$

5. $x + 7 = 4$

6. $\dfrac{1}{3}x = 5$

7. $5x = 13$

8. $x + 5 = 0$

9. $3x - 4 = 11$

10. $3y + 5 = 11$

11. $\dfrac{1}{2}x - 3 = 1$

12. $\dfrac{x}{3} + 2 = 6$

13. $2a + 3 = 8$

14. $5b - 2 = 7$

15. $2c + 7 = 1$

16. $3d + 11 = 4$

17. $2n + 3n = 15$

18. $5k = 6 + 2k$

19. $7 - p = p$

20. $3x = 8 + 5x$

21. $3(x + 2) = 15$

22. $2(y - 5) = 7$

23. $4(7 - x) = 11$

24. $2(2p - 7) = 9$

25. $3 + 2n = 9 + n$

26. $8 - x = 2 + 3x$

27. $17 + 3p = 9 - p$

28. $13 + 2y = 7 - 3y$

29. $3(x - 4) = 2(x - 3)$

30. $5(7 - y) = 3(2 - y)$

31. $4(n - 2) - 2(n + 3) = 11$

32. $7(h + 3) - 2(h - 4) = 9$ **33.** $x(x + 5) = x(x + 2) + 12$

34. $y(y - 6) - y(y - 1) = 8$ **35.** $n(n - 6) = n^2 - 4$

36. $p^2 - p(p + 7) = 14$ **37.** $(x + 6)(x + 1) = (x + 3)(x + 1)$

38. $(y - 7)(y + 3) = (y + 1)(y - 6)$ **39.** $(n - 3)(n - 5) = (n - 1)(n - 4)$

40. $(p - 4)^2 = (p - 3)(p - 2)$

Solving linear equations containing fractions

Method 1   First make all the denominators the same.

Example 1 $\dfrac{x}{3} + \dfrac{x}{4} = 5$

Make each
denominator 12:

$$\frac{4x}{12} + \frac{3x}{12} = \frac{60}{12}$$

Add the fractions on the left:

$$\frac{7x}{12} = \frac{60}{12}$$

$$7x = 60$$

$$x = \frac{60}{7} = 8\frac{4}{7}$$

Example 2 $\dfrac{5}{x} - \dfrac{3}{4} = \dfrac{1}{2x}$

Make each
denominator $4x$:

$$\frac{20}{4x} - \frac{3x}{4x} = \frac{2}{4x}$$

$$\frac{20 - 3x}{4x} = \frac{2}{4x}$$

$$20 - 3x = 2$$

$$18 = 3x$$

$$x = 6$$

Example 3 $\dfrac{x + 3}{6} - \dfrac{x - 3}{2} = \dfrac{2}{3}$

Make each denominator 6:

$$\frac{x + 3}{6} - \frac{3(x - 3)}{6} = \frac{4}{6}$$

$$\frac{x + 3 - 3(x - 3)}{6} = \frac{4}{6}$$

$$x + 3 - 3(x - 3) = 4$$

$$x + 3 - 3x + 9 = 4$$

$$-2x + 12 = 4$$

$$8 = 2x$$

$$4 = x$$

Method 2   First remove the denominators by multiplying each term by a suitable number or expression.

Example 4 $\dfrac{x}{2} - \dfrac{x}{5} = \dfrac{3}{4}$

Multiply each term by 20:

$$\dfrac{20x}{2} - \dfrac{20x}{5} = \dfrac{60}{4}$$

$$10x - 4x = 15$$

$$6x = 15$$

$$x = 2\dfrac{1}{2}$$

Example 5 $\dfrac{x + 1}{3} + \dfrac{x - 1}{4} = \dfrac{1}{2}$

Multiply each term by 12:

$$\dfrac{12(x + 1)}{3} + \dfrac{12(x - 1)}{4} = \dfrac{12}{2}$$

$$4(x + 1) + 3(x - 1) = 6$$

$$4x + 4 + 3x - 3 = 6$$

$$7x = 5$$

$$x = \dfrac{5}{7}$$

Exercise 14.2

Solve the following equations:

1. $\dfrac{x}{3} + \dfrac{x}{2} = 10$

2. $\dfrac{y}{2} - \dfrac{y}{5} = 6$

3. $\dfrac{n}{3} - \dfrac{n}{5} = \dfrac{4}{3}$

4. $\dfrac{a}{4} - \dfrac{a}{8} = 2$

5. $\dfrac{1}{2}x + \dfrac{1}{3}x = 5$

6. $\dfrac{1}{4}n - \dfrac{1}{5}n = 3$

7. $\dfrac{2x}{3} + \dfrac{3x}{4} = 34$

8. $\dfrac{5x}{6} - \dfrac{3x}{4} = \dfrac{2}{3}$

9. $\dfrac{7y}{10} = 6 + \dfrac{2y}{5}$

10. $\dfrac{n}{4} = \dfrac{n}{3} - \dfrac{1}{6}$

11. $\dfrac{x + 3}{5} = \dfrac{x}{2}$

12. $\dfrac{n + 3}{4} = \dfrac{n + 2}{5}$

13. $0.3x + 1.2 = 0.5x$

14. $0.7x - 2 = 1.1x$

15. $\sqrt{\dfrac{x - 1}{3}} = 2$ (Square both sides.)

16. $\sqrt{\dfrac{y + 13}{2}} = 3$

17. $\dfrac{y - 1}{4} + \dfrac{y + 1}{3} = \dfrac{1}{2}$

18. $\dfrac{p + 4}{3} + \dfrac{p + 2}{5} = 2$

19. $\dfrac{4x - 1}{3} = 1 + \dfrac{2x + 1}{2}$

20. $\dfrac{2y + 1}{5} - 1 = \dfrac{y - 3}{2}$

21. $\dfrac{4}{x} - \dfrac{1}{2} = \dfrac{3}{2x}$

22. $\dfrac{2}{3} + \dfrac{1}{y} = \dfrac{11}{3y}$

23. $\dfrac{3}{5} - \dfrac{4}{5a} = \dfrac{2}{a}$

Cross-multiplying

Suppose that $\dfrac{a}{b} = \dfrac{c}{d}$.

Multiplying each side by bd, we have $\dfrac{abd}{b} = \dfrac{cbd}{d}$.

This simplifies to $ad = bc$ (by cancelling b on the left and d on the right.)

We can go from $\dfrac{a}{b} = \dfrac{c}{d}$ to $ad = bc$ in a single step, known as 'cross-multiplying'. a and d are multiplied together, and b and c are multiplied together, like this:

$$\dfrac{a}{b} \diagdown \dfrac{c}{d} \text{ giving } a \times d = b \times c$$

This method can be used to shorten the working, when an equation has just one fraction on each side.

Example 1 $\quad \dfrac{6}{y} = \dfrac{8}{5}$

$6 \times 5 = 8 \times y$

$30 = 8y$

$y = \dfrac{30}{8} = 3\dfrac{3}{4}$

Example 2 $\quad \dfrac{5}{n-2} = \dfrac{7}{n+3}$

$5(n+3) = 7(n-2)$

$5n + 15 = 7n - 14$

$29 = 2n$

$n = 14\dfrac{1}{2}$

Exercise 14.3

Solve the following equations:

1. $\dfrac{3}{x} = \dfrac{2}{5}$ 2. $\dfrac{4}{y} = \dfrac{5}{7}$ 3. $\dfrac{2}{n} = \dfrac{1}{4}$ 4. $\dfrac{5}{x} = 3$

5. $\dfrac{7}{x} - \dfrac{4}{3} = 0$ $\left(\text{First write this as } \dfrac{7}{x} = \dfrac{4}{3}.\right)$

6. $\dfrac{5}{x} + \dfrac{2}{3} = 0$ 7. $\dfrac{x+1}{5} = \dfrac{3}{2}$ 8. $\dfrac{y-2}{5} = \dfrac{3}{4}$

9. $\dfrac{2}{3} = \dfrac{7}{x+5}$ 10. $\dfrac{7}{y-3} = \dfrac{2}{5}$ 11. $\dfrac{n+2}{3} = \dfrac{n+5}{4}$

12. $\dfrac{x+3}{7} = \dfrac{x}{5}$ 13. $\dfrac{5}{n-1} = \dfrac{3}{n-4}$ 14. $\dfrac{3}{y-1} = \dfrac{4}{y+1}$

15. $\dfrac{2}{3p+1} = \dfrac{1}{2p-5}$ 16. $\dfrac{4}{1-3y} = \dfrac{3}{2-y}$ 17. $\sqrt{\dfrac{5x}{x+1}} = 2$

Problems

Exercise 14.4

1. A certain number is multiplied by 5, and 6 is subtracted from the answer. The result is 14. Let the number be x. Write down an equation for x and then solve it.

2. There are 35 coins in a box. Some of them are 20p coins and the rest are 10p coins. The total value is £4. If there are x 20p coins, how many 10p coins are there? Form an equation for x and then solve it.

3. Mary is 4 years older than John. Let Mary's age be x years. Write down an expression for John's age. If three times Mary's age equals four times John's, write down an equation for x and then solve it.

4. The width of a rectangle is $\frac{1}{3}$ of its length. Its perimeter is 40 cm. Let the length be x cm. Write down an equation for x and then solve it.

5. At the start of a game, x counters were shared equally among 6 players. How many did each get? In the next game, the same number of counters were shared equally among 5 players. Each got 2 more than in the previous game. Form an equation for x and then solve it.

6. When cards are dealt to 6 people, each receives 3 more than when the same cards are dealt to 8 people. How many cards are there?

7. In a competition, a prize of £60 was shared equally among x girls so that each girl received £$\frac{60}{x}$. In a similar competition, a prize of £60 was shared equally among $2x$ boys. Write an expression for the amount received by each boy. If each boy received £3 less than each girl, write down an equation for x and then solve it.

8. A boat travelled 3 miles upstream at a speed of x miles per hour. How long did the journey take? On the return journey downstream, its speed was $2x$ miles per hour. If the downstream journey took a quarter of an hour less than the upstream journey, write down an equation for x and then solve it.

Inequalities

$>$ means 'is greater than'. Thus $7 > 5$ and $-2 > -3$.

If $a > b$, then a is to the right of b on the number line.

The number line

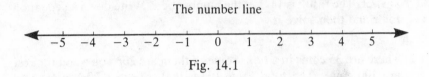

Fig. 14.1

$<$ mean 'is less than'. Thus $6 < 9$ and $-4 < 1$.

If $c < d$, then c is to the left of d on the number line.

$\geqslant$ means 'is greater than or equal to'. Suppose that $x \geqslant 3$.

Then on the number line x is either to the right of 3 or at 3.

$\leqslant$ means 'is less than or equal to'.

The statement $-5 < n < -1$ means that the number n is greater than -5 and less than -1. If n is an integer, then the only possible values for it are -4, -3, and -2. If n is a real number, then there is an infinite number of possible values. Three possible values are -2.38, $-3\frac{11}{17}$ and -4.999.

The inequality $-4 \leqslant k < 2$, where k is a real number, is illustrated in Fig. 14.2. We use ● to show that -4 is a possible value and ○ to show that 2 is not a possible value.

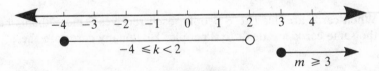

Fig. 14.2

Figure 14.2 also illustrates the inequality $m \geqslant 3$. The arrow shows that the line has no end, since there is no upper limit to the size of m.

Rules for inequalities

1. The same number may be added to or subtracted from each side of an inequality. For example,

    since $1 > -4$ we have $1 + 2 > -4 + 2$, that is $3 > -2$

    and $1 - 3 > -4 - 3$, that is $-2 > -7$.

2. Multiplying or dividing both sides of an inequality by a positive number leaves the inequality sign unchanged. For example,

$$\text{since } 1 > -4 \text{ then } 3 > -12 \text{ and } \frac{1}{2} > -2.$$

3. Multiplying or dividing both sides by a negative number reverses the sign of the inequality. For example,

$$\text{since } 1 > -4 \text{ then } 1 \times (-2) < -4 \times (-2), \text{ that is } -2 < 8$$

$$\text{and also } 1 \div (-2) < -4 \div (-2), \text{ that is } -\frac{1}{2} < 2.$$

Solving inequalities

The steps are the same as for solving equations. But remember rule 3 above.

Example 1 $3x + 7 > 1$

Subtracting 7 from both sides:

$$3x + 7 - 7 > 1 - 7$$

$$3x > -6$$

Dividing both sides by 3:

$$x > -2$$

Example 2 $x - 6 > 3x + 4$

Adding 6 to both sides:

$$x - 6 + 6 > 3x + 4 + 6$$

$$x > 3x + 10$$

Subtracting $3x$ from both sides:

$$x - 3x > 3x + 10 - 3x$$

$$-2x > 10$$

Dividing both sides by -2:

$$x < -5$$

(Notice that the sign is reversed.)

Example 3 If x and y are elements of $\{1, 2, 3\}$, list the elements of $\{(x, y) : y > x\}$.

We have to list all possible pairs of numbers from 1 to 3, such that the second number in the pair is larger than the first.

The pairs are $(1, 2)$, $(1, 3)$ and $(2, 3)$.

For inequalities in the $x - y$ plane, see page 163 in the section on graphs.

Exercise 14.5

1. n is an integer. State all possible values of n such that:

(i) $7 < n < 10$ (ii) $-3 < n < 2$ (iii) $1 \leqslant n \leqslant 3$

(iv) $-5 < n \leqslant -3$ (v) $0 \leqslant n < 3$.

2. k is an integer. State all possible values of k such that:

(i) $1 < 2k < 9$ (ii) $-4 < 3k \leqslant 6$.

3. $p \in \{-3, -2, -1, 0, 1, 2\}$. State the possible values of p if:

(i) $p < -1$ (ii) $p \geqslant 1$ (iii) $-2 < p \leqslant 2$.

4. State the inequalities for a, b and c, as shown in Fig. 14.3.

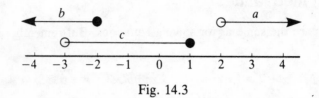

Fig. 14.3

5. Draw a figure to illustrate: (i) $d > -1$ (ii) $e \leqslant 3$

(iii) $-2 < f < 3$ (iv) $1 < g \leqslant 4$ (v) $-4 \leqslant h < 0$.

6. Solve the following inequalities, where x is a real number:

(i) $x - 5 > 2$ (ii) $x + 3 < -4$ (iii) $5x - 2 \leqslant 13$

(iv) $1 + 3x \geqslant 5$ (v) $2 - x < 8$ (vi) $5 - 2x > x + 8$.

7. n is a real number.

(i) Solve $2 < n + 3 < 5$. (Solve separately the inequalities $2 < n + 3$ and $n + 3 < 5$, and then combine the results.)

(ii) Solve $5 < 4 - n < 8$.

8. List the integers, positive and negative, for which $x^2 < 5$.

9. y is real. State the range of values for y for which $y^2 \leqslant 16$.

10. x and y are elements of $\{1, 2, 3, 4\}$. List the elements of:

(i) $\{(x, y) : y < x\}$ (ii) $\{(x, y) : y > x + 1\}$

(iii) $\{(x, y) : x + y \leqslant 3\}$ (iv) $\{(x, y) : x + y \geqslant 7\}$.

11. Find the set of integers which satisfy $2 \leqslant x + 4 < 6$. (Solve separately $2 \leqslant x + 4$ and $x + 4 < 6$ and then combine the results.)

12. Find the set of integers which satisfy $-3 < 2x + 3 \leqslant 1$.

15 Formulae

Constructing formulae

An aircraft travels at 740 km/h for 3 hours. It travels 740 km each hour, so in 3 hours it travels $740 \times 3 = 2220$ km.

A ship travels at v km/h for t hours and covers a distance of d km. It travels v km each hour, so in t hours it travels vt km. Therefore:

$$d = vt$$

This is a formula for the distance travelled in terms of the speed and time. It is the expression in symbols of the statement:

$$\text{distance} = \text{speed} \times \text{time}.$$

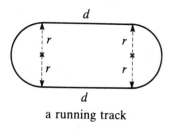

a running track

Fig. 15.1

Figure 15.1 shows a running track, consisting of two semicircles of radius r metres and two straights of length d metres. We can find formulae for: (i) the perimeter of the track, (ii) the area enclosed.

(i) The two semicircles together make a circle of circumference $2\pi r$.
If the perimeter is p metres, then $p = 2\pi r + 2d$.

(ii) The area enclosed can be divided into a rectangle and two semicircles.
The area of the rectangle is $2r \times d = 2rd$ square metres.
The area of the two semicircles (one circle) is πr^2 square metres.
If the area enclosed is A square metres then $A = 2rd + \pi r^2$.

Exercise 15.1

1. (i) I buy 3 oranges at 8p each and 2 lemons at 10p each. How much change do I get from a £1 note?

   (ii) I buy k oranges at x p each and n lemons at y p each. I receive c p change from a £1 note. Express c in terms of k, n, x and y.

2. (i) A hotel charges £15 for the first day and £13 for each additional day. Find the charge for a stay of 6 days.

(ii) A hotel charges £x for the first day and £y for each additional day. The charge for a stay of n days is £c. Express c in terms of x, y and n.

3. (i) An aircraft makes a journey of 2400 km at an average speed of 600 km/h. How long does it take?

   (ii) Write down a formula for the time, t hours, taken for a journey of d km at a speed of s km/h.

   (iii) Use the formula to find the time taken by a ship to travel 70 km at a speed of 20 km/h.

4. (i) A girl has £3.50 in her piggy bank. She puts another 40p in it every week for 6 weeks. How much does she then have in it?

   (ii) She starts with x p in her piggy bank and adds k p to it every week for n weeks. She then has y p. Express y in terms of x, k and n.

5. A lorry weighs x tonnes when empty. When it has a load of y tonnes of sand, it weighs p tonnes.

   (i) Write down a formula for y in terms of x and p.

   (ii) On a weighbridge the weight of the loaded lorry is 12.3 tonnes. Its weight when empty is 4.9 tonnes. Use the formula to find the weight of the load.

6. Write down a formula for each of the following statements.

   (i) The length of a rectangle is found by dividing its area by its width. (Use L, A and W.)

   (ii) If the area of a square is given, the length of a side is found by taking the square root of the area. (Use A and L.)

   (iii) The volume of a sphere is approximately 4.2 times the cube of the radius. (Use V and r.)

7. Write down a formula for each of the following statements.

   (i) Given one of the base angles of an isosceles triangle, the angle at the apex can be found by doubling this base angle and subtracting the answer from 180°.

   (ii) The size of an interior angle of a regular polygon is found by dividing 360° by the number of sides and then subtracting the answer from 180°.

8. A man has a car journey of x km from his home to the factory where he works. He travels to the factory each morning and home again each afternoon. He notices one day that the car's meter shows d km. After n more days it shows f km. Assuming he used the car only for journeys to and from work in those n days, state a formula for f in terms of x, d and n.
   Use the formula to find f if $d = 19273$, $x = 75$ and $n = 6$.

9. The cost of running a coach to the seaside is £x. The coach carries n passengers who each pay y pence. Write down a formula for the profit £p that the coach company makes.

   Use the formula to find the profit for a coach carrying 30 passengers who each pay £1.60, if the coach costs £34 to run.

   What happens if there are only 15 passengers, who each pay £1.60?

10. Find a formula for the area, A m², of a concrete path of width w metres round a circular fish pond of radius r metres.

    Calculate A correct to 2 sig.fig. for r = 8 and w = 1, taking π as 3.14.

Rearranging formulae

To calculate the area of a circle when given its radius, we use the formula $A = \pi r^2$. We can also use this formula to calculate the radius of a circle when given its area. For example, if the area is 58 cm², then $\pi r^2 = 58$,

$$3.14r^2 \approx 58, \qquad r^2 \approx \frac{58}{3.14} \approx 18.47, \qquad r \approx \sqrt{18.47} \approx 4.3.$$

If we have to calculate the radii of several circles from their areas, it is better to rearrange the formula so that r becomes the subject. This

gives $r = \sqrt{\dfrac{A}{\pi}}$. The above calculation then becomes $r \approx \sqrt{\dfrac{58}{3.14}} \approx 4.3$.

The steps involved in rearranging formulae are the same as the steps used in solving equations. This is shown in the following four examples.

Example 1   (i) Solve $7 = 2 + 3x$.
                (ii) Make t the subject of $v = u + at$.

(i) $7 = 2 + 3x$ (ii) $v = u + at$

Subtracting 2 from each side: Subtracting u from each side:

$$7 - 2 = 3x$$ $$v - u = at$$

Dividing both sides by 3:   Dividing both sides by a:

$$\frac{5}{3} = \frac{3x}{3} = x,$$ $$\frac{v - u}{a} = \frac{at}{a} = t$$

$$x = 1\frac{2}{3}$$ $$t = \frac{v - u}{a}$$

Example 2   (i) Solve $76 = 13x^2$.

(ii) Make r the subject of $V = \pi r^2 h$.

(i) $13x^2 = 76$

Dividing both sides by 13:

$$x^2 = \frac{76}{13} \approx 5.85$$

Taking the square root of each side:

$$x \approx \sqrt{5.85} \approx 2.4$$

(ii) $\pi r^2 h = V$

Dividing both sides by πh:

$$r^2 = \frac{V}{\pi h}$$

Taking the square root of each side:

$$r = \sqrt{\frac{V}{\pi h}}$$

Example 3   (i)   Solve $17 = 5\sqrt{\dfrac{x}{4}}$.

(ii) Make l the subject of $T = 2\pi\sqrt{\dfrac{l}{g}}$.

(i) $5\sqrt{\dfrac{x}{4}} = 17$

Dividing both sides by 5:

$$\sqrt{\frac{x}{4}} = \frac{17}{5}$$

Squaring both sides:

$$\frac{x}{4} = \frac{289}{25}$$

Multiplying both sides by 4:

$$x = \frac{289 \times 4}{25}$$

$$\approx 46.2$$

(ii) $2\pi\sqrt{\dfrac{l}{g}} = T$

Dividing both sides by 2π:

$$\sqrt{\frac{l}{g}} = \frac{T}{2\pi}$$

Squaring both sides:

$$\frac{l}{g} = \frac{T^2}{4\pi^2}$$

Multiplying both sides by g:

$$l = \frac{T^2 g}{4\pi^2}$$

Example 4   (i) Solve $3 = \dfrac{x + 7}{x - 1}$.

(ii) Make x the subject of $y = \dfrac{x + 7}{x - 1}$.

(i) Cross-multiplying:

$$3(x - 1) = x + 7$$
$$3x - 3 = x + 7$$
$$3x - x = 3 + 7$$
$$2x = 10$$
$$x = 5$$

(ii) Cross-multiplying:

$$y(x - 1) = x + 7$$
$$yx - y = x + 7$$
$$yx - x = y + 7$$
$$(y - 1)x = y + 7$$
$$x = \frac{y + 7}{y - 1}$$

Exercise 15.2

In each of questions 1 to 12 there is an equation and a formula. First solve the equation, at each step stating what you are doing to the two sides. Then use the same steps to make x the subject of the formula.

1. $x + 3 = 10$; $x + a = b$

2. $9x = 7$; $cx = d$

3. $\frac{1}{4}x = 3$; $\frac{x}{f} = g$

4. $x - 4 = 7$; $x - h = k$

5. $x^2 = 25$; $x^2 = n$

6. $3x^2 = 75$; $px^2 = q$

7. $\sqrt{x} = 9$; $\sqrt{x} = r$

8. $\frac{7}{x} = 3$; $\frac{t}{x} = u$

9. $\frac{2}{5}x = 3$; $\frac{v}{w}x = y$

10. $\frac{x}{5} = \frac{4}{7}$; $\frac{x}{a} = \frac{b}{c}$

11. $5x + 3 = 26$; $dx + e = f$

12. $\frac{1}{3}x - 2 = 5$; $\frac{x}{g} - h = m$

Rearrange the following formulae so that the stated letter is the subject:

13. $n = p + q$; p

14. $h = k - m$; k

15. $c = \pi d$; d

16. $s = \frac{d}{t}$; d

17. $A = h^2$; h

18. $k = \sqrt{p}$; p

19. $v = lbh$; h

20. $A + B + C = 180$; B

21. $A = \frac{1}{2}bh$; h

22. $A = \pi r^2$; r

23. $v = u + at$; a

24. $b = \sqrt{cd}$; d

25. $E = \frac{1}{2}mv^2$; v

26. $I = \frac{PRT}{100}$; R

27. $v^2 = u^2 + 2as$; s

28. $a^2 + b^2 = c^2$; a

29. $v = \frac{4}{3}\pi r^3$; r

30. $t = \frac{1}{3}\sqrt{s}$; s

31. $s = \frac{d}{t}$; t

32. $n = \sqrt{\frac{a}{b}}$; b

33. $T = 2\pi\sqrt{\frac{l}{g}}$; g

34. $y = \frac{2}{3 + x}$; x

35. $y = \frac{x + 5}{x - 2}$; x

36. $y = \frac{3x - 4}{2x - 5}$; x

37. $s = ut + \dfrac{1}{2}at^2$; a **38.** $y = 3x^2 + 5$; x **39.** $y = \dfrac{7}{x} + 3$; x

40. $S = \pi r^2 + 2\pi rh$; h **41.** $\dfrac{1}{u} + \dfrac{1}{v} = \dfrac{1}{f}$; v

Exercise 15.3

1. A rectangular box has a square base of side x cm and a height of 9 cm. Its volume is V cm³.

 (i) Write down a formula for V in terms of x.

 (ii) Rearrange the formula to make x the subject.

 (iii) Calculate x if $V = 900$.

2. A regular polygon has n sides. The size of each angle is $x°$ where
$$x = 180 - \frac{360}{n}.$$

 (i) Calculate x if $n = 10$.

 (ii) Make n the subject of the formula.

 (iii) Calculate n if $x = 150$.

3. From the top of a cliff of height h metres, the approximate distance of the horizon, d kilometres, is given by the formula $d = \sqrt{13h}$.

 (i) Find the distance of the horizon from the top of a cliff of height 208 m.

 (ii) Rearrange the formula to make h the subject.

 (iii) Find the height of a cliff from which the distance of the horizon is 65 km.

4. In triangle ABC, $\hat{B} = \hat{C} = x°$ and $\hat{A} = y°$.

 (i) Write down a formula expressing y in terms of x.

 (ii) If $x = 53$, calculate y.

 (iii) Rearrange the formula to make x the subject.

 (iv) If $y = 112$, calculate x.

5. The formula for converting a temperature of F degrees Fahrenheit to C degrees Celsius is $C = \dfrac{5}{9}(F - 32)$.

 (i) Calculate C when F is: (a) 86 (b) 14.

 (ii) Rearrange the formula to make F the subject.

 (iii) Calculate F when C is: (a) 140 (b) -15.

 (iv) At what temperature is $F = C$?

6. The volume V of a beaker is given by the formula $V =$ $\frac{1}{3}\pi h(2R^2 - r^2)$, where R is the radius of the top, r the radius of the bottom and h the vertical height of the beaker.

(a) Taking π as 3.142, and showing the steps in your calculations, find the volume when $h = 11$, $R = 4$ and $r = 2$.

(b) Make R the subject of the formula.

(c) Given that the volume V can also be expressed by the formula $V = \frac{7}{3}\pi r^2 h$, express R in terms of r.   (L)

16 Simultaneous linear equations

The equation $3x - 2y = 10$ contains two unknown quantities, x and y. It is satisfied by an infinite number of pairs of values, (x, y), such as $(4, 1)$, $(6, 4)$, $(8, 7)$, $\left(5, 2\frac{1}{2}\right)$, $(5.8, 3.7)$, $(-2, -8)$. The equation $4x - 3y = 12$ is also satisfied by an infinite number of pairs of values such as $(0, -4)$, $(3, 0)$, $(6, 4)$, $(9, 8)$. There is only one pair of values, $(6, 4)$, which satisfies both equations. When two such equations are true at the same time, they are called *simultaneous equations*.

Two methods of solution, by substitution and by elimination, are given here.

Solving simultaneous equations by substitution

Example

$$2x + y = 5 \qquad (1)$$
$$8x + 3y = 21 \qquad (2)$$

Equation (1) can be written as:

$$y = 5 - 2x \qquad (3)$$

Substituting $5 - 2x$ for y in (2) gives:

$$8x + 3(5 - 2x) = 21$$
$$8x + 15 - 6x = 21$$
$$2x = 6$$
$$x = 3$$

Substituting for x in (3) gives:

$$y = 5 - 6 = -1$$

The solution is therefore $x = 3$, $y = -1$.

Checking by substitution in (1): $2x + y = 6 + (-1) = 5$

Checking by substitution in (2): $8x + 3y = 24 + 3(-1) = 24 - 3 = 21$

This method is useful when one of the equations has just one x (that is, no number in front of x) or one y.

Solving simultaneous equations by elimination

(a) By adding the equations

Example

$$3x + 4y = 23 \tag{1}$$
$$5x - 4y = 17 \tag{2}$$

Adding the two equations gives:

$$3x + 4y + 5x - 4y = 23 + 17$$
$$8x = 40$$
$$x = 5$$

Substituting 5 for x in (1) gives:

$$15 + 4y = 23$$
$$4y = 8$$
$$y = 2$$

The solution is $x = 5$, $y = 2$.

Checking by substitution in (2): $5x - 4y = 5 \times 5 - 4 \times 2 = 17$

We do not check in (1) because this has been used to obtain the value of y.

(b) By subtracting one equation from the other

Example

$$6x + 5y = 14 \tag{1}$$
$$2x + 5y = 8 \tag{2}$$

Adding does not eliminate y. It gives $8x + 10y = 22$. Subtracting (2) from (1) gives:

$$6x + 5y - 2x - 5y = 14 - 8$$
$$4x = 6$$
$$x = 1\frac{1}{2}$$

Substituting in (1) gives:

$$9 + 5y = 14$$
$$5y = 5$$
$$y = 1$$

The solution is $x = 1\frac{1}{2}$, $y = 1$.

Checking by substitution in (2): $2x + 5y = 2 \times 1\frac{1}{2} + 5 \times 1 = 3 + 5 = 8$

(c) By multiplication before adding or subtracting

Example

$$5x + 2y = 18 \qquad (1)$$

$$7x - 3y = 2 \qquad (2)$$

We multiply equation (1) by 3 and equation (2) by 2 so that we have $6y$ in both equations.

(1) × 3 gives: $15x + 6y = 54$

(2) × 2 gives: $14x - 6y = 4$

Adding, and continuing as in method (a), we obtain the solution $x = 2$, $y = 4$.

(d) By eliminating x

We eliminated y in the three examples above. Sometimes the working is simpler if we eliminate x.

Example

$$3x + 11y = 10 \qquad (1)$$

$$x + 5y = 6 \qquad (2)$$

Multiplying (2) by 3: $3x + 15y = 18$

(1) is: $3x + 11y = 10$

Subtracting, and continuing as in the other examples, we obtain the solution $x = -4$, $y = 2$.

Exercise 16.1

Solve, by substitution:

1. $y = 4x - 3$ **2.** $y = x - 10$ **3.** $y - 2x = 4$ **4.** $5x + 3y = 2$
    $x + y = 7$ $2x + 3y = 5$ $3y - 8x = 10$ $y + 2x = 0$
5. $n = 2p - 1$ **6.** $a = 2 + b$ **7.** $f + 3g = 7$ **8.** $x - 2y = 9$
    $n + 3p = 9$ $4a - 3b = 5$ $9f - 2g = 5$ $3x + 7y = 1$

Solve by elimination:

9. $x + y = 14$ **10.** $3x - 2y = 10$ **11.** $4x + y = 7$
    $x - y = 4$ $x + 2y = 6$ $3x + y = 5$
12. $7x + 3y = 15$ **13.** $x + 5y = 17$ **14.** $5x + 4y = 4$
    $4x + 3y = 6$ $3x - 5y = 11$ $3x + 4y = 8$
15. $x + 2y = 13$ **16.** $2a - 5b = 9$ **17.** $8p - 7q = 13$
    $3x - y = 11$ $3a + 2b = 4$ $3p + 2q = 28$

18. $7k + 3n = 21$
    $5k + 2n = 15$
19. $3c + 4d = 1$
    $4c + 3d = 6$
20. $x + 2y = 7$
    $x + y = 2$
21. $3x + 5y = 5$
    $3x + 9y = -3$
22. $3x + 5y = 11$
    $x + 2y = 5$
23. $3x - 4y + 23 = 0$
    $3x + 5y + 5 = 0$
24. $2x + 5y + 11 = 0$
    $3x + 4y + 6 = 0$

Exercise 16.2

These are harder questions.
Solve:

1. $3x + 7y = 13$
   $2x - 7y = 4$
2. $4x + 2y = 11$
   $3x + 4y = 5$
3. $5x + 7y + 2 = 0$
   $3x + 2y + 5 = 0$
4. $2x - 5y = 7$
   $3x - 4y = 6$

5. The identity $x - 4 = A(x + 2) + B(x + 3)$ is true for all values of x. By putting $x = 0$ and $x = 4$, form two equations for A and B and so find the values of A and B.

6. The identity $5y + 4 = C(y + 1) + D(y + 2)$ is true for all values of y. By putting $y = 0$ and $y = 1$, find the values of C and D.

7. 5 pencils and 3 pens cost 72p; 3 pencils and 5 pens cost 88p. Let each pencil cost x p and each pen cost y p. Write down two equations for x and y. By solving them, find the cost of one pencil and the cost of one pen.

8. For a certain concert there were x seats costing £2 each and y seats costing £1 each. When all the seats were sold the takings were £310. For the next concert in the same hall, the price of the dearer seats was raised to £3 and the price of the others to £2. This time the takings were £530, when all the seats were sold. Write down two equations and use them to find the total number of seats in the hall.

9. The sum of two numbers is 32 and their difference is 6. Use two equations to find the numbers.

10. Adrian is 5 years older than his sister Betty. The sum of their ages is 29 years. Let Adrian's age be x years and Betty's be y years. Write down two equations and use them to find the two ages.

11. In a factory, goods are packed in both large and small boxes. The total mass of 3 large and 7 small boxes is 48.3 kg. The total mass of 4 large and 5 small boxes is 46.2 kg. Find the mass of each type of box.

12. A bag contains 29 coins. Some of them are 5p coins and the rest are 2p coins Let the number of 5p coins be x and the number of 2p coins be y. Write down an equation for x and y. The value of the coins is £1. Write down another equation for x and y. By solving the equations, find the number of each type of coin.

13. The total time taken to walk x km at 6 km/h and to cycle y km at 15 km/h is 3 hours. The total time to walk y km and cycle x km at the same speeds is 4 hours. Find x and y.

14. Dave is x years old and Jill is y years old. The ratio of their ages is $5:7$ $\left(\text{that is, } \dfrac{x}{y} = \dfrac{5}{7}\right)$. In three years' time the ratio of their ages will be $3:4$. Use two equations to find x and y.

17 Quadratic equations

An equation of the form $ax^2 + bx + c = 0$ where a, b and c are numbers, is called a *quadratic equation*. The values of x which satisfy the equation are called its *roots*.

Solving quadratic equations

Equations which have rational roots can be solved by factorisation. The basis of the method is as follows:

If $ab = 0$ then either $a = 0$ or $b = 0$ (or both are 0)

Thus if $(x - 5)(x + 3) = 0$

then either $(x - 5) = 0$ or $(x + 3) = 0$

from which $x = 5$ or $x = -3$

Example 1 $x^2 + 3x - 10 = 0$

Factorising the left-hand side:

$(x + 5)(x - 2) = 0$

Either $x + 5 = 0$ or $x - 2 = 0$
$$x = -5 \text{ or } x = 2$$

Example 2 $6x^2 + 10 = 19x$

Subtracting $19x$ from each side, we have:

$6x^2 - 19x + 10 = 0$

$(2x - 5)(3x - 2) = 0$

Either $2x - 5 = 0$ or $3x - 2 = 0$

$2x = 5$ or $3x = 2$

$$x = 2\frac{1}{2} \text{ or } x = \frac{2}{3}$$

Example 3 $5x^2 + 4x = 0$

(This equation has only two terms, because $c = 0$.)

Factorising: $x(5x + 4) = 0$

Either $x = 0$ or $5x + 4 = 0$

$$x = 0 \text{ or } x = -\frac{4}{5} \text{ (or } -0.8)$$

(In an equation of this type, students often 'lose' the solution $x = 0$ because they divide $5x^2 + 4x = 0$ by x.)

Example 4 $16x^2 = 25$

$$16x^2 - 25 = 0$$

$$(4x + 5)(4x - 5) = 0$$

Either $4x + 5 = 0$ or $4x - 5 = 0$

$$x = -\frac{5}{4} \text{ or } x = \frac{5}{4}$$

Exercise 17.1

Solve the following equations:

1. $(x - 6)(x - 2) = 0$
2. $(x - 8)(x + 5) = 0$
3. $x(x - 3) = 0$
4. $x(x + 1) = 0$
5. $x^2 + 5x + 6 = 0$
6. $x^2 - 7x + 10 = 0$
7. $x^2 - 4x + 3 = 0$
8. $x^2 + 8x + 7 = 0$
9. $x^2 - 2x = 0$
10. $x^2 + 5x = 0$
11. $x^2 - 2x - 3 = 0$
12. $x^2 + 4x - 5 = 0$
13. $x^2 + 6x = 7$
14. $x^2 = x + 2$
15. $x^2 - 9 = 0$
16. $x^2 = 36$
17. $(3x + 5)(2x - 7) = 0$
18. $(2x + 1)(5x + 2) = 0$
19. $2x^2 + 7x + 5 = 0$
20. $3x^2 - 7x + 2 = 0$
21. $6x^2 + x - 1 = 0$
22. $10x^2 - 7x + 1 = 0$
23. $5x^2 + 3x - 2 = 0$
24. $2x^2 - 3x - 2 = 0$
25. $6x^2 + 7x + 2 = 0$
26. $6x^2 - 11x + 3 = 0$
27. $9x^2 - 4 = 0$
28. $4x^2 - 25 = 0$
29. $3x^2 - 7x + 4 = 0$
30. $3x^2 + 13x + 4 = 0$
31. $x(2x + 3) = 2$
32. $3(x^2 - 1) = 8x$
33. $3x(x + 2) = x$
34. $2x(x - 1) + x^2 = 0$
35. $(x - 1)(x - 2) = 2$
36. $(x - 3)(x + 4) = 18$

Constructing quadratic equations with given numbers as roots

Example 1   Construct an equation having 4 and -7 as roots.

We reverse the working for solving an equation.

If $x = 4$ or $x = -7$

then either $x - 4 = 0$ or $x + 7 = 0$

Hence $(x - 4)(x + 7) = 0$

Expanding: $x^2 + 3x - 28 = 0$

Check: when $x = 4$, $x^2 + 3x - 28 = 16 + 12 - 28 = 0$

when $x = -7$, $x^2 + 3x - 28 = 49 - 21 - 28 = 0$

Example 2   Construct an equation having $-1\frac{1}{2}$ and $1\frac{1}{4}$ as roots.

$$x = -1\frac{1}{2} \text{ or } x = 1\frac{1}{4}$$

$$2x = -3 \text{ or } 4x = 5$$

$$2x + 3 = 0 \text{ or } 4x - 5 = 0$$

Hence $(2x + 3)(4x - 5) = 0$

Expanding: $8x^2 + 2x - 15 = 0$

Check: when $x = -1\frac{1}{2}$,

$$8x^2 + 2x - 15 = 8 \times \frac{9}{4} - 3 - 15 = 18 - 18 = 0$$

when $x = 1\frac{1}{4}$,

$$8x^2 + 2x - 15 = 8 \times \frac{25}{16} + \frac{5}{2} - 15 = 12\frac{1}{2} + 2\frac{1}{2} - 15 = 0$$

Exercise 17.2

Construct equations with the following pairs of numbers as roots. Your equations should not contain any fractions. Check each one as in the examples above.

1. 2 and 3   2. 5 and 1   3. 3 and −5   4. −2 and 7

5. −4 and −1   6. −3 and −6   7. $\frac{1}{2}$ and 2   8. $\frac{2}{3}$ and $\frac{1}{3}$

9. $1\frac{1}{2}$ and −2   10. $-\frac{2}{5}$ and 5   11. $-\frac{1}{3}$ and $-\frac{4}{3}$ 12. $2\frac{1}{2}$ and $-1\frac{1}{2}$

A quick way to construct equations

Suppose that an equation has α and β as its roots.

Then $x = \alpha$ or $x = \beta$, and so $x - \alpha = 0$ or $x - \beta = 0$.

Hence $(x - \alpha)(x - \beta) = 0$

Expanding: $x^2 - \alpha x - \beta x + \alpha\beta = 0$

$$x^2 - (\alpha + \beta)x + \alpha\beta = 0$$

That is: $x^2 - \text{(sum of roots)}x + \text{(product of roots)} = 0$

This can be used to write down an equation having two given numbers as its roots.

Example 1   Construct an equation having 9 and -13 as its roots.

Sum of roots $= 9 - 13 = -4$
Product of roots $= 9 \times (-13) = -117$

The equation is $x^2 - (-4)x - 117 = 0$
$$x^2 + 4x - 117 = 0$$

Example 2   Construct an equation having $-\dfrac{1}{2}$ and $-\dfrac{1}{3}$ as its roots.

Sum of roots $= -\dfrac{1}{2} - \dfrac{1}{3} = -\dfrac{5}{6}$

Product of roots $= \left(-\dfrac{1}{2}\right) \times \left(-\dfrac{1}{3}\right) = \dfrac{1}{6}$

The equation is $x^2 + \dfrac{5}{6}x + \dfrac{1}{6} = 0$

Multiplying by 6: $6x^2 + 5x + 1 = 0$

Checking roots

We have shown above that for the equation $x^2 + bx + c = 0$, $-b =$ sum of roots and $c =$ product of roots. This can be used for checking roots that have already been calculated.

Example   A student gives 5 and -2 as the roots of $x^2 + 3x - 10 = 0$. Are these the correct roots?

Product of his roots $= 5 \times (-2) = -10$ which is correct.

Sum of his roots $= 5 + (-2) = 3$. It should be -3.

His answer is wrong.

Exercise 17.3

Write down equations that have as roots:

1. 3 and 4   **2.** 8 and 1   **3.** -2 and 5   **4.** -6 and 4
5. -5 and -2 **6.** -3 and -1 **7.** -2 and 2   **8.** 0 and 3

Find, in the form $ax^2 + bx + c = 0$ (where a, b and c are integers), the equations that have as roots:

9. $\dfrac{1}{2}$ and 1   **10.** $\dfrac{1}{3}$ and $\dfrac{2}{3}$ **11.** $-\dfrac{1}{3}$ and 1   **12.** $-\dfrac{1}{2}$ and $-1\dfrac{1}{2}$

13. Each part of this question has a quadratic equation and two values of x. State whether or not these two values of x are the roots of the equation.

(i) $x^2 - 7x + 6 = 0$; $x = 6$ or 1

(ii) $x^2 - 11x + 10 = 0$; $x = 5$ or 2

(iii) $x^2 - 5x + 4 = 0$; $x = 3$ or 2

(iv) $x^2 + 4x + 3 = 0$; $x = -3$ or -1

(v) $x^2 + 6x - 7 = 0$; $x = 7$ or -1

(vi) $x^2 - 4x - 5 = 0$; $x = 5$ or -1

(vii) $x^2 + 7x + 10 = 0$; $x = -4$ or -3

(viii) $x^2 + 6x + 8 = 0$; $x = -4$ or -2.

Equations with irrational roots

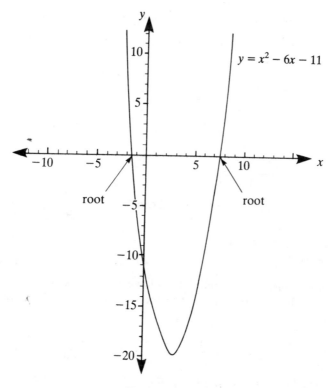

Fig. 17.1

The expression $x^2 - 6x - 11$ cannot be factorised, so we cannot solve the equation $x^2 - 6x - 11 = 0$ by the method used above. But there *are* values of x which satisfy this equation. Approximations to these values can be found by drawing the graph of $y = x^2 - 6x - 11$ and reading from the graph, as accurately as possible, the values of x for which $y = 0$. (See page 158.)

Better approximations can be obtained by the following method.

Solution by formula

The standard quadratic equation is $ax^2 + bx + c = 0$ where a, b and c are numbers ($a \neq 0$). The two roots can be found using this formula:

$$x = \frac{-b \pm \sqrt{b^2 - 4ac}}{2a}$$

One root is $\dfrac{-b + \sqrt{b^2 - 4ac}}{2a}$; the other is $\dfrac{-b - \sqrt{b^2 - 4ac}}{2a}$

Example   Solve $3x^2 - 7x - 2 = 0$, giving the roots correct to 2 d.p.

Here $a = 3, b = -7$ and $c = -2$, so:

$$x = \frac{-(-7) \pm \sqrt{(-7)^2 - 4 \times 3(-2)}}{2 \times 3}$$

$$= \frac{7 \pm \sqrt{49 + 24}}{6} = \frac{7 \pm \sqrt{73}}{6} = \frac{7 \pm 8.544}{6} \text{ (to 3 d.p.)}$$

$$= \frac{15.544}{6} \text{ or } \frac{-1.544}{6}$$

$$= 2.59 \text{ or } -0.26 \text{ (to 2 d.p.)}$$

Exercise 17.4

The equations in questions 1 to 6 have rational roots, so they can be solved by the factors method. For practice, solve them here by using the formula:

1. $x^2 - 6x + 5 = 0$ **2.** $x^2 + 2x - 8 = 0$ **3.** $x^2 - 10x + 21 = 0$
4. $6x^2 + 7x + 2 = 0$ **5.** $2x^2 + 5x - 3 = 0$ **6.** $6x^2 - 11x - 10 = 0$

Solve the following equations, giving your answers correct to 2 d.p.:

7. $x^2 + 6x + 4 = 0$ **8.** $x^2 + 5x + 3 = 0$ **9.** $x^2 + 2x - 5 = 0$
10. $3x^2 + 7x + 3 = 0$ **11.** $2x^2 - 3x - 1 = 0$ **12.** $5x^2 + 9x + 2 = 0$
13. $2x^2 - 11x + 7 = 0$ **14.** $3y^2 + 6y - 7 = 0$ **15.** $4p^2 + 7p - 6 = 0$
16. $3t^2 - 5t - 8 = 0$ **17.** $2k^2 + 4k = 3$ **18.** $2x^2 + 5 = 9x$

Problems

Example   The length of a rectangle is 6 m greater than its width.

The area is 160 m². Find the width.

Let the width be x m.

Then the length is $(x + 6)$ m and the area is $x(x + 6)$ m².

Therefore $x(x + 6) = 160$

So $x^2 + 6x - 160 = 0$

$$(x + 16)(x - 10) = 0$$

$$x = -16 \text{ or } 10$$

As the width must be positive, the answer is 10 metres.

Exercise 17.5

1. I write down two numbers. One of them is 5 larger than the other. I square each number and add the squares together, which gives the answer 73. Let the smaller number be n. Form an equation for n and simplify it. Then solve the equation to find the two original numbers.

2. When a number is added to its square, the answer is 56. Let the number be x. Write down an equation for x and solve it.

3. Two consecutive numbers are squared and added together. The answer is 61. Taking the numbers as n and $n + 1$, form an equation and simplify it. Solve the equation and hence find the two numbers.

4. Two consecutive even numbers are squared and added together. The answer is 340. Find the two numbers.

5. The length of a rectangle is 5 m greater than its width. The area is 104 m². Find the width.

6. The perimeter of a rectangle is 28 cm and its area is 45 cm². If the width is x cm, what is the length? Write down an equation for x and solve it.

7. A piece of wire of length 44 cm is cut into two parts. Each part is bent to form the four sides of a square. If the length of a side of one square is x cm, what is the length of the side of the other square? The sum of the areas of the squares is 65 cm². Form an equation for x and solve it.

8. A rectangle has a length of x cm and its width is $(x - 7)$ cm. The diagonals are each $(x + 1)$ cm. Form an equation for x, using Pythagoras' Theorem, and then solve it. Hence state the length of the rectangle.

9. Triangle ABC has a right-angle at A. AC is 7 cm longer than AB; BC is 2 cm longer than AC. Let AB = y cm. Form an equation for y, using Pythagoras' Theorem. Solve the equation and hence state the lengths of the three sides of the triangle.

10. The straight line $y = x + 2$ cuts the curve $y = 14 - x^2$ at two points A and B. At these points $x + 2 = 14 - x^2$. Solve this equation and so state the co-ordinates of A and B.

11. Show that at the points of intersection of $y = x + 3$ and $xy = 10$, the equation $x^2 + 3x - 10 = 0$ is satisfied. Solve the equation and so find the co-ordinates of the two points.

12. Find, correct to 2 d.p., the values of x at the points of intersection of
$$y = \frac{2}{x} - 2 \text{ and } y = 2x + 4.$$

Equations with fractions

Example $\dfrac{5}{x-2} - \dfrac{3}{x+2} = 2$

Multiplying by $(x - 2)(x + 2)$:
$$\frac{5(x-2)(x+2)}{(x-2)} - \frac{3(x-2)(x+2)}{(x+2)} = 2(x-2)(x+2)$$

$$5(x+2) - 3(x-2) = 2(x^2 - 4)$$
$$5x + 10 - 3x + 6 = 2x^2 - 8$$
$$0 = 2x^2 - 2x - 24$$
$$0 = x^2 - x - 12$$
$$0 = (x-4)(x+3)$$
$$x = 4 \text{ or } -3$$

Exercise 17.6

Solve:

1. $\dfrac{2}{x} - \dfrac{1}{x+2} = \dfrac{1}{3}$

2. $\dfrac{4}{y-3} - \dfrac{4}{y} = \dfrac{3}{10}$

3. $\dfrac{6}{x} + \dfrac{3}{x+1} = 4$

4. $\dfrac{7}{x-1} = 2 + \dfrac{6}{x+1}$

5. $\dfrac{1}{n-3} - \dfrac{1}{n-1} = \dfrac{1}{4}$

6. $\dfrac{2}{x-1} - \dfrac{1}{x+2} = \dfrac{1}{2}$

7. $\dfrac{3}{n-2} - \dfrac{1}{2} = \dfrac{1}{n+3}$

8. $\dfrac{8}{x+2} + \dfrac{7}{2x-1} = 3$

9. $\dfrac{2}{x+1} + \dfrac{1}{3x} = \dfrac{3}{x}$

10. $\dfrac{1}{x-1} - \dfrac{1}{2x} = \dfrac{1}{x}$

11. $\dfrac{3}{n} + \dfrac{5}{n-2} = \dfrac{4}{n+1}$ 12. $\dfrac{2}{x+1} + \dfrac{5}{x+3} = \dfrac{3}{x-3}$

13. In a competition a prize of £30 was shared equally among x people, so that each received £$\dfrac{30}{x}$. In another competition a prize of £30 was shared equally among $x + 4$ people. How much did each receive? If this second amount is £2 less than the first amount, write down an equation for x and solve it.

14. The cost of hiring a minibus for a certain journey was £60. One Saturday x people shared the cost. How much did each pay? The following Wednesday $x - 2$ people shared the cost for the same journey and each person paid £1 more than was paid by each person on the Saturday. Form an equation for x and solve it.

15. A motor boat travelled 8 km up a river at a speed of x km/h. Write down an expression for the time taken. On the return journey downstream the speed was $(x + 4)$ km/h. Write down an expression for the time taken on the return journey. If the difference in time was 10 minutes, write down an equation for x, and show that it reduces to $x^2 + 4x - 192 = 0$. Solve this equation and state the speed of the boat upstream.

16. A racing cyclist completes the uphill section of a mountainous course of 75 km at an average speed of V km/h. Write down, in terms of V, the time taken for the journey.

    He then returns downhill along the same route at an average speed of $(V + 20)$ km/h. Write down the time taken for the return journey.

    Given that the difference between the times is one hour, form an equation in V and show that it reduces to $V^2 + 20V - 1500 = 0$.

    Solve this equation and calculate the time to complete the uphill section of the course.

    Calculate also the cyclist's average speed over the 150 km.   (C)

17. A housewife bought n oranges for 70p. Write down an expression for the price of one orange. Later she bought $(n + 1)$ smaller oranges for 60p. Each larger orange cost 4p more than each smaller orange. Write down an equation for n and solve it.

18. A packet contains b biscuits and has a mass of 400 g. Another packet contains $(b + 5)$ biscuits and has a mass of 450 g. It is found that each biscuit from the second packet is 1 g lighter than each from the first packet. Write down an equation for b and solve it.

Simultaneous equations — one linear and one quadratic

We use the method of substitution. We make x or y the subject of the linear equation and then substitute in the quadratic equation. There are usually two solutions for x and two for y. You must make clear which value of y goes with which value of x.

Example

$$2x^2 + xy - y^2 = 0 \qquad (1)$$
$$x + 4y = 9 \qquad (2)$$

From (2), $x = 9 - 4y.$ $\qquad (3)$

Substituting in (1), $2(9 - 4y)^2 + (9 - 4y)y - y^2 = 0$
$$2(81 - 72y + 16y^2) + 9y - 4y^2 - y^2 = 0$$
$$162 - 144y + 32y^2 + 9y - 5y^2 = 0$$
$$27y^2 - 135y + 162 = 0$$
$$y^2 - 5y + 6 = 0$$
$$(y - 3)(y - 2) = 0$$
$$y = 3 \text{ or } 2.$$

From (3), if $y = 3$, $x = 9 - 12 = -3$.
and   if $y = 2$, $x = 9 - 8 = 1$.

The solutions are $\begin{Bmatrix} x = -3 \\ y = 3 \end{Bmatrix}$ and $\begin{Bmatrix} x = 1 \\ y = 2 \end{Bmatrix}$.

Exercise 17.7

Solve the following pairs of equations:

1. $y = x + 5, x^2 + y^2 = 13$
2. $x^2 - xy = 8, 2x - y = 6$
3. $xy = 8, y = 2 + x$
4. $xy = 2, x + 3y = 5$
5. $5x^2 - 3y^2 = 5, x - y + 1 = 0$
6. $3x^2 + 4y^2 = 19, x + 2y = 5$
7. $2x + y = 7, x^2 + 13xy + y^2 = 91$
8. $x - 3y + 5 = 0, 2x^2 - 3y^2 = 5$
9. $x^2 + 4y^2 = 5, 3x + 2y = 5$
10. $8x^2 + 3y^2 = 35, 4x - 3y = 5$
11. $x^2 - xy + y^2 = 7, 2x = 3y - 7$
12. $3x^2 - 2y^2 + 5 = 0, 3x - 2y = 1$

13. Find the points of intersection of the graphs of $y = x + 3$ and $x^2 - 5x + y = 0$.

14. Find the points of intersection of the graphs of $3x^2 - y^2 = 3$ and $x + y = 1$.

15. The perimeter of a rectangle is 31 cm. If the length is decreased by 3 cm and the breadth is increased by 1 cm, the area becomes 45 cm^2. Express these statements as equations using x and y for the original length and breadth. Solve the equations to find the original dimensions.

18 Functions

Function, domain and range

'Take a number, double it and then add three.' If we apply this instruction to each element of the set $\{-2, -1, 0, 1, 2\}$, we obtain the set $\{-1, 1, 3, 5, 7\}$. This instruction is an example of a *function* or *mapping*. A function maps each element of one set to one and only one element of another set. The first set is called the *domain* and the second set is called the *range*.

Three ways of illustrating the above function are shown in Fig. 18.1.

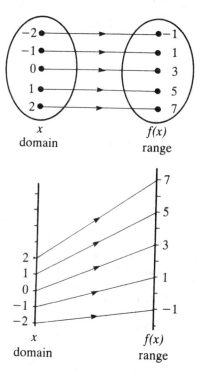

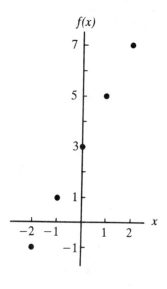

Fig. 18.1

Let the above function be denoted by f. Let x be any element of the domain. Doubling x gives $2x$. Adding 3 gives $2x + 3$. Thus the function f is such that x is mapped to $2x + 3$. This can be written as:

$$f : x \rightarrow 2x + 3$$

The element -2 becomes $2(-2) + 3 = -1$. -1 is called the *image* of -2.

If the image of x is denoted by $f(x)$, then $f(x) = 2x + 3$. Thus $f(-2) = -1$ and $f(2) = 7$.

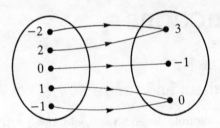

Fig. 18.2

Fig. 18.2 illustrates the function $f : x \rightarrow x^2 - 1$ for the domain $\{-2, -1, 0, 1, 2\}$.

Here the two elements -2 and 2 are both mapped to 3, and the two elements -1 and 1 are both mapped to 0. This is an example of a *many–one mapping*.

The Cartesian graph for this mapping is shown in Fig. 18.3.

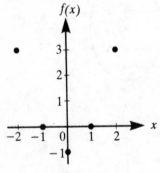

Fig. 18.3

If the domain for this function is {real numbers from -2 to 2}, the Cartesian graph is a continuous curve, since x can have any value from -2 to 2.

For example, taking x as 1.4, we have $f(1.4) = 1.4^2 - 1 = 0.96$.

The graph is shown in Fig. 18.4.

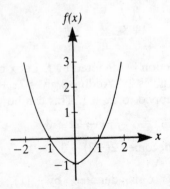

Fig. 18.4

Exercise 18.1

1. The function f is given by $f : x \to 2x - 5$. Find the values of $f(4)$, $f(1)$, $f(0)$ and $f(-2)$.

2. If f is given by $f : x \to 10 - x^2$, find the values of $f(3)$, $f(-3)$, $f(0)$ and $f\left(\frac{1}{2}\right)$.

3. $f : x \to 3(x - 1)$. Find $f(3)$, $f(1)$, $f(0)$ and $f(-2)$.

4. $f : x \to \dfrac{20}{x}$. Find $f(5)$, $f(2)$, $f(-4)$ and $f\left(\frac{1}{2}\right)$.

5. $f : x \to 3x + 1$. Find the range for the domain $\{0, 1, 2, 3\}$.

6. $f : x \to x^2 + 2$. Find the range for the domain $\{-1, 0, 1\}$.

7. $f : x \to \dfrac{1}{2}x - 3$. Find the range for the domain $\{0, 1, 2, 3, 4\}$.

8. $f : x \to 2^x$. Find the range for the domain $\{-2, -1, 0, 1, 2\}$.

9. $f : x \to x + 4$.
   (i) Find $f(3)$ and $f(-3)$.
   (ii) If $f(a) = 9$ and $f(b) = 1$ find a and b.

10. $f : x \to x^2$.
   (i) Find $f(5)$ and $f(-5)$.
   (ii) Find values of c such that $f(c) = 36$.

11. $f : x \to \dfrac{12}{x}$.
   (i) Find the images of 2 and 9.
   (ii) What numbers have 3 and 2 as images?

12. Sketch the Cartesian graph for $f : x \to 6 - 2x$ where the domain is:
   (i) $\{0, 1, 2, 3\}$
   (ii) {real numbers from 0 to 3}.

13. State the range of each of the following where x is real:
   (i) $f : x \to 3x$, domain $\{x : 0 \leqslant x \leqslant 2\}$
   (ii) $g : x \to x + 3$, domain $\{x : -5 \leqslant x \leqslant 5\}$
   (iii) $h : x \to x^2 - 2$, domain $\{x : -3 \leqslant x \leqslant 3\}$
   (iv) $k : x \to \dfrac{1}{x + 2}$, domain $\{x : 0 \leqslant x \leqslant 1\}$.

14. $f : x \to x^2 + 1$. Find:
   (i) the range when the domain is $\{-2, 0, 2\}$
   (ii) the domain when the range is $\{1, 10\}$.

15. $f : x \to x^2 - 3x$. Find:
   (i) the range for the domain $\{-1, 0, 1\}$
   (ii) the domain for the range $\{0, -2\}$.

Inverse of a function

If $f: x \to 3x + 5$, the calculation for $f(x)$ can be shown in a flow chart, like this:

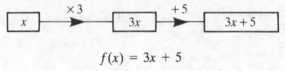

$$f(x) = 3x + 5$$

For $f(-4)$ we have

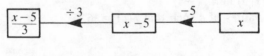

$$f(-4) = -7$$

We can get from the answer back to the number we started with, by reversing the instructions in the flow chart, like this:

$$\boxed{-4} \xleftarrow{\div 3} \boxed{-12} \xleftarrow{-5} \boxed{-7}$$

−7 is mapped back to −4

Note how the instructions have changed.
Replacing −7 with x, we have

$$\boxed{\dfrac{x-5}{3}} \xleftarrow{\div 3} \boxed{x-5} \xleftarrow{-5} \boxed{x}$$

x is mapped back to $\dfrac{x-5}{3}$

The function which reverses f is called the *inverse* of f and is written as f^{-1}. Thus

$$f^{-1}: x \to \frac{x-5}{3}$$

We can obtain the inverse of $f: x \to 3x + 5$ without using a flow chart. Let $y = 3x + 5$, so that $f: x \to y$. For the inverse of f we have to find the value of x that corresponds to a given value of y. That is, we have to rearrange the formula $y = 3x + 5$ to read $x = \ldots\ldots$ in terms of y.

From $y = 3x + 5$, we have $y - 5 = 3x$ so $\dfrac{y-5}{3} = x$.

Thus, given a number y, we calculate $\dfrac{y-5}{3}$ to obtain the starting number. In the same way, given a number x, we calculate $\dfrac{x-5}{3}$ to obtain the starting number. That is, $f^{-1}: x \to \dfrac{x-5}{3}$, which is what we obtained from the flow chart.

Now consider $g : x \rightarrow x^2 - 2$.

Let $y = x^2 - 2$, so that $g : x \rightarrow y$.

If $y = x^2 - 2$, then $y + 2 = x^2$ so $x = \pm \sqrt{y + 2}$.

Thus, given y, we calculate $\pm \sqrt{y + 2}$ to obtain the starting number.
In the same way given x, we calculate $\pm \sqrt{x + 2}$, to obtain the starting number. That is, $g^{-1} : x \rightarrow \pm \sqrt{x + 2}$.

For example, $g(5) = 5^2 - 2 = 23$ and $g(-5) = (-5)^2 - 2 = 23$.

$$g^{-1}(23) = \pm \sqrt{23 + 2} = \pm \sqrt{25} = \pm 5,$$

It was stated earlier that, for a function, the image must have only one value. So g^{-1} is not a function, since the image has two values. However, we have assumed above that the domain of g is {all real numbers}. If we restrict the domain of g to {positive real numbers}, then g^{-1} is a function, since now only the positive square roots can be considered.

Exercise 18.2

1. State the inverse of: (i) $f : x \rightarrow 2x$ (ii) $f : x \rightarrow x - 5$

   (iii) $f : x \rightarrow \dfrac{1}{4}x$ (iv) $f : x \rightarrow \dfrac{3}{5}x$.

2. (i) State the images of 1, 2 and 4, using the function $f : x \rightarrow 3x + 2$.
   (ii) Find f^{-1} in the form $f^{-1} : x \rightarrow$ ....
   (iii) Apply f^{-1} to the numbers you obtained as answers in (i). You should get back to 1, 2 and 4.

3. (i) Given $f : x \rightarrow 5 - 2x$, find $f(0)$, $f(1)$ and $f(2)$.
   (ii) Find f^{-1} in the form $f^{-1} : x \rightarrow$ ....
   (iii) Find $f^{-1}(5)$, $f^{-1}(3)$ and $f^{-1}(1)$.

4. Find f^{-1} in the form $f^{-1} : x \rightarrow$ ...... if:
   (i) $f : x \rightarrow 5x - 3$ (ii) $f : x \rightarrow 3(x + 2)$ (iii) $f : x \rightarrow \dfrac{1}{2}x + 7$

5. (i) Given $f : x \rightarrow \dfrac{12}{x}$, find $f(2)$, $f(3)$ and $f(4)$.

   (ii) Find f^{-1} in the form $f^{-1} : x \rightarrow$ ....
   (iii) Apply f^{-1} to the numbers you obtained in (i). Do you get back to 2, 3, and 4?

6. (i) Given $f : x \rightarrow 7 - x$, find $f(0)$, $f(2)$ and $f(4)$.
   (ii) Find f^{-1} in the form $f^{-1} : x \rightarrow$ ....
   (iii) Apply f^{-1} to the numbers you obtained as answers in (i).

7. State the inverses of $f : x \rightarrow x - 6$, $g : x \rightarrow 6 - x$ and $h : x \rightarrow \dfrac{6}{x}$.

   If $f(2) = a$, $g(2) = b$ and $h(2) = c$, state the values of a, b and c.
   Check that $f^{-1}(a) = 2$, $g^{-1}(b) = 2$ and $h^{-1}(c) = 2$.

8. Using the set of real numbers from 0 to 6 as the domain, sketch the graph of each function and its inverse:

(i) $f : x \rightarrow 2x + 3$ (ii) $f : x \rightarrow \dfrac{1}{3}x - 4$.

9.   (i) Given that $f(x) = (x - 1)^3$, find $f(4)$ and $f(-1)$.

(ii) Find the inverse of f.

(iii) Find $f^{-1}(27)$ and $f^{-1}(-8)$.

10.   (i) Given $f : x \rightarrow x^2 + 1$, find $f(2)$ and $f(-2)$.

(ii) Write down the inverse of f and find $f^{-1}(5)$.

(iii) Explain why f^{-1} is not a function.

Composite functions

Let f and g be two functions, such that $f : x \rightarrow 3x$ and $g : x \rightarrow x - 2$. The expression gf means use function f first and then apply g to the result. In other words, first multiply by 3 and then subtract 2. For example:

$$5 \xrightarrow{\ f\ } 15 \xrightarrow{\ g\ } 13.$$

Here f maps 5 to 15 and g maps 15 to 13.

$f(5) = 15$ and then $gf(5) = g(15) = 15 - 2 = 13$.

In general $f(x) = 3x$ and then $g(3x) = 3x - 2$. We can combine these as $gf(x) = g(3x) = 3x - 2$.

Hence we can write $gf : x \rightarrow 3x - 2$.

Usually fg gives a different result from gf.

For example $g(5) = 5 - 2 = 3$ so $fg(5) = f(3) = 3 \times 3 = 9$.

In general $fg(x) = f(x - 2) = 3(x - 2)$.

Hence we can write $fg : x \rightarrow 3(x - 2)$.

Example   Given $f : x \rightarrow \dfrac{6}{x}$ and $g : x \rightarrow 4 + x$, find the values of $f(2)$,

$g(2)$, $gf(2)$, $fg(2)$, $gf(x)$ and $fg(x)$.

$$f(2) = \frac{6}{2} = 3, \quad g(2) = 4 + 2 = 6, \quad gf(2) = g(3) = 4 + 3 = 7,$$

$$fg(2) = f(6) = \frac{6}{6} = 1, \quad gf(x) = g\left(\frac{6}{x}\right) = 4 + \frac{6}{x}$$

$$fg(x) = f(4 + x) = \frac{6}{4 + x}$$

Exercise 18.3

1. $f : x \rightarrow x + 3$ and $g : x \rightarrow 2x$.

Find $f(5)$, $g(5)$, $gf(5)$, $fg(5)$, $gf(x)$ and $fg(x)$.

2. $f : x \rightarrow \dfrac{x}{3}$ and $g : x \rightarrow x - 4$.

Find $f(12)$, $g(12)$, $gf(12)$, $fg(12)$, $gf(x)$ and $fg(x)$.

3. $f : x \rightarrow x^2$ and $g : x \rightarrow 10 - x$.

Find $f(3)$, $g(3)$, $gf(3)$, $fg(3)$, $gf(x)$ and $fg(x)$.

4. $f : x \rightarrow 2 - x$ and $g : x \rightarrow x^2 + 1$.

Find $f(-1)$, $g(-1)$, $gf(-1)$, $fg(-1)$, $gf(x)$ and $fg(x)$.

5. $f : x \rightarrow x + 3$.

(i) Find $f(4)$ and $ff(4)$ and express $ff(x)$ in its simplest form.

(ii) Find $f^{-1}(2)$ and $f^{-1}f^{-1}(2)$ and express $f^{-1}f^{-1}(x)$ in its simplest form.

6. $f : x \rightarrow 2x + 1$. Find $f(3)$, $f(-3)$, $ff(3)$ and $ff(-3)$.

Show that $ff(x) = 4x + 3$.

7. $f : x \rightarrow 3x - 2$. Find $ff(4)$ and $ff(-1)$.

Find $ff(x)$ in its simplest form and check your result by using it to find $ff(4)$ and $ff(-1)$.

8. Two simple functions f and g are such that $gf : x \rightarrow x^2 + 3$.

What are f and g? Find fg.

9. Two simple functions f and g are such that $gf : x \rightarrow \dfrac{1}{x + 1}$.

What are f and g? Find fg. Find $fg(2)$ and $gf(2)$.

Harder questions

Example   Find three simple functions f, g and h so that
$fgh : x \rightarrow 2 - 4x - x^2$.

$$2 - 4x - x^2 = 6 - 4 - 4x - x^2 = 6 - (4 + 4x + x^2)$$
$$= 6 - (2 + x)^2$$

Hence $f : x \rightarrow 2 + x$, $g : x \rightarrow x^2$ and $h : x \rightarrow 6 - x$.

Exercise 18.4

1. $f : x \rightarrow 1 - 3x$ and $g : x \rightarrow \dfrac{6}{x}$.

(i) Find $f(0.2)$, $g\left(\dfrac{1}{2}\right)$ and $gf(-3)$.

(ii) Express fg in the form $fg : x \rightarrow$ ...... and hence find x such that $fg(x) = -3.5$.

2. $f : x \to x^2$, $g : x \to 3x$ and $h : x \to x - 1$.

 Express $x \to 3x^2 - 6x + 3$ in terms of f, g and h. Find hgf.

3. f, g and h are simple functions such that $fgh : x \to 2x^2 + 12x + 18$.
 What are f, g and h?. Find hgf and gfh.

4. Find f in the form $f : x \to$ ...... for each of the following:
   (i) $f(1) = 4$, $f(2) = 5$ and $f(3) = 6$
   (ii) $f(1) = 1$, $f(2) = 3$ and $f(3) = 5$
   (iii) $f(1) = 1$, $f(2) = 4$ and $f(3) = 9$
   (iv) $f(1) = 6$, $f(2) = 3$ and $f(3) = 2$.

5. Find the inverse of (i) $f : x \to 20 - 3x$ (ii) $f : x \to \dfrac{12}{x} + 5$.

 Test your answers using the number 6.

6. $f : x \to 2x + 1$ and $g : x \to x^2 - 1$.
   (i) Find two values of x such that $fg(x) = 17$.
   (ii) Find two values of x such that $gf(x) = 8$.

7. $f(x) = x + 2$, $g(x) = 2x$ and $h(x) = \dfrac{2}{x}$.

 Express in terms of x: $gf(x)$, $fg(x)$ and $gh(x)$.
 Show that there is no value of x for which $gf(x) = fg(x)$.
 Find the values of x for which $fg(x) = gh(x)$.

8. $f : x \to x^2 - 2x$.
   (i) Find the range set whose domain is $\{3, 0, -3\}$.
   (ii) Find the domain set whose range is $\{0, 3\}$.

9. The mappings f and g are defined as follows:
 $f : x \to x - 3$, $g : x \to x^2$.

 Express in the form $x \to$ ...... :
   (i) the inverse mapping f^{-1}
   (ii) the inverse mapping g^{-1} (This is a 1-to-2 mapping as also are (v)
       and (vi) below.)
   (iii) gf (iv) fgf (v) $(gf)^{-1}$ (vi) $(fgf)^{-1}$.
 Find a mapping h such that $hgf : x \to x^2 - 6x + 3$.
 Find $(hgf)^{-1}$, and hence, or otherwise, solve the equation
 $x^2 - 6x + 3 = 2$, giving each answer correct to 2 decimal places. (L)

19 Graphs

Coordinates

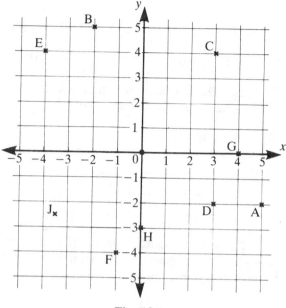

Fig. 19.1

Figure 19.1 shows two lines at right-angles to each other, with a scale on each. The lines are called *axes* and the point of intersection, O, is called the *origin*. The line across the page is the *x-axis* and the line up the page is the *y-axis*. Distances from O to the right and up the page are positive; those to the left and down from O are negative.

To go from O to A we move 5 units to the right and 2 units down. We say that the *x*-coordinate of A is 5 and the *y*-coordinate of A is −2. A is the point (5, −2). The *x*-coordinate must alway be placed first in the brackets. Notice that (−2, 5) is the point B, not A.

The origin O is (0, 0).

Exercise 19.1

1. State in the form (x, y) the coordinates of the points C, D, E, F, G, H, J and O in Fig. 19.1.

2. Using paper with squares of side 5 mm, draw axes O*x* and O*y* with numbers from −5 to +5, as in Fig. 19.1. Mark the points P(3, 2), Q(−1, 4), R(−3, 0) and S(1, −2). What shape is quadrilateral PQRS? Draw the diagonals PR and QS to intersect at T. State the coordinates of T.

3. Using axes numbered as in Fig. 19.1, mark the points K(5, 3), L(−1, 1) and M(−3, −5). Mark point N so that KLMN is a parallelogram. State the coordinates of N. Also state the coordinates of V, the mid-point of LM, and of W, the point of intersection of KM and LN.

Straight line graphs: standard equation $y = mx + c$

The equation $y = mx + c$, where m and c are constants or fixed numbers, is an equation of the first degree since it contains no powers of x or y higher than the first. The graphs of such equations are straight lines so the equations are often called *linear equations*.

Example Draw the graph of $y = 2x + 3$.
Taking the values −2, 0 and 2 for x we obtain the following table:

x	−2	0	2
y	−1	3	7

The points (−2, −1), (0, 3) and (2, 7) have been plotted in Fig. 19.2, and a straight line drawn through them. Of course, two points are sufficient to fix a straight line, but it is better to use three points, the third one being a check.

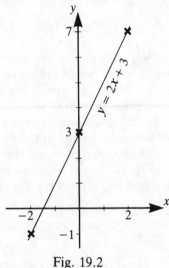

Fig. 19.2

Alternative notation

The graph in Fig. 19.2 has been described as the graph of $y = 2x + 3$. It can also be described as the graph of the function $f : x \rightarrow 2x + 3$. We read this as 'the function f such that x is mapped to $2x + 3$'. Using this notation, the axes can be labelled x and $f(x)$.

Gradient and intercept

The table below shows some pairs of values of x and y, for the equation $y = \frac{4}{3}x + 5$.

x	0	3	6	9
y	5	9	13	17

For each increase of 3 in x, there is an increase of 4 in y. These increases are shown in the graph of $y = \frac{4}{3}x + 5$ in Fig. 19.3.

The *gradient* or slope of a straight line is $\dfrac{\text{vertical change}}{\text{horizontal change}}$ or $\dfrac{y \text{ step}}{x \text{ step}}$.

So for the graph in Fig. 19.3, the gradient is $\frac{4}{3}$. Notice that this is also the number in front of x in the equation $y = \frac{4}{3}x + 5$. Notice also that the graph cuts the y-axis at $(0, 5)$ and that 5 appears in $y = \frac{4}{3}x + 5$.

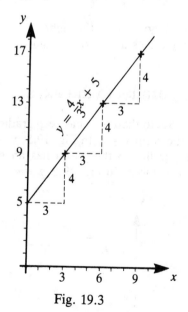

Fig. 19.3

Similarly, the graph of $y = \frac{2}{7}x - 3$ has a gradient of $\frac{2}{7}$ and cuts the y-axis at $(0, -3)$. (When $x = 0$ in the equation, $y = 0 - 3$.)

The graph of $y = 3x + 7$ has a gradient of $\frac{3}{1}$, since 3 can be written as $\frac{3}{1}$.

The graph of $y = -\frac{3}{5}x + 8$ cuts the y-axis at $(0, 8)$ and has a gradient of $-\frac{3}{5}$. In Fig. 19.4, each step consists of 5 units to the right and 3 units downwards.

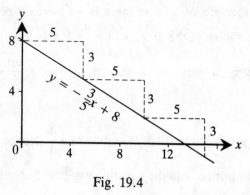

Fig. 19.4

In general, the graph of $y = mx + c$ cuts the y-axis at $(0, c)$ and has a gradient of m.
If $m > 0$, the graph goes up to the right.
If $m < 0$, the graph goes down to the right.

Straight lines parallel to the axes

A straight line parallel to the x-axis has zero gradient. Hence $m = 0$ and the equation has the form $y = c$. In Fig. 19.5, $y = 3$.

A straight line parallel to the y-axis has an equation of the form $x = d$, where d is a number. In Fig. 19.5, $x = 4$.

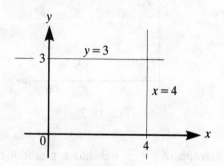

Fig. 19.5

Solving simultaneous linear equations

The values of x and y which satisfy two linear equations, such as $x - 3y + 2 = 0$ and $x + y - 4 = 0$, can be found by drawing the graphs of the equations and reading off the coordinates of the point of intersection. See Exercise 19.2, questions 12 and 13.

Exercise 19.2

On squared paper, draw the graphs of:

1. $y = x + 2$, taking $x = -2$, 0 and 2.

2. $y = 3x - 2$, taking $x = -1$, 1 and 3.

3. $y = \frac{1}{2}x - 1$, taking $x = -4$, 0 and 4.

4. $y = -\frac{2}{3}x + 4$, taking $x = -3$, 0 and 3.

5. For each of the graphs in Fig. 19.6, state the point at which it cuts the y-axis, its gradient and its equation.

6. State the gradient of:
   (i) $y = 5x + 7$ (ii) $y = x - 3$
   (iii) $y = -2x - 5$ (iv) $y = -x + 6$.

7. Make sketches, as in Fig. 19.6, to show the graphs of the following equations. In each case show clearly where the graph cuts the y-axis. Also show two steps as in the second part of Fig. 19.6:
   (i) $y = \frac{2}{3}x + 4$ (ii) $y = -\frac{2}{3}x + 7$
   (iii) $y = \frac{3}{2}x - 5$ (iv) $y = -\frac{3}{2}x - 1$.

8. On a single diagram, sketch and label the graphs of:
   (i) $y = 2$ (ii) $y = 5$ (iii) $y = -3$ (iv) $x = 4$.

9. Write each of the following equations in the form $y = mx + c$, then state its gradient and the point where it cuts the y-axis:
   (i) $3x + 7y = 21$ (ii) $2x + y = 5$
   (iii) $4x - y = 8$ (iv) $2x - 5y - 9 = 0$.

10. On a diagram mark the points A $(0, 6)$ and B $(-6, 2)$. The straight line through A and B has the equation $y = mx + c$. State:
    (i) the value of c (ii) the value of m
    (iii) the point at which the line crosses the x-axis.

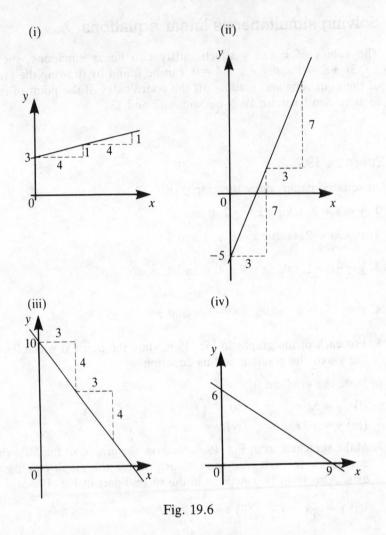

Fig. 19.6

11. (i) The line $y = 2x + k$ passes through the point $(4, 11)$. By substituting $x = 4$ and $y = 11$, find the value of k.

(ii) Find c so that $y = \frac{2}{3}x + c$ passes through $(6, -1)$.

12. The equations $x - 3y + 6 = 0$ and $x + y - 4 = 0$ can be rearranged as $y = \frac{1}{3}x + 2$ and $y = -x + 4$. Using axes numbered from 0 to 6, draw the graphs of these equations. State the coordinates of the point of intersection of the graphs. Check that these values of x and y satisfy both equations simultaneously.

13. For values of x from 0 to 5, draw the graphs of $4x - 5y - 15 = 0$ and $2x + y - 4 = 0$. From your graphs find the values of x and y which satisfy both equations simultaneously.

Graphs of quadratic functions

The graph of $y = ax^2 + bx + c$, where a, b and c are constants $(a \neq 0)$, is a curve called a *parabola*. It has one of the shapes shown in Fig. 19.7.

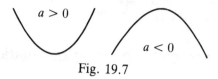

Fig. 19.7

Example   Draw the graph of $y = x^2 - 3x - 6$ for values of x from -2 to 5.

The table below shows the values of y for various values of x.

x	-2	-1	0	1	2	3	4	5
x^2	4	1	0	1	4	9	16	25
$-3x$	6	3	0	-3	-6	-9	-12	-15
-6	-6	-6	-6	-6	-6	-6	-6	-6
y	4	-2	-6	-8	-8	-6	-2	4

Figure 19.8 shows the graph. The points $(-2, 4)$, $(-1, -2)$ and so on were plotted, then joined with a smooth curve.

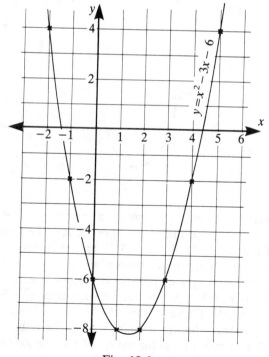

Fig. 19.8

Every point on the curve (not just those plotted) represents a pair of values of x and y which satisfy $y = x^2 - 3x - 6$. For example, the point $(2.3, -7.61)$ lies on the curve. Its values satisfy the equation since $2.3^2 - 3 \times 2.3 - 6 = 5.29 - 6.9 - 6 = -7.61$.

Using alternative notation (page 152) we can refer to the above graph as 'the graph of the function $f : x \rightarrow x^2 - 3x - 6$'.

Using graphs to solve quadratic equations

Example   Use the graph of $y = x^2 - 3x - 6$ to find approximate sol-
              utions of the equations $x^2 - 3x - 6 = 0$, $x^2 - 3x - 9 = 0$ and
              $x^2 - 3x - 1 = 0$.

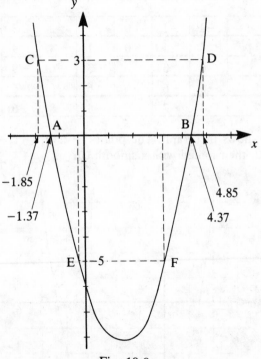

Fig. 19.9

If we put $y = 0$ in the equation $y = x^2 - 3x - 6$, we have $0 = x^2 - 3x - 6$. $y = 0$ on the graph where it cuts the x-axis, that is at A and B in Fig. 19.9. At A, $x = -1.37$ approximately and at B, $x = 4.37$ approximately. Since these values of x make y zero, they also make $x^2 - 3x - 6$ zero, so they are approximate solutions of $x^2 - 3x - 6 = 0$.

The equation $x^2 - 3x - 9 = 0$ can be written as $x^2 - 3x - 6 = 3$ by adding 3 to each side. Compare this with $x^2 - 3x - 6 = y$. They are the same if $y = 3$, so the solutions to $x^2 - 3x - 9 = 0$ are the values of x on the graph where $y = 3$, that is the values of x at the points C and D in Fig. 19.9. These are approximately -1.85 and 4.85.

The equation $x^2 - 3x - 1 = 0$ can be written as $x^2 - 3x - 6 = -5$ by subtracting 5 from each side. For the solutions of $x^2 - 3x - 1 = 0$ we find the values of x where $y = -5$, that is at E and F. They are approximately -0.30 and 3.30.

Remember, the values of x which satisfy an equation are called its roots.

Hints for drawing graphs

1. Draw up a table of values for y for the given values of x. (Choose suitable values if you are not given any.) Do not write the table on the graph paper, since you will not know how much space is needed for the graph.

2. Examine the table to find the range of values of y which must be used.

3. Choose suitable scales. Draw and number the axes.

4. Plot points for the values in your table.

5. If there are any large gaps between points, choose some other values of x, add them to your table and plot the extra points.

6. Draw a smooth curve through the points and write the equation of the graph beside it.

7. If there is room, now copy the table of values onto your graph paper.

Exercise 19.3

1. Copy and complete this table for the graph of $y = x^2 - 20$:

x	-6	-4	-2	0	2	4	6
x^2	36	16		0			36
$x^2 - 20$	16	-4		-20			16

Number the x-axis from -6 to 6 using 1 cm to each unit and the y-axis from -20 to 20 using 2 cm to 10 units. Plot the points and draw the graph.

State the values of x where $y = 0$. These are the solutions of the equation $0 = x^2 - 20$, that is, of $x^2 = 20$. So they are the two square roots of 20.

Also state the values of x where $y = -14$. What equation has these values as roots? Of what number are these values the square roots?

2. Draw up a table to show the values of $x^2 - 5x + 3$ for $x = 0, 1, 2, 3,$
4, 5. Draw the graph of $y = x^2 - 5x + 3$ for values of x from 0 to 5.

From your graph obtain approximate solutions to the equation
$x^2 - 5x + 3 = 0$.

State the values of x where $y = 2$. What equation has these values
as roots?

Write the equation $x^2 - 5x + 5 = 0$ in the form $x^2 - 5x + 3 = n$ and
hence find approximations to the roots of $x^2 - 5x + 5 = 0$.

3. Draw the graph of $y = x^2 - x - 7$ for values of x from -3 to 4.

From your graph, obtain approximate values for the roots of:

(i) $x^2 - x - 7 = 0$ (ii) $x^2 - x - 10 = 0$ (iii) $x^2 - x - 3 = 0$.

Are there any values of x for which $y = -8$? Does the equation
$x^2 - x - 7 = -8$ (that is $x^2 - x + 1 = 0$) have any roots?

4. By drawing the graph of $y = x^2 - 6x + 4$ for values of x from 0 to 6,
find approximations to the roots of:

(i) $x^2 - 6x + 4 = 0$ (ii) $x^2 - 6x + 7 = 0$ (iii) $x^2 - 6x + 1 = 0$.

What happens if you try to solve $x^2 - 6x + 11 = 0$?

What is the smallest possible value of $x^2 - 6x + 4$ (that is, of y) on
your graph?

Intersection of graphs

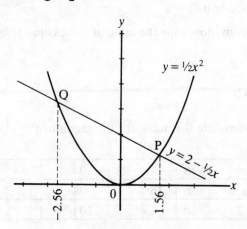

Fig. 19.10

Figure 19.10 shows the graphs of $y = \dfrac{1}{2}x^2$ and $y = 2 - \dfrac{1}{2}x$. At P the value
of y is the same for both graphs. So the value of x at P must make
$\dfrac{1}{2}x^2$ equal to $2 - \dfrac{1}{2}x$. That is, the value of x at P satisfies the equation

$\frac{1}{2}x^2 = 2 - \frac{1}{2}x$, which can be rearranged in the simpler form of $x^2 + x - 4 = 0$.

Similarly, the value of x at Q satisfies this equation. These two values of x, at P and Q, are the solutions (or roots) of the equation $x^2 + x - 4 = 0$. From the graphs we find that they are approximately 1.56 and −2.56.

Exercise 19.4

1. Show that the values of x at the points of intersection of $y = x^2$ and $y = 5x - 6$ satisfy the equation $x^2 - 5x + 6 = 0$. Solve the equation by the factors method and hence find the coordinates of the two points of intersection of the graphs.

2. Find the coordinates of the points of intersection of the curve $y = x^2$ and the straight line $y = 2x + 3$.

3. Draw the graphs of $y = \frac{1}{2}x^2$ and $y = \frac{1}{2}x + 3$. Use scales of 2 cm to 1 unit on each axis. Number the x-axis from −3 to +3 and the y-axis from 0 to 10. Mark the points of intersection of the graphs as R and S and state the values of x at these points. These values satisfy the equation $\frac{1}{2}x^2 = \frac{1}{2}x + 3$. Show that this equation can be rearranged to give $x^2 - x - 6 = 0$. Check that the values of x at R and S satisfy this equation.

4. Copy and complete the table:

x	−2	−1½	−1	−½	0	½	1	1½	2
x^2		2¼		¼					

Using 2 cm to represent 1 unit on both axes, draw the graph of $y = x^2$. Draw also the graph of $y = \frac{1}{2}x + 3$. State the values of x at the points of intersection of the graphs. Show that the equation satisfied by these values is $2x^2 - x - 6 = 0$.

Using the same axes, draw the graph of $y = x + 1$. What equation is satisfied by the values of x at the points of intersection of this graph with the graph of $y = x^2$? From your graph find approximate values for the roots of this equation.

5. Write down the two values missing from the following table, which gives values of $\frac{1}{5}x^3 - 3x + 2$ for values of x from −5 to +4:

x	-5	-4	-3	-2	-1	0	1	2	3	4
$\frac{1}{5}x^3 - 3x + 2$	-8	1.2		6.4	4.8	2	-0.8		-1.6	2.8

Draw the graph of $y = \frac{1}{5}x^3 - 3x + 2$ for values of x from -5 to $+4$,

taking 2 cm to represent 1 unit on the x-axis and 2 cm to represent 2 units on the y-axis.

(i) Use your graph to estimate the solutions of the equation $\frac{1}{5}x^3 - 3x + 2 = 0$.

(ii) Using the same axes, draw the graph of $y = 2 - x$.

(iii) Write down and simplify the equation which has, as the members of its solution set, the values of x at the points of intersection of the two graphs.
(W)

6. Functions f and g are defined by $f(x) = \frac{1}{x}$, $g(x) = 4 - x$.

(a) Using a scale of 4 cm to 1 unit on each axis, plot and draw with the same axes the graphs of these two functions for $0.2 \leqslant x \leqslant 4.0$.

(b) From your graphs, estimate the two values of x for which $f(x) = g(x)$.

(c) Write down an equation of which these two values are the roots, and simplify it.
(L)

7. (a) Given that $y = 7x - 2x^2$, copy and complete the following table.

x	-2	-1	0	1	2	3	4	-5
y	-22			5	6		-4	-15

(b) Using a scale of 2 cm to represent 1 unit on the x-axis and a scale of 2 cm to represent 5 units on the y-axis, draw the graph of $y = 7x - 2x^2$ from $x = -2$ to $x = 5$ inclusive.

(c) Using the same axes and the same scales draw the graph of the straight line $y = 2x - 5$.

(d) Find, in the form $f(x) = 0$, the equation whose solutions are the values of x at the points where the line meets the curve. Estimate as accurately as you can from your graphs the smallest value of x which satisfies this equation.
(EA-C Joint GCE/CSE)

Inequalities in the x–y plane

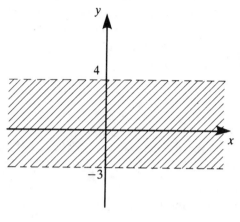

Fig. 19.11

For all points in the shaded area of Fig. 19.11, the y-coordinate is greater than -3 and less than 4. That is, the shaded area shows the set of points such that $-3 < y < 4$. This is often written as $\{(x, y) : -3 < y < 4\}$. In Fig. 19.11, the shaded area is bounded by broken lines because points on the boundaries are not included in the set. If boundary points are included in the set we use a continuous line, as in Fig. 19.12.

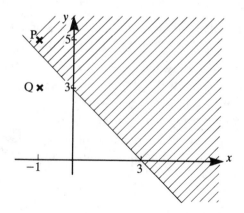

Fig. 19.12

Figure 19.12 shows the set of points (x, y) such that $x + y \geqslant 3$. This can be written as $\{(x, y) : x + y \geqslant 3\}$. Point P $(-1, 5)$ is in the set, because $-1 + 5 = 4$ which is greater than 3. Point Q $(-1, 3)$ is not in the set, because $-1 + 3 = 2$ which is less than 3.

Exercise 19.5

1. For each of the following, make a sketch to show the set of points in the x–y plane which satisfies the inequality:

    (i) $y \geqslant 2$ (ii) $y < -1$ (iii) $x > -3$

    (iv) $-2 < y \leqslant 3$ (v) $1 \leqslant x < 4$.

2. Make a sketch to show the area for which $-1 < y < 2$ and $1 < x < 4$.

3. Make a sketch showing the graph of $x + y = 4$. This divides the plane into three sets of points: $A = \{(x, y) : x + y = 4\}$, $B = \{(x, y) : x + y > 4\}$ and $C = \{(x, y) : x + y < 4\}$. On your sketch label the sets B and C.

4. Make a sketch of the graph of $y = x + 2$. Label as G the area containing $\{(x, y) : y > x + 2\}$ and as L the area containing $\{(x, y) : y < x + 2\}$.

5. On a diagram shade the area for which $y > x$, $x > -2$ and $y < 3$.

6. $A = \{(x, y) : x + y > 4\}$ and $B = \{(x, y) : y > 2\}$. On a diagram show, by shading, $A \cap B$.

7. On squared paper, draw the graphs of $x = 3$, $y = 1$ and $x + y = 8$ for $0 \leqslant x \leqslant 8$, $0 \leqslant y \leqslant 8$. Shade the region in which $x + y > 8$. Next shade the region in which $x < 3$. Then shade the region in which $y < 1$. The region now unshaded contains the points (x, y) for which $x > 3$, $y > 1$ and $x + y < 8$. From this region write down all pairs of integers (x, y) which satisfy all three inequalities.

8. Use the method of question 7 to find all pairs of integers (x, y) which satisfy the inequalities $x > 1$, $x < 4$, $y < 5$ and $x + 2y > 6$.

9. On squared paper, draw the graphs of $y = x$, $y = 4 + \frac{1}{3}x$, $2x + y = 6$ and $x = 6$ for $0 \leqslant x \leqslant 6$, $0 \leqslant y \leqslant 6$.

    $P = \{(x, y): x < 6, y < x, 2x + y > 6\}$

    $Q = \{(x, y): y > x, y < 4 + \frac{1}{3}x, 2x + y < 6\}$

    On your graph, label the regions containing these sets of points. From your graph, write down all pairs of integers for which $y > x$, $y > 6 - 2x$ and $y < 4 + \frac{1}{3}x$.

Travel graphs: distance–time

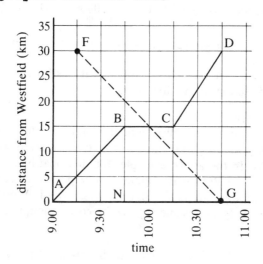

Fig. 19.13

The continuous line in Fig. 19.13 is the *distance–time graph* for a river boat which left Westfield at 9.00 and travelled downstream to Eastley, stopping at Middleton on the way.

The graph was drawn from the following table:

Time	9.00	9.45	10.15	10.45
Distance from Westfield	0	15	15	30
Point on the graph	A	B	C	D

Between 9.00 and 9.45 the boat travelled 15 km.

Its speed was $\dfrac{\text{distance}}{\text{time}} = \dfrac{15\,\text{km}}{\frac{3}{4}\,\text{h}} = 15 \times \dfrac{4}{3} = 20\,\text{km/h}$.

Notice that on the graph $\dfrac{\text{distance}}{\text{time}} = \dfrac{BN}{AN}$ which is the gradient of the line AB.

From 9.45 to 10.15 the boat was stationary at Middleton. The corresponding part of the graph is the horizontal line BC.

From 10.15 to 10.45 the speed was $\dfrac{15\,\text{km}}{\frac{1}{2}\,\text{h}} = 30\,\text{km/h}$. On the graph this is the gradient of the line CD.

The broken line in Fig. 19.13 is the graph for another boat. Point F shows that at 9.15 this boat was 30 km from Westfield, that is, at Eastley. Point G shows that at 10.45 it reached Westfield. In $1\frac{1}{2}$ h it travelled

30 km, so its speed was $\dfrac{30 \text{ km}}{1\frac{1}{2} \text{ h}} = 30 \times \dfrac{2}{3} = 20 \text{ km/h}$. The gradient of FG

is $\dfrac{-30}{1\frac{1}{2}} = -20$. The negative sign is used because the line FG slopes

down to the right (see page 154). Notice that this boat is travelling in the opposite direction to the first boat.

Speed and velocity
Velocity is speed in a stated direction. If we take a velocity of 60 km/h northwards as positive ($+60$) then a velocity of 60 km/h southward is negative (-60). In the above example, the first boat had a velocity of 20 km/h away from Westfield. Taking this as positive, the velocity of the second boat was -20 km/h, since it was towards Westfield. Each velocity therefore has the same sign as the corresponding gradient.

Summary
(a) The gradient of a distance–time graph represents the velocity.
(b) The larger the gradient, the greater the velocity.
(c) A positive gradient represents a positive velocity (speed in one direction); a negative gradient represents a negative velocity (speed in the opposite direction).
(d) A horizontal line (such as BC) represents zero velocity—the object is stationary.

Travel graphs: velocity–time

An underground train accelerated steadily from rest to 12 m/s in 15 seconds, stayed at that speed for 20 seconds and then slowed steadily to rest in 10 seconds. Figure 19.14 shows the *velocity–time graph* for the train.
  The velocity increased by 12 m/s in 15 s, that is at a rate of $\dfrac{12}{15}$ m/s

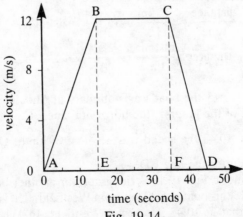

Fig. 19.14

each second, or 0.8 m/s per second. We say that the *acceleration* was 0.8 m/s per second and we write it as 0.8 m/s^2. Notice that the gradient of AB is $\dfrac{12}{15}$ so the gradient represents the acceleration.

The horizontal line BC represents the constant velocity of 12 m/s.

Over the final 10 seconds the velocity decreased from 12 m/s to zero. The *deceleration* or *retardation* was $\dfrac{12}{10}$ m/s^2, or 1.2 m/s^2. This is represented by the negative gradient of CD.

During the first 15 seconds, the average velocity was $\dfrac{1}{2}(0 + 12)$ m/s = 6 m/s. Hence the distance travelled in the 15 seconds was 15 × 6 m = 90 m. Notice that the area of triangle AEB is calculated in the same way $\left(\dfrac{1}{2} \times \text{base} \times \text{height} = \dfrac{1}{2} \times \text{AE} \times \text{BE} = \dfrac{1}{2} \times 15 \times 12\right)$ so it represents the distance travelled.

During the next 20 seconds, the train travelled 20 × 12 m = 240 m which is represented by the area of the rectangle BCFE.

Over the final 10 seconds the average speed was 6 m/s and the distance travelled was 10 × 6 m = 60 m. This is represented by the area of triangle CFD.

The total distance travelled was 390 m and it is represented by the total area of the quadrilateral ABCD.

Summary
(a) The gradient of a velocity–time graph represents acceleration. A positive gradient represents a positive acceleration; a negative gradient represents a negative acceleration, that is a deceleration or retardation. Zero gradient means constant velocity.
(b) The area under a velocity–time graph represents the distance travelled.

Exercise 19.6

Questions 1–3 are on distance–time graphs. The remaining questions are on velocity–time graphs.

1. A man went for a walk. His distance–time graph is shown in Fig. 19.15.

   (i) If he left home at 9.40, when did he arrive back?
   (ii) How far did he walk altogether?
   (iii) How long did he rest?
   (iv) How fast did he walk on the way home?
   (v) What was his slowest speed?

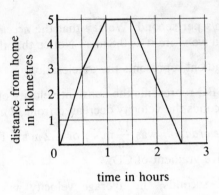

Fig. 19.15

2. Draw graphs to illustrate the following:
   (i) A man cycles for 1 h at 15 km/h, stops for 1 h and then continues in the same direction for 2 h at 10 km/h.
   (ii) A boat cruises at 20 km/h for $1\frac{1}{2}$ h and returns to its starting point in 2 h.

3. A cyclist leaves P at 10.00 and cycles at 15 km/h towards Q which is 25 km away. At what time does he arrive?
   Draw a distance–time graph for his journey using 6 cm for 1 hour on the x-axis and 2 cm for 5 km on the y-axis.
   A car starts from P at 10.40 and travels towards Q at 75 km/h. It stays at Q for 20 minutes and then returns to P at the same speed. Draw its distance–time graph using the same axes.
   At what times, and how far from P, does the car pass the cyclist?

4. (i) A train accelerated steadily from rest to 14 m/s in 20 seconds. What was its acceleration in m/s²?
   (ii) Another train took 20 seconds to accelerate from a speed of 12 m/s to 30 m/s. What was its acceleration?
   (iii) A car accelerated from 40 km/h to 100 km/h in 15 seconds. What was its acceleration in km/h per second?

5. Figure 19.16 shows the velocity–time graph of a train.

   (i) For how long was it travelling at 15 m/s?
   (ii) What was its maximum velocity?
   (iii) State the two accelerations.
   (iv) State the retardation.
   (v) Find the distance travelled in the first 30 seconds.

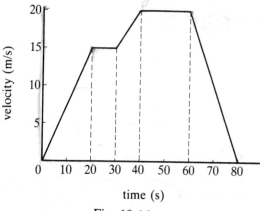

time (s)

Fig. 19.16

6. An electric milk van accelerated steadily from 0 to 8 m/s in 20 s, remained at that speed for 10 s and then slowed steadily to rest in 16 s. Draw its velocity–time graph. Suitable scales are 2 cm to 10 s and 1 cm to 10 m/s. State the acceleration for the first 20 s and the deceleration for the last 16 s. Also calculate the distance travelled in the 46 s.

7. A car accelerated steadily from rest to 24 m/s in 30 seconds, stayed at that speed for 40 seconds and then slowed steadily to rest in 20 seconds. Draw its velocity–time graph using 1 cm for 10 s and 1 cm for 5 m/s.

(i) When was the speed 12 m/s? (ii) What was the acceleration?
(iii) What was the retardation? (iv) How far did the car travel?

8. Figure 19.17 shows a velocity–time graph. If the total distance travelled in the 70 seconds was 880 metres, calculate:

(i) u (ii) the acceleration (iii) the retardation.

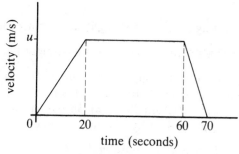

time (seconds)

Fig. 19.17

For curved travel graphs, see page 179 and 180.

20 Variation

Direct variation

The table below shows the distance travelled by a certain aircraft in various times.

Time (t minutes)	2	4	6	8	10	12
Distance (s kilometres)	26	52	78	104	130	156
$s \div t$	13	13	13	13	13	13

For all values of s and t in the table, $\dfrac{s}{t} = 13$. The aircraft is travelling at a constant speed of 13 km/min.

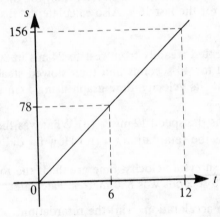

Fig. 20.1

The graph of s against t is a straight line through the origin. Since $\dfrac{s}{t} = 13$, $s = 13t$. This is the equation of the straight line.

From the table we notice that if t is doubled (for example from 4 to 8) then s is doubled (from 52 to 104); if t is trebled (for example from 2 to 6) then s is trebled (from 26 to 78). In fact, if t is multiplied by any number, then s is multiplied by the same number.

We say that s is *directly proportional* to t
or that s *varies directly* as t.

The symbol $\propto$ is used for 'is proportional to'. We write: $s \propto t$.
In general, if $y \propto x$ (if y is proportional to x), then:

(i) $\dfrac{y}{x}$ is the same for all pairs of values of x and y, that is, $\dfrac{y}{x} = k$, a constant. So $y = kx$.

(ii) the graph of y against x is a straight line through the origin.

(iii) if x is multiplied (or divided) by a certain number, then y is multiplied (or divided) by the same number.

Example $y \propto x$ and $y = 15$ when $x = 6$. Find y when $x = 8$.

If $y \propto x$, then $y = kx$ or $\dfrac{y}{x} = k$.

Since $y = 15$ when $x = 6$, then $k = \dfrac{15}{6} = \dfrac{5}{2}$.

The formula is therefore $y = \dfrac{5}{2}x$.

When $x = 8$, $y = \dfrac{5}{2} \times 8 = 20$.

Consider a cube with an edge of x cm. The area of each face is x^2 cm^2, and the cube has 6 faces. So the total surface area is $6x^2$ cm^2. Denoting this total surface area by y cm^2, we have the formula $y = 6x^2$.

Some values of x and y are shown in the table.

x	1	2	3	4	5	6
y	6	24	54	96	150	216
$y \div x$	6	12	18	24	30	36
$y \div x^2$	6	6	6	6	6	6

Notice that $y \div x$ is not constant, so y is not proportional to x in this case.

The graph of $y = 6x^2$ is not a straight line; it is a parabola (Fig. 20.2).

The bottom line of the table shows that $y \div x^2$ is always 6 and Fig. 20.3 shows that the graph of y against x^2 is a straight line through the origin.

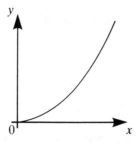

Fig. 20.2

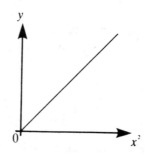

Fig. 20.3

We say that y is directly proportional to x^2 or that y varies directly as x^2.

We write: $y \propto x^2$, or $y = kx^2$, where k is a constant.

Similarly, if x and y are two variables and $y \div x^3$ is constant, then $y \propto x^3$, and it follows that:

(i) there is a formula $y = kx^3$ where k is a constant.

(ii) the graph of y against x^3 is a straight line through the origin.

Example $y \propto x^3$ and $y = 128$ when $x = 8$. Find:

(i) y when $x = 12$ (ii) x when $y = 54$.

Since $y \propto x^3$, we can write $y = kx^3$.

Since $y = 128$ when $x = 8$, then $128 = k \times 8^3$ so

$$k = \frac{128}{8^3} = \frac{128}{512} = \frac{1}{4}$$

The formula is therefore $y = \frac{1}{4}x^3$.

(i) When $x = 12$, $y = \frac{1}{4} \times 12 \times 12 \times 12 = 432$

(ii) When $y = 54$, $54 = \frac{1}{4}x^3$ so $x^3 = 4 \times 54 = 216$

$x = 6$

Exercise 20.1

1. If y varies directly as the cube of x, then we can write $y \propto x^3$ and $y = kx^3$. Make two such statements for each of the following:

(i) A varies directly as the square of r

(ii) d varies directly as v

(iii) w is directly proportional to the cube of h.

2.

x	2	3	4	5
y	12	27	48	75

For the table of values above, decide whether $y \propto x$, $y \propto x^2$ or $y \propto x^3$. Then write down the formula for y in terms of x.

3.

x	3	4	5	6	7
y	1.5	2.0	2.5	3.0	3.5

For the table of values above, decide whether $y \propto x$, $y \propto x^2$ or $y \propto x^3$. Then write down the formula for y in terms of x.

4. Copy and complete the table so that $y \propto x$.

x	2	4	6			20
y	3			12	15	

5. $y \propto x$ and $y = 10$ when $x = 15$.

(i) Find y when $x = 9$ and when $x = 21$.

(ii) Find x when $y = 8$.

6. $y \propto x^2$ and $y = 80$ when $x = 4$.

(i) Find y when $x = 3$ and when $x = 5$.

(ii) Find the positive value of x when $y = 20$.

7. $y \propto x$ and $y = 3$ when $x = 5$.

Find y when $x = 4$ and when $x = 6$.

8. $y \propto x^2$ and $y = 3.6$ when $x = 3$.

Find y when $x = 2$ and when $x = 5$.

9. $y \propto x^3$ and $y = 54$ when $x = 3$.

(i) Find y when $x = 4$ and when $x = 5$.

(ii) Find x when $y = 16$.

10. $p \propto n^2$ and $p = 2.5$ when $n = 5$.

Find p when $n = 6$ and find n when $p = 4.9$.

11. $v \propto h^3$ and $v = 40$ when $h = 2$.

(i) State the formula for v in terms of h and find v when $h = 3$.

(ii) Express h in terms of v and find h when $v = 320$.

12. If $y \propto \sqrt{x}$ and $y = 15$ when $x = 25$, find the formula for y in terms of x and use it to find y when $x = 4$. Also state the formula for x in terms of y and use it to find x when $y = 12$.

Inverse variation

The table shows the average speed (v km/h) for vehicles which take various times (t h) for a journey of 60 km.

Time, t hours	1	2	3	4	5	6
Average speed, v km/h	60	30	20	15	12	10

As t increases, v decreases. If t is multiplied by 3 (for example from 2 to 6) then v is divided by 3 (30 to 10).

We say that v is *inversely* proportional to t or that v *varies inversely* as t.

We write: $v \propto \dfrac{1}{t}$.

The product vt is constant and has the value 60.

We can therefore write: $vt = 60$ or $v = \dfrac{60}{t}$.

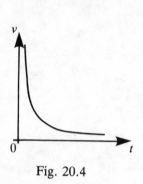

Fig. 20.4   Fig. 20.5

The graph of v against t is not a straight line; it is a curve called a *hyperbola* (Fig. 20.4). But the graph of v against $\dfrac{1}{t}$ is a straight line (Fig. 20.5).

In general, if $y \propto \dfrac{1}{x}$ then:

(i) $yx = k$, a constant, or $y = \dfrac{k}{x}$

(ii) the graph of y against $\dfrac{1}{x}$ is a straight line through the origin

(iii) if x is multiplied (or divided) by a certain number then y is divided (or multiplied) by the same number.

Similarly, if $y \propto \dfrac{1}{x^2}$ (if y varies inversely as the square of x), then:

(i) $yx^2 = k$, a constant, or $y = \dfrac{k}{x^2}$

(ii) the graph of y against $\dfrac{1}{x^2}$ is a straight line through the origin.

Example y varies inversely as the square of x and $y = 9$ when $x = 2$.
Find y when $x = 4$ and find x when $y = 4$.

Since $y \propto \dfrac{1}{x^2}$, we can write $y = \dfrac{k}{x^2}$ or $yx^2 = k$.

Since $y = 9$ when $x = 2$, $k = 9 \times 2^2 = 36$.

The formula is therefore $y = \dfrac{36}{x^2}$.

When $x = 4$, $y = \dfrac{36}{4^2} = \dfrac{36}{16} = \dfrac{9}{4} = 2\dfrac{1}{4}$.

Also $yx^2 = 36$, so when $y = 4$, $4x^2 = 36$, $x^2 = 9$ and $x = \pm 3$.

Exercise 20.2

1. $y \propto \dfrac{1}{x}$. (a) Explain how y will change if x is:
   (i) multiplied by 3   (ii) divided by 2.
   (b) $y = 12$ when $x = 5$. Find y when x is:
   (i) 15 (ii) $2\frac{1}{2}$.

2. If $y = \dfrac{20}{x}$, state the relation between x and y in words.

   (i) Find y when $x = 5$.   (ii) Find x when $y = 10$.

3. If $y = \dfrac{100}{x^2}$, state the relation between x and y in words.

   (i) Find y when $x = 4$.   (ii) Find x when $y = 4$.

4. If $y \propto \dfrac{1}{x}$ and $y = 12$ when $x = 4$, find y when $x = 6$.

5. If $y \propto \dfrac{1}{x^2}$ and $y = 2$ when $x = 5$, find y when $x = 2$.

6. If $y \propto \dfrac{1}{x}$ and $y = 4$ when $x = 6$, find:

   (i) y when $x = 3$ (ii) x when $y = 2$.

7. If $y \propto \dfrac{1}{x^2}$ and $y = 18$ when $x = 2$, find:

   (i) y when $x = 3$ (ii) x when $y = 2$.

8. When a given mass of gas is kept at a constant temperature, its pressure, p pascals, varies inversely as its volume, v cm^3. $p = 300$ when $v = 1.6$. Find p when $v = 2.4$ and find v when $p = 50$.

9. Make separate sketches to show the following relationships (for positive values of x only):

   (i) $y \propto x$ (ii) $y \propto x^2$ (iii) $y \propto \dfrac{1}{x}$.

10. The intensity of illumination on a surface varies inversely as the square of its distance from a bulb. The illumination is 4 units at a distance of 3 m. Find the illumination at:

    (i) 2 m   (ii) 5 m.

Joint variation

In joint variation, three or more variables are involved.

For a cylinder, the formula for the volume is $V = \pi r^2 h$. The three variables V, r and h are involved. π is a constant. We can write $V \propto r^2 h$ and say that V varies directly as the height and as the square of the radius.

Sum of two parts

If $y = 5x + \dfrac{3}{x}$ then y is the sum of two parts. One of these parts is directly proportional to x and the other is inversely proportional to x.

Exercise 20.3

1. Express each as a formula with a constant c:
    (i) E varies directly as m and as the square of v
    (ii) p varies directly as q and inversely as r
    (iii) d varies directly as m and inversely as the cube of x.

2. $m \propto r^2 h$. If $m = 5$ when $r = 3$ and $h = 2$, find the formula for m in terms of r and h, and hence find the value of m when $r = 2$ and $h = 3$.

3. $F \propto \dfrac{m}{d^2}$. If $F = 18$ when $m = 20$ and $d = 4$, find the formula for F in terms of m and d, and hence find the value of F when $m = 25$ and $d = 3$.

4. Express each as a formula with constants a and b:
    (i) y is the sum of two parts, one part directly proportional to x and the other directly proportional to x^2.
    (ii) h is the sum of two parts, one part inversely proportional to p and the other directly proportional to p.
    (iii) y is the sum of two parts, one a constant and the other directly proportional to the square of n.

5. y is the sum of two parts. One part is directly proportional to x and the other is inversely proportional to x. When $x = 2$, $y = 16$ and when $x = 3$, $y = 19$. Find the formula for y in terms of x and hence find the value of y when $x = 4$.

6. S is the sum of two parts. One part is directly proportional to n and the other is directly proportional to n^2. When $n = 3$, $S = 6$ and when $n = 5$, $S = 15$. Find the formula for S and hence the value of S when $n = 7$.

21 Gradients

Gradient of a straight line

To obtain the gradient of a straight line, we mark two points on the line and then find the change in y and the change in x between the two points.

$$\text{Gradient of straight line} = \frac{\text{change in } y}{\text{change in } x}.$$

In Fig. 21.1, A is $(3, 5)$ and B is $(7, 8)$. Gradient of the straight line AB is $\dfrac{\text{NB}}{\text{AN}} = \dfrac{8-5}{7-3} = \dfrac{3}{4}$. In Fig. 21.1 y increases as x increases, so the gradient is positive.

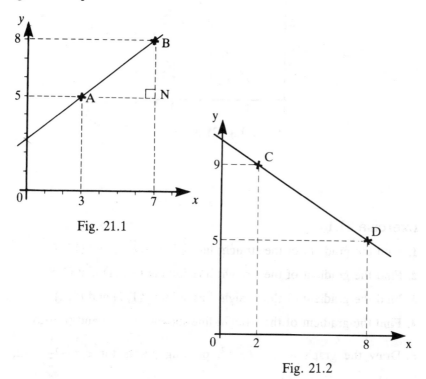

Fig. 21.1

Fig. 21.2

Now look at the line CD in Fig. 21.2. y changes from 9 to 5, as x changes from 2 to 8. So the change in y is -4, while the change in x is 6.

$$\text{Gradient of CD} = -\frac{4}{6} = -\frac{2}{3}.$$

Since y decreases as x increases, the gradient of the line CD is negative.

Gradient of a curve

The gradient of a curve at a point is taken to be the gradient of the tangent to the curve at that point.

Figure 21.3 shows the curve $y = x^2$. The tangent at $(2, 4)$ is drawn. Notice that it passes through the point B $(3, 8)$.

Gradient of the tangent at A is $\dfrac{NB}{AN} = \dfrac{8 - 4}{3 - 2} = \dfrac{4}{1} = 4$.

Thus the gradient of the curve at $(2, 4)$ is 4.

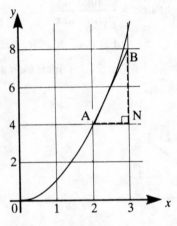

Fig. 21.3

Exercise 21.1

1. Find the gradient of the straight line joining $(2, 3)$ and $(12, 7)$.

2. Find the gradient of the straight line joining $(-2, 1)$ and $(2, 8)$.

3. Find the gradient of the straight line joining $(1, 7)$ and $(5, 2)$.

4. Find the gradient of the straight line joining $(-7, 6)$ and $(8, -3)$.

5. Draw the graph of $y = 4 - x^2$, plotting points for $x = -2\frac{1}{2}$, -2, $-1\frac{1}{2}$, -1, ... 2, $2\frac{1}{2}$, and using 2 cm to represent 1 unit, on each axis. Carefully draw the tangent at $(-2, 0)$. It should pass through $(-1, 4)$. Find the gradient of this tangent and hence the gradient of the curve at $(-2, 0)$.

   Also draw the tangent at $(1, 3)$. It should pass through $(2, 1)$. Find the gradient of the curve at $(1, 3)$.

6. Draw the graph of $y = \frac{1}{2}x^2$ for $x = 0, \frac{1}{2}, 1, 1\frac{1}{2}, \ldots\ldots 3\frac{1}{2}$ using a scale

of 2 cm to 1 unit, on each axis. Label the points P $\left(1, \frac{1}{2}\right)$, Q (2, 2)

and R $\left(3, 4\frac{1}{2}\right)$. Draw tangents to the curve at P, Q and R. Using

suitable points on the tangents, find the gradient of the curve at each
of the points P, Q and R.

Distance–time graphs

In the summary on page 166 it was stated that the gradient of a distance–
time graph represents velocity. The graphs in that section were straight
lines because the velocities were constant. We now consider distance–
time graphs which are curves.

A ball is allowed to roll down a sloping road. After rolling for

t seconds it is x metres from its starting point, where $x = \frac{1}{10}t^2$.

When $t = 4, x = \frac{1}{10} \times 16 = 1.6$ and when $t = 8, x = \frac{1}{10} \times 64 = 6.4$. So,

in the four seconds from $t = 4$ to $t = 8$, the ball travels $6.4 - 1.6 = 4.8$
metres. Its average velocity over this four seconds is $4.8 \div 4 = 1.2$ m/s.

Figure 21.4 shows the graph of $x = \frac{1}{10}t^2$. The point A (4, 1.6)

represents the time 4 s and the distance 1.6 m; the point B (8, 6.4)
represents the time 8 s and the distance 6.4 m. The gradient of the

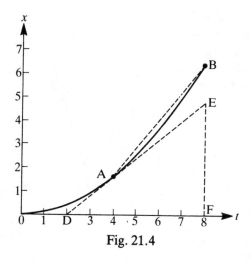

Fig. 21.4

straight line AB is $\dfrac{\text{change in } x}{\text{change in } t} = \dfrac{4.8}{4} = 1.2$. Therefore the gradient of
AB represents the average velocity over the four seconds from $t = 4$
to $t = 8$. By taking B nearer to A, we can get the average velocity over a
shorter interval of time. On the graph, if B approaches A, the line AB
approaches the tangent at A. Thus the gradient of the tangent at A
represents the velocity at 4 seconds. From triangle DEF, we estimate
the gradient of the tangent at A as $\dfrac{4.8}{6} = 0.8$, so the estimated velocity at
time 4 s is 0.8 m/s. As the gradient of this graph increases from O to B,
the velocity increases as t increases. If the gradient of a graph decreases,
the velocity decreases.

Summary

For a distance–time graph, the gradient of the tangent at a point
represents the velocity at this instant.

Velocity–time graphs

In the summary on page 167, it was stated that the gradient of a
velocity–time graph represents acceleration. The graphs in that section
were straight lines.

For a curved velocity–time graph, the gradient of the tangent at
a point represents the acceleration at that instant.

Exercise 21.2

1. Figure 21.5 shows the
   distance–time graph for a
   cyclist. By considering how
   the gradient of the graph
   changes over the 30 seconds,
   describe how the velocity
   changes.

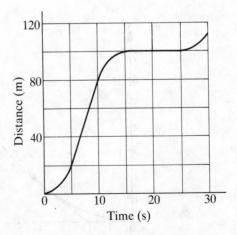

Fig. 21.5

2. A bus starts from rest, increases its speed gradually, runs at a constant speed, comes to rest at traffic lights, starts again, runs at a constant speed and comes to rest at a bus stop. Sketch its distance–time graph.

3. A goods train moved slowly from rest to a signal at red. The table shows the distance from the starting point at various times.

Time (minutes)	0	1	2	3	4	5
Distance (metres)	0	20	120	420	680	740

Draw the distance–time graph using 2 cm to represent 1 minute and 200 metres.
  (i) Find the average velocity over the first two minutes and then over the whole five minutes.
  (ii) By drawing tangents, estimate the velocity at 2 minutes and at 4 minutes.
  (iii) By drawing a suitable tangent, estimate the greatest velocity and the time at which this was reached.

4. A stone is hurled into the air. Its height is h metres at time t seconds where $h = 30t - 5t^2$. Using 2 cm to represent 1 s and 10 m, draw the graph of h for values of t from 0 to 6.
  By drawing tangents, estimate the velocity: (i) at 2 s (ii) at 5 s (iii) initially.
  When was the stone momentarily at rest?

5. Figure 21.6 shows the velocity–time graph for a car which took 10 seconds to reduce its speed from 11 m/s to zero.
  (i) State the average retardation over the 10 seconds.
  (ii) Using the triangle, estimate the retardation at 4 seconds.
  (iii) When was the retardation greatest?

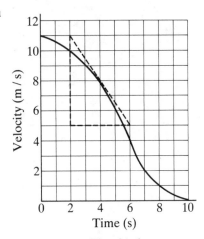

Fig. 21.6

6. The table shows the velocity of a car at various times.

Time (s)	0	1	2	3	4	5	6
Velocity (m/s)	0	0.6	1.7	4.2	10	12.8	13.2

Using 2 cm to represent 1 s and 2 m/s, plot the points and draw a smooth curve through them.

  (i) Find the average acceleration over the six seconds.

 (ii) Estimate the acceleration at 2 seconds.

(iii) Estimate the greatest acceleration and the time at which it occurs.

22 Areas under graphs

1. By counting squares

An approximate value for the area under a graph can be found by counting the squares under the graph. If more than half of a square is in the required area, count it as a whole square; if less than half a square is in the required area, ignore it.

Attention must be given to the *scales* on the axes. For example, in question 2 of Exercise 22.1, 2 cm on the x-axis represents $\frac{1}{2}$ unit and 2 cm on the y-axis represents 1 unit, so $2 \times 2 = 4$ cm^2 on the graph paper represents $\frac{1}{2} \times 1 = \frac{1}{2}$ unit^2 of area.

2. Using the trapezium method

Example Find the approximate value of the area bounded by $y = x^2 + 1$, $x = 1$, $x = 2$ and the x-axis, using strips of width $\frac{1}{2}$.

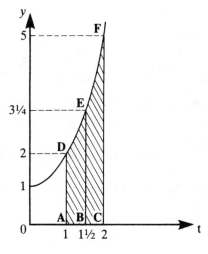

Fig. 22.1

The required area is shaded in Fig. 22.1. It is approximately the sum of the areas of the trapezia ABED and BCFE. We have the following table of values:

x	1	$1\frac{1}{2}$	2
y	2	$3\frac{1}{4}$	5

Hence AD = 2, BE = $3\frac{1}{4}$ and CF = 5.

Area of a trapezium

$= \frac{1}{2}$(sum of parallel sides) × (distance between them)

Area of ABED $= \frac{1}{2}\left(2 + 3\frac{1}{4}\right) \times \frac{1}{2} = 5\frac{1}{4} \times \frac{1}{4} = \frac{21}{16}$

Area of BCFE $= \frac{1}{2}\left(3\frac{1}{4} + 5\right) \times \frac{1}{2} = 8\frac{1}{4} \times \frac{1}{4} = \frac{33}{16}$

Sum of areas of trapezia $= \frac{54}{16} = 3\frac{3}{8} \approx 3.4$

Exercise 22.1

1. Copy and complete the table for $y = \frac{1}{2}x^2$ and draw the graph, using 2 cm to represent 1 unit on each axis:

x	0	$\frac{1}{2}$	1	$1\frac{1}{2}$	2	$2\frac{1}{2}$	3
y	0	0.125		1.125		3.125	4.5

Now find an approximate value for the area bounded by the curve, and the lines $x = 1$, $x = 3$ and the x-axis, by:

(i) counting squares

(ii) the trapezium method, using two trapezia each of 1 unit width.

2. Draw up a table to show the values of $y = 4 - x^2$, for $x = -1$, $-\frac{1}{2}$, $0, \frac{1}{2}, 1, 1\frac{1}{2}, \ldots\ldots, 3$. Then plot the graph of the function, using 2 cm to represent $\frac{1}{2}$ unit on the x-axis and 2 cm to represent 1 unit on the y-axis. Now find an approximate value for the area bounded by the curve, the positive x-axis and the positive y-axis by:

(i) counting squares

(ii) calculating the area of a trapezium and a triangle.

3. Draw a graph of $y = \dfrac{12}{x}$, for $x = 1, 2, 3, 4, 5$. Use 2 cm to represent 1 unit on each axis. Now find an approximate value for the area bounded by $y = \dfrac{12}{x}$, $x = 1$, $x = 5$ and the x-axis, by;

(i) counting squares
(ii) the trapezium method, using two trapezia each of 2 units width.

4. Draw up a table showing the values of $y = x(6 - x)$ for $x = 1, 2, 3, 4, 5$. Make a sketch of the graph for $1 \leqslant x \leqslant 5$.
Use 2 trapezia each of 2 units width to estimate the area between the graph, the x-axis and the lines $x = 1$ and $x = 5$.

Velocity–time graphs

The area under a velocity–time graph, between two values for the time, represents the distance travelled between those two times (see p. 167).

The gradient of the tangent at a point represents the acceleration at that instant (see p. 180).

Example The velocities of a train at various times are shown in the table:

Time in seconds	0	1	2	3	4	5	6	7	8	9	10
Velocity in m/s	0	2	3.4	4.4	5.2	6	6.5	7	7.4	7.7	8

Find: (i) the distance travelled in the 10 seconds
(ii) the acceleration at 2 s and at 6 s.

The velocity–time graph for the table is shown in Fig. 22.2.

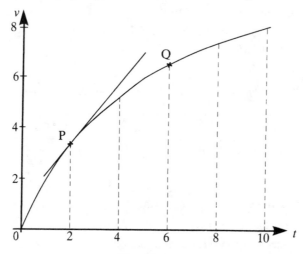

Fig. 22.2

(i) Using the five trapezia, we find the area is approximately 53.0 units. Since each unit on the t-axis represents 1 s and each unit on the v-axis represents 1 m/s, each unit of area represents 1 m. So the distance travelled is approximately 53 m.

(ii) At P the gradient of the tangent is 1.2, so the acceleration at 2 s is 1.2 m/s^2.

Similarly, from the tangent at Q, the acceleration at 6 s is 0.6 m/s^2.

Exercise 22.2

1. The table shows the speedometer readings of a car at various times:

Time from start in seconds	0	5	10	15	20	25	30	35	40
Velocity in m/s	0	6	10	12	$12\frac{1}{2}$	$13\frac{1}{2}$	15	18	20

Using 2 cm to represent 5 s and 5 m/s, plot the points and draw a smooth curve through them.
Estimate: (i) the acceleration at 10 s (ii) the distance travelled.

2. The velocity of a boat at certain times is given below:

Time in seconds	0	1	2	3	4	5	6
Velocity in m/s	0	$\frac{1}{6}$	$\frac{2}{3}$	$\frac{3}{2}$	$\frac{8}{3}$	$\frac{25}{6}$	6

Draw the velocity–time graph.
 (i) Estimate the acceleration after 3 s and after 5 s.
 (ii) Estimate the distance travelled.

3. The table shows the velocity of a certain falling object.

Time in seconds	0	1	2	3	4	5	6
Velocity in m/s	0	4.6	7.6	9.1	9.6	9.9	10

Using 2 cm to 1 s on the time axis and 1 cm to 1 m/s on the velocity axis, draw the velocity–time graph.
Use your graph to estimate
 (i) the acceleration after 2 seconds.
 (ii) the distance travelled in the 6 seconds.

4. The table shows the velocity of a two stage space rocket at 20 second intervals from blast-off.

Time in seconds	0	20	40	60	80	100	120	140	160	180	200
Velocity in km/s	0	0.4	1.0	1.8	2.8	3.6	4.0	4.8	6.4	7.8	8.2

Using 2 cm to represent 20 seconds and 1 km/s, draw the velocity–time graph.

From your graph estimate

  (i) the time at which the second rocket was fired

 (ii) the acceleration, in m/s^2, at 150 seconds

(iii) the distance travelled in the 200 seconds.

23 Matrices

Matrices can be used for storing information. Here are three examples.

Example 1 The results of 10 matches played by 4 football teams can be shown in a table, and in a matrix:

Table

Team	Wins	Draws	Losses
Ayeton	5	3	2
Binford	2	4	4
Canhill	3	6	1
Deemouth	0	2	8

Matrix

$$\begin{pmatrix} 5 & 3 & 2 \\ 2 & 4 & 4 \\ 3 & 6 & 1 \\ 0 & 2 & 8 \end{pmatrix}$$

Example 2 Three points have coordinates as follows: P(2, 1), Q(5, 7), R(−6, 10). The coordinates can be shown in table form, and also in a matrix:

Table

Point	P	Q	R
x	2	5	−6
y	1	7	10

Matrix

$$\begin{pmatrix} 2 & 5 & -6 \\ 1 & 7 & 10 \end{pmatrix}$$

Example 3 The coefficients of x and y in linear equations can be shown in a matrix:

Equations   Matrix

$$\begin{cases} 4x - 2y = 8 \\ 3x + 5y = 1 \end{cases} \qquad \begin{pmatrix} 4 & -2 \\ 3 & 5 \end{pmatrix}$$

Order of a matrix

The matrix in Example 1 above has 4 rows and 3 columns. It is a *4 × 3 matrix*.

The matrix in Example 2 has 2 rows and 3 columns. It is a *2 × 3 matrix*.

The matrix in Example 3 has 2 rows and 2 columns. It is a *2 × 2 matrix*.

A *row matrix* has only one row, for example $(6 \quad 0 \quad -4)$.

A *column matrix* has only one column, for example $\begin{pmatrix} 3 \\ 8 \end{pmatrix}$.

A *square matrix* has the same number of rows and columns. Example 3 above shows a square matrix.

A *zero (or null) matrix* is a matrix where every element is 0, for example $\begin{pmatrix} 0 & 0 \\ 0 & 0 \end{pmatrix}$.

A *unit (or identity) matrix* is a square matrix where every element in the leading diagonal (top left to bottom right) is 1, and every other element is 0. Examples are

$$\begin{pmatrix} 1 & 0 \\ 0 & 1 \end{pmatrix} \quad \text{and} \quad \begin{pmatrix} 1 & 0 & 0 \\ 0 & 1 & 0 \\ 0 & 0 & 1 \end{pmatrix}.$$

Addition of matrices

Suppose that the football teams in Example 1 played 5 more matches, with these results:

$$\begin{pmatrix} 2 & 2 & 1 \\ 1 & 1 & 3 \\ 2 & 1 & 2 \\ 0 & 2 & 3 \end{pmatrix}$$

Then the matrix for the total fifteen matches is:

$$\begin{pmatrix} 5+2 & 3+2 & 2+1 \\ 2+1 & 4+1 & 4+3 \\ 3+2 & 6+1 & 1+2 \\ 0+0 & 2+2 & 8+3 \end{pmatrix} \quad \text{or} \quad \begin{pmatrix} 7 & 5 & 3 \\ 3 & 5 & 7 \\ 5 & 7 & 3 \\ 0 & 4 & 11 \end{pmatrix}$$

This shows that to add two matrices of the same order, we add the corresponding elements.

$$\text{So if } \mathbf{A} = \begin{pmatrix} 3 & 7 \\ 2 & 1 \end{pmatrix} \text{ and } \mathbf{B} = \begin{pmatrix} 2 & 4 \\ 0 & 5 \end{pmatrix}$$

$$\text{then } \mathbf{A} + \mathbf{B} = \begin{pmatrix} 3+2 & 7+4 \\ 2+0 & 1+5 \end{pmatrix} = \begin{pmatrix} 5 & 11 \\ 2 & 6 \end{pmatrix}.$$

Subtraction of matrices

To subtract one matrix from another, we must subtract the corresponding elements.

$$\text{So} \quad \mathbf{A} - \mathbf{B} = \begin{pmatrix} 3-2 & 7-4 \\ 2-0 & 1-5 \end{pmatrix} = \begin{pmatrix} 1 & 3 \\ 2 & -4 \end{pmatrix}$$

Multiplication of a matrix by a scalar

A scalar is an ordinary number such as 5, 9 or 21. To multiply a matrix by a scalar, we must multiply each element of the matrix by the scalar:

$$\text{So} \quad 5\mathbf{A} = 5\begin{pmatrix} 3 & 7 \\ 2 & 1 \end{pmatrix} = \begin{pmatrix} 15 & 35 \\ 10 & 5 \end{pmatrix}$$

Multiplication of two matrices

Consider the following tables for the teams from Example 1 above:

Team	Wins	Draws	Losses
A	5	3	2
B	2	4	4
C	3	6	1
D	0	2	8

Number of points awarded
3 for a win
2 for a draw
1 for a loss

Total points awarded
A $5 \times 3 + 3 \times 2 + 2 \times 1 = 23$
B $2 \times 3 + 4 \times 2 + 4 \times 1 = 18$
C $3 \times 3 + 6 \times 2 + 1 \times 1 = 22$
D $0 \times 3 + 2 \times 2 + 8 \times 1 = 12$

In matrix form this is written as:

$$\begin{pmatrix} 5 & 3 & 2 \\ 2 & 4 & 4 \\ 3 & 6 & 1 \\ 0 & 2 & 8 \end{pmatrix} \times \begin{pmatrix} 3 \\ 2 \\ 1 \end{pmatrix} = \begin{pmatrix} 5 \times 3 + 3 \times 2 + 2 \times 1 \\ 2 \times 3 + 4 \times 2 + 4 \times 1 \\ 3 \times 3 + 6 \times 2 + 1 \times 1 \\ 0 \times 3 + 2 \times 2 + 8 \times 1 \end{pmatrix} = \begin{pmatrix} 23 \\ 18 \\ 22 \\ 12 \end{pmatrix}$$

Similarly, matrices **A** and **B** above can be multiplied together like this:

$$\mathbf{AB} = \begin{pmatrix} 3 & 7 \\ 2 & 1 \end{pmatrix}\begin{pmatrix} 2 & 4 \\ 0 & 5 \end{pmatrix} = \begin{pmatrix} 3 \times 2 + 7 \times 0 & 3 \times 4 + 7 \times 5 \\ 2 \times 2 + 1 \times 0 & 2 \times 4 + 1 \times 5 \end{pmatrix} = \begin{pmatrix} 6 & 47 \\ 4 & 13 \end{pmatrix}$$

Notice that the *top row* of the first matrix and the *left column* of the second matrix give the *top left element* in the product:

$$\begin{pmatrix} 3 & 7 \\ \ldots & \ldots \end{pmatrix}\begin{pmatrix} 2 & \ldots \\ 0 & \ldots \end{pmatrix} = \begin{pmatrix} 3 \times 2 + 7 \times 0 & \ldots\ldots \\ \ldots\ldots\ldots\ldots & \ldots\ldots \end{pmatrix} = \begin{pmatrix} 6 & \ldots \\ \ldots & \ldots \end{pmatrix}$$

Similarly, the *top row* of the first matrix and the *right column* of the second matrix give the *top right element* in the product; the *bottom row* of the first matrix and the *left column* of the second matrix give the *bottom left* element in the product, and so on. Now look at the multiplication **BA**:

$$\mathbf{BA} = \begin{pmatrix} 2 & 4 \\ 0 & 5 \end{pmatrix}\begin{pmatrix} 3 & 7 \\ 2 & 1 \end{pmatrix} = \begin{pmatrix} 2 \times 3 + 4 \times 2 & 2 \times 7 + 4 \times 1 \\ 0 \times 3 + 5 \times 2 & 0 \times 7 + 5 \times 1 \end{pmatrix} = \begin{pmatrix} 14 & 18 \\ 10 & 5 \end{pmatrix}$$

Notice that the answer for **BA** is not the same as the answer for **AB**. In general, if **P** and **Q** are two matrices, **PQ** and **QP** are usually different. We say that *matrix multiplication is not commutative*.

Exercise 23.1

Simplify:

1. $\begin{pmatrix} 7 & 4 \\ 3 & 2 \end{pmatrix} + \begin{pmatrix} 2 & 5 \\ 3 & 4 \end{pmatrix}$
   2. $\begin{pmatrix} 2 & 7 & 4 \\ 3 & 0 & 1 \end{pmatrix} + \begin{pmatrix} 2 & 0 & 1 \\ 1 & 3 & 4 \end{pmatrix}$

3. $\begin{pmatrix} 8 & 4 \\ 9 & 6 \end{pmatrix} - \begin{pmatrix} 5 & 1 \\ 7 & 4 \end{pmatrix}$
   4. $\begin{pmatrix} 9 & 9 & 6 & 2 \\ 5 & 3 & 8 & 6 \end{pmatrix} - \begin{pmatrix} 9 & 7 & 6 & 1 \\ 5 & 2 & 3 & 2 \end{pmatrix}$

5. $3\begin{pmatrix} 1 & 3 \\ 2 & 4 \end{pmatrix}$
   6. $5\begin{pmatrix} 2 & 6 & 1 \\ 3 & 0 & 4 \end{pmatrix}$

7. $\dfrac{1}{3}\begin{pmatrix} 12 & 15 \\ 9 & 6 \end{pmatrix}$
   8. $\begin{pmatrix} 2 & 3 \\ 1 & 2 \end{pmatrix}\begin{pmatrix} 5 \\ 2 \end{pmatrix}$
   9. $\begin{pmatrix} 3 & 4 \\ 2 & 5 \end{pmatrix}\begin{pmatrix} 2 \\ 6 \end{pmatrix}$

10. $\begin{pmatrix} 3 & 2 \\ 4 & 1 \end{pmatrix}\begin{pmatrix} 3 \\ 0 \end{pmatrix}$
   11. $\begin{pmatrix} 4 & 3 & 1 \\ 2 & 0 & 5 \end{pmatrix}\begin{pmatrix} 2 \\ 4 \\ 3 \end{pmatrix}$
   12. $\begin{pmatrix} 5 & 2 & 2 \\ 4 & 1 & 7 \end{pmatrix}\begin{pmatrix} 3 \\ 1 \\ 2 \end{pmatrix}$

13. $\begin{pmatrix} 3 & 1 \\ 1 & 4 \end{pmatrix}\begin{pmatrix} 2 & 5 \\ 1 & 3 \end{pmatrix}$
   14. $\begin{pmatrix} 5 & 2 \\ 3 & 1 \end{pmatrix}\begin{pmatrix} 2 & 1 \\ 1 & 4 \end{pmatrix}$
   15. $\begin{pmatrix} 0 & 4 \\ 1 & 2 \end{pmatrix}\begin{pmatrix} 2 & 5 \\ 3 & 1 \end{pmatrix}$

16. $\begin{pmatrix} 3 & 4 \\ 2 & 1 \end{pmatrix}\begin{pmatrix} 5 & 3 \\ 4 & 6 \end{pmatrix}$
   17. $\begin{pmatrix} 1 & 3 \\ 4 & 2 \end{pmatrix}\begin{pmatrix} 3 & 5 \\ 1 & 2 \end{pmatrix}$
   18. $\begin{pmatrix} 2 & 1 \\ 0 & 3 \end{pmatrix}\begin{pmatrix} 7 & 2 \\ 0 & 5 \end{pmatrix}$

19. $\mathbf{I} = \begin{pmatrix} 1 & 0 \\ 0 & 1 \end{pmatrix}$ and $\mathbf{A} = \begin{pmatrix} 3 & 7 \\ 9 & 4 \end{pmatrix}$. Find **IA** and **AI** and comment on the results.

20. $\mathbf{J} = \begin{pmatrix} 0 & 1 \\ 1 & 0 \end{pmatrix}$ and $\mathbf{B} = \begin{pmatrix} 6 & 2 \\ 5 & 3 \end{pmatrix}$. Find **JB** and **BJ** and comment on the results.

Simplify:

21. $\begin{pmatrix} 5 & -3 \\ -2 & 2 \end{pmatrix} + \begin{pmatrix} -3 & 3 \\ -4 & 1 \end{pmatrix}$

22. $\begin{pmatrix} 6 & -2 & -4 \\ 5 & 0 & 4 \end{pmatrix} + \begin{pmatrix} 3 & 0 & -1 \\ -3 & -7 & -2 \end{pmatrix}$

23. $\begin{pmatrix} -4 & 0 \\ 1 & -5 \end{pmatrix} - \begin{pmatrix} -2 & 1 \\ 3 & 2 \end{pmatrix}$

24. $\begin{pmatrix} 1 & -1 & 0 \\ -1 & 0 & 1 \end{pmatrix} - \begin{pmatrix} 1 & 0 & -1 \\ -1 & 1 & 0 \end{pmatrix}$

25. $\begin{pmatrix} 2 & -1 \\ -1 & 3 \end{pmatrix} \begin{pmatrix} 3 \\ 4 \end{pmatrix}$

26. $\begin{pmatrix} 3 & 1 \\ 2 & 0 \end{pmatrix} \begin{pmatrix} -4 \\ 2 \end{pmatrix}$

27. $\begin{pmatrix} -1 & 1 \\ 1 & -1 \end{pmatrix} \begin{pmatrix} 5 \\ -2 \end{pmatrix}$

28. $\begin{pmatrix} 4 & 2 \\ 3 & 1 \end{pmatrix} \begin{pmatrix} 1 & 2 \\ -1 & -3 \end{pmatrix}$

29. $\begin{pmatrix} 5 & 6 \\ 2 & 3 \end{pmatrix} \begin{pmatrix} 2 & -2 \\ -2 & 4 \end{pmatrix}$

30. $\begin{pmatrix} 1 & -2 \\ -3 & 2 \end{pmatrix} \begin{pmatrix} 3 & 2 \\ -1 & -2 \end{pmatrix}$

31. $\begin{pmatrix} 2 & 0 & 1 \\ 1 & 3 & 1 \end{pmatrix} \begin{pmatrix} 3 & 2 \\ 1 & 3 \\ 2 & 1 \end{pmatrix}$

32. $\begin{pmatrix} 1 & 3 \\ 4 & 1 \\ 2 & 0 \end{pmatrix} \begin{pmatrix} 2 & 1 & 2 \\ 1 & 3 & 1 \end{pmatrix}$

33. If $A = \begin{pmatrix} 2 & 3 \\ 4 & 1 \end{pmatrix}$ and $B = \begin{pmatrix} 1 & 2 \\ 3 & 0 \end{pmatrix}$, find $A + B$, $A - B$, $3A$, AB and BA.

34. If $C = \begin{pmatrix} 3 & -2 \\ 0 & 2 \end{pmatrix}$ and $D = \begin{pmatrix} 2 & 0 \\ 4 & 1 \end{pmatrix}$, find $C + D$, $C - D$, $5D$, CD and DC.

35. Using A and B from question 33, find $2B + 3B$, A^2, B^2 and $A^2 - B^2$.

36. Find a, b, c and d so that:

$$\begin{pmatrix} a & 1 \\ 2 & b \end{pmatrix} \begin{pmatrix} 3 \\ 2 \end{pmatrix} = \begin{pmatrix} 14 \\ 0 \end{pmatrix} \text{ and } \begin{pmatrix} 1 & 2 & -1 \\ 3 & 0 & c \end{pmatrix} \begin{pmatrix} 2 \\ 1 \\ d \end{pmatrix} = \begin{pmatrix} 7 \\ 0 \end{pmatrix}$$

37. If $\begin{pmatrix} 3 & x \\ 2x & 1 \end{pmatrix} \begin{pmatrix} 1 \\ 2 \end{pmatrix} = \begin{pmatrix} 7 \\ y \end{pmatrix}$, find x and y.

38. Find p, q, r and s so that $\begin{pmatrix} 3 & 1 \\ r & 2 \end{pmatrix} \begin{pmatrix} 1 & q \\ p & 2 \end{pmatrix} = \begin{pmatrix} 6 & 17 \\ 10 & s \end{pmatrix}$.

39. $A = \begin{pmatrix} 4 & 2 \\ 1 & 3 \end{pmatrix}$, $B = \begin{pmatrix} 6 & 5 \\ 3 & 4 \end{pmatrix}$ and $C = \begin{pmatrix} 2 \\ 3 \end{pmatrix}$.

(a) Simplify, where possible, $A + B$, $B + C$, BC, CA, AB and BA.

(b) Solve (i) $X + A = B$ (ii) $Y + B = 3A$.

40. $P = \begin{pmatrix} 4 & 2 \\ -2 & 1 \end{pmatrix}$, $Q = \begin{pmatrix} -3 & 1 \\ 3 & -4 \end{pmatrix}$ and $R = \begin{pmatrix} 2 \\ -1 \end{pmatrix}$.

(a) Simplify, where possible, $P + Q$, $P - Q$, $Q + R$, QR, PQ, QP and Q^2.

(b) Solve $X + P = Q$ and $Y + Q = 4P$.

41. Tim and Sarah are on holiday in Solaria. The table below shows the number of large postcards, small postcards and stamps they bought.

	Large cards	Small cards	Stamps
Sarah	4	2	7
Tim	2	3	8

Show this as a 2×3 matrix, P.
A large card costs 7 sols, a small card 5 sols and a stamp 3 sols. Show this data as a 3×1 column matrix, C.
Form the product matrix PC, and explain the meaning of the numbers in this matrix.

42. A factory makes three products A, B and C. The table shows the units of labour, materials and other items needed to produce one of each product.

	Labour	Materials	Other items
A	3	2	1
B	5	1	2
C	2	3	3

Represent this data by a matrix P. Labour costs £8 per unit, materials £3 per unit and other items £5 per unit. Represent this by a column matrix Q.
Find PQ and hence state the total cost of each product.

43. In an athletics match between two clubs, Binsbury A.C. obtained 5 first places, 3 seconds and 6 thirds; Camptown A.C. obtained 4 firsts, 6 seconds and 3 thirds. Show this as a 2×3 matrix R.

Binsbury wanted to award 3 points for each first place, 2 for each second and 1 for each third; Camptown wanted to award 5 for each first, 3 for each second and 1 for each third. Show this as a 3×2 matrix T.

Form the product matrix RT. What was the result of the match using:

(i) Binsbury's scoring system (ii) Camptown's scoring system?

The zero matrix

Let $\mathbf{O} = \begin{pmatrix} 0 & 0 \\ 0 & 0 \end{pmatrix}$, the 2×2 zero matrix, and $\mathbf{M} = \begin{pmatrix} a & b \\ c & d \end{pmatrix}$.

$$\mathbf{O} + \mathbf{M} = \begin{pmatrix} 0 & 0 \\ 0 & 0 \end{pmatrix} + \begin{pmatrix} a & b \\ c & d \end{pmatrix} = \begin{pmatrix} 0+a & 0+b \\ 0+c & 0+d \end{pmatrix} = \begin{pmatrix} a & b \\ c & d \end{pmatrix} = \mathbf{M}$$

That is, $\mathbf{O} + \mathbf{M} = \mathbf{M}$. Similarly $\mathbf{M} + \mathbf{O} = \mathbf{M}$.

So the matrix $\mathbf{O}$ acts just like zero does in the addition of numbers. $(0 + 5 = 5$ and $5 + 0 = 5)$.

The unit matrix

Let $\mathbf{I} = \begin{pmatrix} 1 & 0 \\ 0 & 1 \end{pmatrix}$, the 2×2 unit matrix, and $\mathbf{M} = \begin{pmatrix} a & b \\ c & d \end{pmatrix}$.

$$\mathbf{IM} = \begin{pmatrix} 1 & 0 \\ 0 & 1 \end{pmatrix} \begin{pmatrix} a & b \\ c & d \end{pmatrix} = \begin{pmatrix} 1 \times a + 0 \times c & 1 \times b + 0 \times d \\ 0 \times a + 1 \times c & 0 \times b + 1 \times d \end{pmatrix} = \begin{pmatrix} a & b \\ c & d \end{pmatrix} = \mathbf{M}$$

That is, $\mathbf{IM} = \mathbf{M}$. Similarly $\mathbf{MI} = \mathbf{M}$.

Compare $\mathbf{IM} = \mathbf{M}$ and $\mathbf{MI} = \mathbf{M}$ with the statements $1 \times n = n$ and $n \times 1 = n$, where n is any number.

The matrix $\mathbf{I}$ corresponds to the number 1 (unity).

The determinant of a 2 × 2 matrix

The *determinant* of $\begin{pmatrix} a & b \\ c & d \end{pmatrix}$ is the number $ad - bc$. For example,

the determinant of $\begin{pmatrix} 5 & 6 \\ 2 & 3 \end{pmatrix}$ is $5 \times 3 - 2 \times 6 = 15 - 12 = 3$

and the determinant of $\begin{pmatrix} 3 & 8 \\ 1 & 2 \end{pmatrix}$ is $3 \times 2 - 8 \times 1 = 6 - 8 = -2$.

Inverse matrices

If a and b are two numbers such that $ab = 1$, then b is the *inverse* of a for multiplication, and a is the inverse of b.

For example $\frac{7}{3}$ is the inverse of $\frac{3}{7}$ since $\frac{7}{3} \times \frac{3}{7} = 1$.

Similarly, if $\mathbf{A}$ and $\mathbf{B}$ are two matrices such that $\mathbf{AB} = \mathbf{I}$, the unit matrix, then $\mathbf{B}$ is the inverse of $\mathbf{A}$ for multiplication, and $\mathbf{A}$ is the inverse of $\mathbf{B}$.

For example $\begin{pmatrix} 5 & -7 \\ -2 & 3 \end{pmatrix}$ is the inverse of $\begin{pmatrix} 3 & 7 \\ 2 & 5 \end{pmatrix}$ because

$$\begin{pmatrix} 3 & 7 \\ 2 & 5 \end{pmatrix}\begin{pmatrix} 5 & -7 \\ -2 & 3 \end{pmatrix} = \begin{pmatrix} 1 & 0 \\ 0 & 1 \end{pmatrix}.$$

The inverse of $\begin{pmatrix} a & b \\ c & d \end{pmatrix}$ can be obtained as follows:

(i) switch the positions of a and d

(ii) change the signs of b and c

(iii) divide each number by the determinant of $\begin{pmatrix} a & b \\ c & d \end{pmatrix}$.

Example Find the inverse of $\begin{pmatrix} 5 & 6 \\ 2 & 3 \end{pmatrix}$.

We interchange 5 and 3, and change the signs of 6 and 2,

which gives $\begin{pmatrix} 3 & -6 \\ -2 & 5 \end{pmatrix}$.

The determinant of $\begin{pmatrix} 5 & 6 \\ 2 & 3 \end{pmatrix}$ is $5 \times 3 - 6 \times 2 = 3$ so we divide

all numbers in $\begin{pmatrix} 3 & -6 \\ -2 & 5 \end{pmatrix}$ by 3, which gives $\begin{pmatrix} 1 & -2 \\ -\frac{2}{3} & \frac{5}{3} \end{pmatrix}$.

Check: $\begin{pmatrix} 5 & 6 \\ 2 & 3 \end{pmatrix}\begin{pmatrix} 1 & -2 \\ -\frac{2}{3} & \frac{5}{3} \end{pmatrix} = \begin{pmatrix} 5-4 & -10+10 \\ 2-2 & -4+5 \end{pmatrix} = \begin{pmatrix} 1 & 0 \\ 0 & 1 \end{pmatrix}$

Also $\begin{pmatrix} 1 & -2 \\ -\frac{2}{3} & \frac{5}{3} \end{pmatrix}\begin{pmatrix} 5 & 6 \\ 2 & 3 \end{pmatrix} = \begin{pmatrix} 5-4 & 6-6 \\ -\frac{10}{3}+\frac{10}{3} & -4+5 \end{pmatrix} = \begin{pmatrix} 1 & 0 \\ 0 & 1 \end{pmatrix}.$

The symbol $\mathbf{M}^{-1}$ is used for the inverse of $\mathbf{M}$, so we can write $\mathbf{M}\mathbf{M}^{-1} = \mathbf{I}$ and $\mathbf{M}^{-1}\mathbf{M} = \mathbf{I}$.

Singular matrices

A matrix with zero determinant has no inverse. It is called a *singular matrix*. For example, $\begin{pmatrix} 8 & 6 \\ 4 & 3 \end{pmatrix}$ is singular, since $8 \times 3 - 6 \times 4 = 0$.

Another example is $\begin{pmatrix} 9 & 3 \\ 6 & 2 \end{pmatrix}$.

Exercise 23.2

1. Find the determinants of the following matrices:

(i) $\begin{pmatrix} 4 & 5 \\ 2 & 3 \end{pmatrix}$ (ii) $\begin{pmatrix} 7 & 5 \\ 5 & 4 \end{pmatrix}$ (iii) $\begin{pmatrix} 1 & 2 \\ 5 & 6 \end{pmatrix}$ (iv) $\begin{pmatrix} 3 & 5 \\ 4 & 7 \end{pmatrix}$

(v) $\begin{pmatrix} -1 & 2 \\ 0 & 1 \end{pmatrix}$ (vi) $\begin{pmatrix} 4 & 2 \\ 8 & 4 \end{pmatrix}$ (vii) $\begin{pmatrix} -3 & 5 \\ -4 & 2 \end{pmatrix}$ (viii) $\begin{pmatrix} -1 & 1 \\ 1 & -1 \end{pmatrix}$.

2. Show that each matrix is the inverse of the other. In each case, form the product **AB** and also the product **BA**.

(i) $\begin{pmatrix} 7 & 3 \\ 2 & 1 \end{pmatrix}$ and $\begin{pmatrix} 1 & -3 \\ -2 & 7 \end{pmatrix}$ (ii) $\begin{pmatrix} 0.2 & -0.2 \\ 0.6 & 0.4 \end{pmatrix}$ and $\begin{pmatrix} 2 & 1 \\ -3 & 1 \end{pmatrix}$.

3. The determinant of each of the following matrices is 1. Write down the inverse of each matrix and check your answer by multiplying it by the original matrix.

(i) $\begin{pmatrix} 3 & 5 \\ 1 & 2 \end{pmatrix}$ (ii) $\begin{pmatrix} 7 & 2 \\ 10 & 3 \end{pmatrix}$ (iii) $\begin{pmatrix} 4 & 7 \\ 5 & 9 \end{pmatrix}$ (iv) $\begin{pmatrix} 7 & 3 \\ -5 & -2 \end{pmatrix}$

Find the inverse of each matrix and check your answer by multiplication.

4. $\begin{pmatrix} 3 & 2 \\ 5 & 4 \end{pmatrix}$ 5. $\begin{pmatrix} 5 & 2 \\ 6 & 3 \end{pmatrix}$ 6. $\begin{pmatrix} 7 & 2 \\ 4 & 1 \end{pmatrix}$ 7. $\begin{pmatrix} 4 & 3 \\ 6 & 4 \end{pmatrix}$

8. $\begin{pmatrix} -4 & -3 \\ 7 & 5 \end{pmatrix}$ 9. $\begin{pmatrix} -3 & 0 \\ 5 & -2 \end{pmatrix}$

10. Show that each of the following matrices is its own inverse:

(i) $\begin{pmatrix} 1 & 0 \\ 0 & -1 \end{pmatrix}$ (ii) $\begin{pmatrix} 0 & 1 \\ 1 & 0 \end{pmatrix}$ (iii) $\begin{pmatrix} 1 & 1 \\ 0 & -1 \end{pmatrix}$ (iv) $\begin{pmatrix} 0 & -1 \\ -1 & 0 \end{pmatrix}$.

11. The determinant of $\begin{pmatrix} 5 & 7 \\ 1 & x \end{pmatrix}$ is 8. Find x.

12. State, in its simplest form, the determinant of $\begin{pmatrix} k+6 & k \\ 5 & 2 \end{pmatrix}$. Find the value of k for which the matrix is singular.

13. State, in its simplest form, the determinant of $\begin{pmatrix} n-2 & 5 \\ 3 & n \end{pmatrix}$. Find two values of n for which the matrix is singular.

24   Vectors

To fully describe a displacement (or movement) we must state both the distance moved and the direction of the movement. For example, an aircraft might fly 80 km in the direction 065°. Quantities which have direction as well as magnitude are called *vector quantities* or just *vectors*. Other examples of vectors are velocity and force.

A quantity which has magnitude only is called a *scalar quantity* or just a *scalar*. Examples are distance, mass, time, cost and numbers.

In Fig. 24.1, P is displaced to Q. A short way of writing 'the displacement from P to Q is $\overrightarrow{PQ}$.

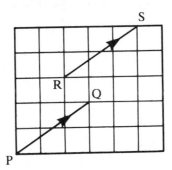

Fig. 24.1

Since Q is 3 squares to the right of P (the *x* direction) and 2 squares up (the *y* direction), we can write $\overrightarrow{PQ} = \begin{pmatrix} 3 \\ 2 \end{pmatrix}$. This is called a *column vector*.

Notice that $\overrightarrow{RS}$ is also $\begin{pmatrix} 3 \\ 2 \end{pmatrix}$. $\overrightarrow{PQ}$ and $\overrightarrow{RS}$ are equivalent. We can write $\overrightarrow{PQ} = \overrightarrow{RS}$.

The displacement $\begin{pmatrix} 3 \\ 2 \end{pmatrix}$ can be denoted by a single small letter such as **a** or a̱. **a** (heavy type) is used in print, but a̱ (a letter with a line under it) is easier to use in written work. We thus have $\mathbf{a} = \begin{pmatrix} 3 \\ 2 \end{pmatrix}$.

Multiplication of a vector by a scalar

3**a** is a vector in the same direction as **a** but with 3 times its magnitude.

$3a = 3\binom{3}{2} = \binom{9}{6}$. In Fig. 24.2, $\mathbf{b} = 3\mathbf{a}$ and $\mathbf{c} = \dfrac{1}{2}\mathbf{a} = \dfrac{1}{2}\binom{3}{2} = \binom{1\frac{1}{2}}{1}$.

In general, if k is a scalar, $k\mathbf{a}$ is a vector in the same direction as $\mathbf{a}$ but with k times its magnitude. If k is negative, $k\mathbf{a}$ is in the opposite direction to $\mathbf{a}$. In Fig. 24.2, $\mathbf{d} = -2\mathbf{a} = -2\binom{3}{2} = \binom{-6}{-4}$.

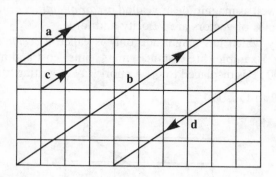

Fig. 24.2

Addition of vectors

In Fig. 24.3, $\overrightarrow{AB} = \overrightarrow{DC} = \mathbf{p}$ and $\overrightarrow{AD} = \overrightarrow{BC} = \mathbf{q}$.

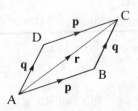

Fig. 24.3

The displacement $\overrightarrow{AC}$ has the same result as the displacement $\overrightarrow{AB}$ followed by the displacement $\overrightarrow{BC}$.

We write $\overrightarrow{AC} = \overrightarrow{AB} + \overrightarrow{BC}$ or $\mathbf{r} = \mathbf{p} + \mathbf{q}$.

Since displacement $\overrightarrow{AD}$ followed by displacement $\overrightarrow{DC}$ also has the result $\overrightarrow{AC}$, we can write $\overrightarrow{AC} = \overrightarrow{AD} + \overrightarrow{DC}$ or $\mathbf{r} = \mathbf{q} + \mathbf{p}$.

Thus $\mathbf{p} + \mathbf{q} = \mathbf{q} + \mathbf{p}$, so vector addition is commutative.

If $\mathbf{p} = \binom{5}{1}$ and $\mathbf{q} = \binom{3}{4}$, then $\mathbf{r} = \mathbf{p} + \mathbf{q} = \binom{5}{1} + \binom{3}{4} = \binom{8}{5}$.

To subtract vector **q** from vector **p**, we add −**q** to **p**. Thus

$$\mathbf{p} - \mathbf{q} = \mathbf{p} + (-\mathbf{q}) = \begin{pmatrix} 5 \\ 1 \end{pmatrix} + \begin{pmatrix} -3 \\ -4 \end{pmatrix} = \begin{pmatrix} 2 \\ -3 \end{pmatrix}.$$

It is simpler to write $\mathbf{p} - \mathbf{q} = \begin{pmatrix} 5 \\ 1 \end{pmatrix} - \begin{pmatrix} 3 \\ 4 \end{pmatrix} = \begin{pmatrix} 2 \\ -3 \end{pmatrix}$.

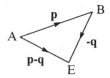

Fig. 24.4

Magnitude or modulus of a vector

By Pythagoras' Theorem, the magnitude of **a** in Fig. 24.2 is $\sqrt{3^2 + 2^2} = \sqrt{9 + 4} = \sqrt{13}$. We can write the magnitude of **a** as $|\mathbf{a}|$. If $\mathbf{t} = \begin{pmatrix} x \\ y \end{pmatrix}$, then $|\mathbf{t}| = \sqrt{x^2 + y^2}$.

Direction of a vector

If **a** makes an angle θ with the positive x direction, then $\tan \theta = \dfrac{2}{3}$ so $\theta \approx 33.7°$.

Fig. 24.5

Example If $\mathbf{h} = \begin{pmatrix} 6 \\ 5 \end{pmatrix}$ and $\mathbf{k} = \begin{pmatrix} 2 \\ 8 \end{pmatrix}$, express in the form $\begin{pmatrix} x \\ y \end{pmatrix}$, $\mathbf{h} + \mathbf{k}$,

$\mathbf{h} - \mathbf{k}$, $2\mathbf{h}$, $\dfrac{1}{2}\mathbf{k}$, $2\mathbf{h} - \mathbf{k}$ and $\mathbf{h} - 3\mathbf{k}$. Also find $|\mathbf{h} - \mathbf{k}|$ and $|2\mathbf{h} - \mathbf{k}|$.

$$\mathbf{h} + \mathbf{k} = \begin{pmatrix} 6 \\ 5 \end{pmatrix} + \begin{pmatrix} 2 \\ 8 \end{pmatrix} = \begin{pmatrix} 8 \\ 13 \end{pmatrix}, \quad \mathbf{h} - \mathbf{k} = \begin{pmatrix} 6 \\ 5 \end{pmatrix} - \begin{pmatrix} 2 \\ 8 \end{pmatrix} = \begin{pmatrix} 4 \\ -3 \end{pmatrix},$$

$$2\mathbf{h} = 2\begin{pmatrix} 6 \\ 5 \end{pmatrix} = \begin{pmatrix} 12 \\ 10 \end{pmatrix}, \quad \frac{1}{2}\mathbf{k} = \frac{1}{2}\begin{pmatrix} 2 \\ 8 \end{pmatrix} = \begin{pmatrix} 1 \\ 4 \end{pmatrix},$$

$$2\mathbf{h} - \mathbf{k} = \binom{12}{10} - \binom{2}{8} = \binom{10}{2},$$

$$\mathbf{h} - 3\mathbf{k} = \binom{6}{5} - 3\binom{2}{8} = \binom{6}{5} - \binom{6}{24} = \binom{0}{-19}$$

$$|\mathbf{h} - \mathbf{k}| = \sqrt{4^2 + (-3)^2} = \sqrt{16 + 9} = \sqrt{25} = 5$$

$$|2\mathbf{h} - \mathbf{k}| = \sqrt{10^2 + 2^2} = \sqrt{104}$$

Unit base vectors

A unit vector has a magnitude of 1. In the x—y plane, $\mathbf{i}$ and $\mathbf{j}$ are the unit vectors in the directions Ox and Oy (Fig. 24.6). $\mathbf{i} = \binom{1}{0}$ and $\mathbf{j} = \binom{0}{1}$.

Any vector in the x—y plane can be expressed in terms of $\mathbf{i}$ and $\mathbf{j}$.

In Fig. 24.6, $\overrightarrow{BL} = \binom{3}{0} = 3\binom{1}{0} = 3\mathbf{i}$

and $\overrightarrow{LC} = \binom{0}{2} = 2\binom{0}{1} = 2\mathbf{j}$.

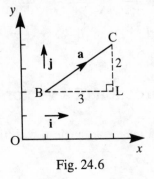

$\overrightarrow{BC} = \overrightarrow{BL} + \overrightarrow{LC}$ so $\mathbf{a} = 3\mathbf{i} + 2\mathbf{j}$.

Similarly, $\mathbf{t} = \binom{7}{-5} = 7\mathbf{i} - 5\mathbf{j}$.

$\mathbf{a} + \mathbf{t} = (3\mathbf{i} + 2\mathbf{j}) + (7\mathbf{i} - 5\mathbf{j}) = 10\mathbf{i} - 3\mathbf{j}$

$\mathbf{a} - \mathbf{t} = (3\mathbf{i} + 2\mathbf{j}) - (7\mathbf{i} - 5\mathbf{j}) = -4\mathbf{i} + 7\mathbf{j}$.

Fig. 24.6

Position vectors

The *position vector* of point P in the x—y plane is the displacement vector from the origin O to P. In Fig. 24.7, the position vector of P is

$$\overrightarrow{OP} = \mathbf{r} = \binom{4}{3}$$

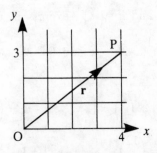

Fig. 24.7

Exercise 24.1

1. In Fig. 24.8, $\vec{PR} = \vec{PS} + \vec{SR} = \mathbf{b} + \mathbf{c}$.

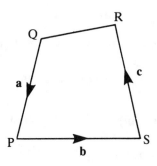

Fig. 24.8

Express in terms of **a**, **b** and **c**: (i) $\vec{QS}$ (ii) $\vec{SQ}$ (iii) $\vec{QR}$.

2. From Fig. 24.9, express in terms of **d** and **e**:

(i) $\vec{VX}$ (ii) $\vec{WY}$ (iii) $\vec{XY}$.

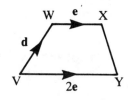

Fig. 24.9

3. Vectors **a**, **b** and **c** are shown in Fig. 24.10. On squared paper show vectors 2**a**, −**a**, **a** + **b**, **a** − **b** and **b** − **a**.

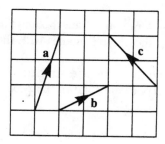

Fig. 24.10

4. Using the vectors in Fig. 24.10, show on squared paper:

$2a + 3c$, $a + b + c$ and $a - b - c$.

5. In Fig. 24.10, $a = \begin{pmatrix} 1 \\ 3 \end{pmatrix}$. Express as column vectors in this way:

(i) b (ii) c (iii) $b + c$ (iv) $b - c$ (v) $2a$ (vi) $2a - c$.

6. $f = \begin{pmatrix} 3 \\ -2 \end{pmatrix}$ and $g = \begin{pmatrix} -2 \\ 4 \end{pmatrix}$. Express as column vectors:

(i) $f + g$ (ii) $f - g$ (iii) $2f$ (iv) $3g$ (v) $2f + 3g$ (vi) $\frac{1}{2}g$.

7. On squared paper show a single vector for each of the following:

(i) $f + g$ (ii) $f - g$ (iii) $2f$ (iv) $3g$ (v) $2f + 3g$ (vi) $\frac{1}{2}g$.

8. Calculate the magnitude of each of the following vectors:

(i) $\begin{pmatrix} 3 \\ 4 \end{pmatrix}$ (ii) $\begin{pmatrix} 12 \\ 5 \end{pmatrix}$ (iii) $\begin{pmatrix} -8 \\ 6 \end{pmatrix}$ (iv) $\begin{pmatrix} 7 \\ -5 \end{pmatrix}$.

9. Calculate the angle between each of the following vectors and the positive x direction:

(i) $\begin{pmatrix} 2 \\ 5 \end{pmatrix}$ (ii) $\begin{pmatrix} 4 \\ 2 \end{pmatrix}$ (iii) $\begin{pmatrix} -5 \\ 3 \end{pmatrix}$.

10. $e = \begin{pmatrix} 2 \\ 5 \end{pmatrix}$ and $f = \begin{pmatrix} -4 \\ 3 \end{pmatrix}$. Express in terms of i and j:

e, $2e$, f, $e + f$ and $e - f$.

11. Express in the form $\begin{pmatrix} x \\ y \end{pmatrix}$: $4i + 2j$, $5i - j$, $-3i + 3j$.

12. $v = 2i - j$ and $w = -i + 5j$. Show that $v + 2w$ is parallel to the y-axis and find the direction of $2v + w$.

13. $p = 4i + 3j$ and $q = 6i - 6j$. State the magnitude of p. Show q in a diagram and state the angle it makes with the positive direction of the x-axis.

$r = kp + q$. Find the value of k so that: (i) r is in the direction of the x-axis   (ii) r is perpendicular to the x-axis.

14. State the position vectors of each of the points in Fig. 24.11.

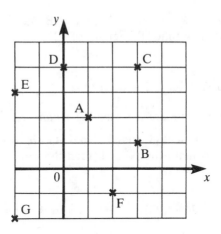

Fig. 24.11

15. Draw a diagram to show the points P, Q, R and S which have

position vectors $\begin{pmatrix} 2 \\ 3 \end{pmatrix}$, $\begin{pmatrix} 4 \\ 2 \end{pmatrix}$, $\begin{pmatrix} -2 \\ 3 \end{pmatrix}$ and $\begin{pmatrix} 4 \\ 0 \end{pmatrix}$.

Express in the form $\begin{pmatrix} x \\ y \end{pmatrix}$ the vectors $\overrightarrow{PQ}$ and $\overrightarrow{RS}$.

$\overrightarrow{RS} = k\overrightarrow{PQ}$. What is the value of k?

16. Express in the form $\begin{pmatrix} x \\ y \end{pmatrix}$:

(i) $\begin{pmatrix} 3 \\ 1 \end{pmatrix} + \begin{pmatrix} 2 \\ 6 \end{pmatrix}$ (ii) $\begin{pmatrix} 1 \\ 5 \end{pmatrix} + \begin{pmatrix} -3 \\ 1 \end{pmatrix}$ (iii) $3\begin{pmatrix} 2 \\ 4 \end{pmatrix}$ (iv) $\begin{pmatrix} 6 \\ 8 \end{pmatrix} - \begin{pmatrix} 4 \\ 5 \end{pmatrix}$

(v) $\begin{pmatrix} 5 \\ -2 \end{pmatrix} - \begin{pmatrix} -1 \\ 1 \end{pmatrix}$ (vi) $\begin{pmatrix} 2 \\ -3 \end{pmatrix} + 3\begin{pmatrix} 2 \\ 1 \end{pmatrix}$.

17. Solve the following equations:

(i) $\begin{pmatrix} x \\ y \end{pmatrix} = \begin{pmatrix} 6 \\ 3 \end{pmatrix} + \begin{pmatrix} 2 \\ -2 \end{pmatrix}$ (ii) $\begin{pmatrix} 5 \\ 7 \end{pmatrix} + \begin{pmatrix} x \\ y \end{pmatrix} = \begin{pmatrix} 8 \\ 3 \end{pmatrix}$

(iii) $2\begin{pmatrix} x \\ y \end{pmatrix} + \begin{pmatrix} 7 \\ 8 \end{pmatrix} = \begin{pmatrix} 1 \\ 2 \end{pmatrix}$ (iv) $\begin{pmatrix} -9 \\ -2 \end{pmatrix} - 3\begin{pmatrix} x \\ y \end{pmatrix} = 2\begin{pmatrix} 3 \\ -4 \end{pmatrix}$.

Proving lines to be parallel

If we find that $\overrightarrow{AB} = k\overrightarrow{CD}$, then the lines AB and CD are parallel and the length of AB is k times the length of CD.

Proving points to be collinear

Points in a straight line are described as *collinear*.

If we find that $\overrightarrow{EF} = k\overrightarrow{FG}$, then $\overrightarrow{EF}$ and $\overrightarrow{FG}$ are in the same direction. Since they both contain the point F then the three points E, F and G are collinear.

Example   In Fig. 24.12, $\overrightarrow{GH} = 3\overrightarrow{HE}$. Express in terms of **a** and **b**:

(i) $\overrightarrow{GE}$ (ii) $\overrightarrow{HE}$ (iii) $\overrightarrow{DH}$ (iv) $\overrightarrow{GH}$ (v) $\overrightarrow{HF}$.

Using the expressions for $\overrightarrow{DH}$ and $\overrightarrow{HF}$, show that the points D, H and F are collinear.

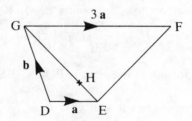

Fig. 24.12

(i) $\overrightarrow{GE} = \overrightarrow{GD} + \overrightarrow{DE} = -\mathbf{b} + \mathbf{a} = \mathbf{a} - \mathbf{b}$

(ii) $\overrightarrow{HE} = \frac{1}{4}\overrightarrow{GH} = \frac{1}{4}(\mathbf{a} - \mathbf{b}) = \frac{1}{4}\mathbf{a} - \frac{1}{4}\mathbf{b}$

(iii) $\overrightarrow{DH} = \overrightarrow{DE} + \overrightarrow{EH} = \overrightarrow{DE} - \overrightarrow{HE}$
$= \mathbf{a} - \left(\frac{1}{4}\mathbf{a} - \frac{1}{4}\mathbf{b}\right) = \frac{3}{4}\mathbf{a} + \frac{1}{4}\mathbf{b}$

(iv) $\overrightarrow{HG} = \frac{3}{4}\overrightarrow{EG} = -\frac{3}{4}\overrightarrow{GE} = -\frac{3}{4}\mathbf{a} + \frac{3}{4}\mathbf{b}$

(v) $\overrightarrow{HF} = \overrightarrow{HG} + \overrightarrow{GF} = -\frac{3}{4}\mathbf{a} + \frac{3}{4}\mathbf{b} + 3\mathbf{a} = 2\frac{1}{4}\mathbf{a} + \frac{3}{4}\mathbf{b}$

$\overrightarrow{HF} = 2\frac{1}{4}\mathbf{a} + \frac{3}{4}\mathbf{b} = 3\left(\frac{3}{4}\mathbf{a} + \frac{1}{4}\mathbf{b}\right) = 3\overrightarrow{DH}$

Hence $\overrightarrow{HF}$ and $\overrightarrow{DH}$ are in the same direction. Since H is common to both vectors, D, H and F are collinear.

Exercise 24.2

1. If $\mathbf{a} = \begin{pmatrix} 2 \\ 3 \end{pmatrix}$, which of the following are parallel to **a**:

(i) $\begin{pmatrix} 3 \\ 2 \end{pmatrix}$ (ii) $\begin{pmatrix} 6 \\ 9 \end{pmatrix}$ (iii) $\begin{pmatrix} 4 \\ -6 \end{pmatrix}$ (iv) $\begin{pmatrix} 1 \\ 1\frac{1}{2} \end{pmatrix}$ (v) $\begin{pmatrix} -10 \\ -15 \end{pmatrix}$?

Express in the form $k\mathbf{a}$ those which are parallel to **a**.

2. P, Q, R and S are points such that $\vec{PQ} = \begin{pmatrix} 5 \\ -1 \end{pmatrix}$, $\vec{PR} = \begin{pmatrix} 2 \\ 3 \end{pmatrix}$, $\vec{PS} = \begin{pmatrix} -1 \\ 7 \end{pmatrix}$. Find $\vec{QR}$ and $\vec{RS}$. What follows for Q, R and S?

3. (i) If $\vec{HK} = 5\,\vec{LM}$, what follows for HK and LM?

(ii) If $\vec{PQ} = 2\,\vec{QR}$, what follows for P, Q and R?

4. Draw diagrams to show the geometric meaning of:

(i) $\vec{AB} = \vec{CD}$ (ii) $\vec{EF} = 2\,\vec{GH}$ (iii) $\vec{KL} = -\,\vec{MN}$
(iv) $\vec{PQ} = \vec{QR}$ (v) $\vec{ST} = 2\,\vec{TV}$ (vi) $\vec{WX} + \vec{XY} + \vec{YW} = 0$.

5. In Fig. 24.13, $\vec{AB} = \vec{BD} = \mathbf{p}$ and $\vec{AC} = \vec{CE} = \mathbf{q}$.

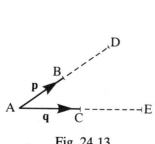

Fig. 24.13

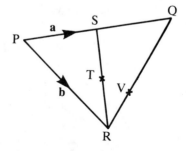

Fig. 24.14

Express in terms of **p** and **q**:

(i) $\vec{AD}$ (ii) $\vec{AE}$ (iii) $\vec{DE}$ (iv) $\vec{BC}$.

From (iii) and (iv), what can you say about BC and DE?

6. In Fig. 24.14, $\vec{PS} = \mathbf{a}$, $\vec{PR} = \mathbf{b}$, S is the mid-point of PQ, T is the mid-point of SR, QV = 2 VR.

Express the following in terms of **a** and **b**:

(i) $\vec{SR}$ (ii) $\vec{ST}$ (iii) $\vec{PT}$ (iv) $\vec{PQ}$ (v) $\vec{QR}$ (iv) $\vec{VR}$ (vii) $\vec{PV}$.

From the expressions for $\vec{PT}$ and $\vec{PV}$, what follows for P, T and V?

7. In the triangle OAB, the mid-point of AB is P.

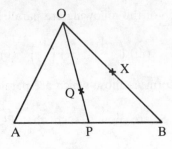

Fig. 24.15

(i) Given that $\overrightarrow{OA}$ = **a** and $\overrightarrow{OB}$ = **b**, express in terms of **a** and **b** the vectors $\overrightarrow{AB}$, $\overrightarrow{AP}$, $\overrightarrow{OP}$.

(ii) The point Q is taken on OP such that $\overrightarrow{OQ} = \frac{2}{3}\overrightarrow{OP}$. Express in terms of **a** and **b** the vectors $\overrightarrow{OQ}$ and $\overrightarrow{AQ}$.

(iii) Given that X is the mid-point of OB, express $\overrightarrow{AX}$ in terms of **a** and **b**. Hence deduce that $\overrightarrow{AX} = k\,\overrightarrow{AQ}$, where k is a scalar, and find the value of k.   (JMB)

8. The figure OABC is a parallelogram and M is the mid-point of AB.

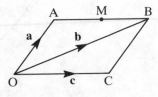

Fig. 24.16

(a) It is given that $\overrightarrow{OA}$ = **a**, $\overrightarrow{OB}$ = **b** and $\overrightarrow{OC}$ = **c**. Write down in terms of **a** and **b** expressions for (i) $\overrightarrow{AB}$ (ii) $\overrightarrow{AM}$ (iii) $\overrightarrow{OM}$.

(b) Express $\overrightarrow{CM}$ in terms of **a**, **b** and **c**.

(c) It is given that P is a point (not shown in the figure) such that $\overrightarrow{OP}$ = **p** and $\overrightarrow{CP} = \frac{2}{3}\overrightarrow{CM}$.

   (i) Find **p** in terms of **a**, **b** and **c** and hence in terms of **b**.

   (ii) Hence, or otherwise, state a result about the position of the point P.   (EA–C Joint 16+)

9. OABC is a quadrilateral. P, Q, R are points on OA, OB, OC respectively such that $\overrightarrow{OA} = 3\overrightarrow{OP}$, $\overrightarrow{OB} = 5\overrightarrow{OQ}$ and $\overrightarrow{OC} = 2\overrightarrow{OR}$.

Taking $\overrightarrow{OP} = \mathbf{p}$, $\overrightarrow{OQ} = \mathbf{q}$ and $\overrightarrow{OR} = \mathbf{r}$, express in terms of $\mathbf{p}$, $\mathbf{q}$ and $\mathbf{r}$ the vectors $\overrightarrow{AB}$, $\overrightarrow{BC}$ and $\overrightarrow{CA}$. Given further that OABC is a parallelogram, show that $3\mathbf{p} - 5\mathbf{q} + 2\mathbf{r} = 0$.

Hence express $\overrightarrow{PQ}$ and $\overrightarrow{QR}$ in terms of $\mathbf{p}$ and $\mathbf{q}$ only.

(a) Show that PQR is a straight line.

(b) Find the ratio PQ:QR. (L)

10. In Fig. 24.17, $\overrightarrow{OP} = \mathbf{a}$ and $\overrightarrow{OS} = \mathbf{b}$.

(i) Express $\overrightarrow{SP}$ in terms of $\mathbf{a}$ and $\mathbf{b}$.

(ii) Given that $\overrightarrow{SX} = h\,\overrightarrow{SP}$, show that $\overrightarrow{OX} = h\mathbf{a} + (1 - h)\mathbf{b}$.

(iii) Given that $\overrightarrow{OQ} = 3\mathbf{a}$ and $\overrightarrow{QR} = 2\mathbf{b}$, write down an expression for $\overrightarrow{OR}$ in terms of $\mathbf{a}$ and $\mathbf{b}$.

(iv) Given that $\overrightarrow{OX} = k\,\overrightarrow{OR}$ use the results of parts (ii) and (iii) to find the values of h and k.

(v) Find the numerical value of the ratio $\dfrac{PX}{XS}$. (C)

(Hint for part (iv). In part (iii) you should obtain $\overrightarrow{OR} = 3\mathbf{a} + 2\mathbf{b}$.

So $\overrightarrow{OX} = k\,\overrightarrow{OR} = 3k\mathbf{a} + 2k\mathbf{b}$. Now from (ii) $\overrightarrow{OX} = h\mathbf{a} + (1 - h)\mathbf{b}$. These two expressions for OX must be the same. Therefore $3k = h$ and $2k = 1 - h$. These are simultaneous linear equations from which you can find values of h and k.)

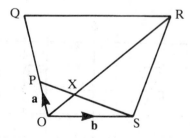

Fig. 24.17

25 Transformations

Translations

In a *translation* the object is moved along a straight line for a certain distance. The translation in Fig. 25.1 can be described by the column vector $\begin{pmatrix} 3 \\ -1 \end{pmatrix}$. The x-coordinate of each point is changed by $+3$ and the y-coordinate by -1. T' is called the *image* of T.

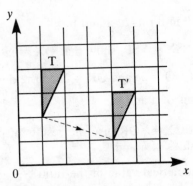

Fig. 25.1

Reflections

In Fig. 25.2, T' is the image of T under *reflection* in the mirror line m.

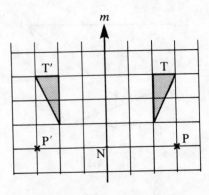

Fig. 25.2

Notice that the image and object face opposite ways.

P' is the image of P. Notice that:

(i) PP' is perpendicular to *m*

(ii) PN = P'N, that is, the object and image are at equal distances from the mirror line.

These two facts are used to construct the position of the image, when an object is reflected in a mirror line.

Example 1   Given the object ABC and the mirror line *n*, in Fig. 25.3, construct the image A'B'C'.

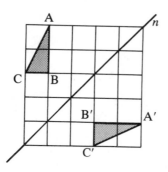

Fig. 25.3

AA' must be perpendicular to *n* and A and A' must be equal distances from *n*. The perpendicular from A to *n* lies along the diagonals of the squares in the grid. A is two diagonals from *n*, so we mark A' at two diagonals from *n* on the opposite side. B' and C' are marked in a similar way.

Example 2   Given PQ and its image P'Q', in Fig. 25.4, construct the position of the mirror line.

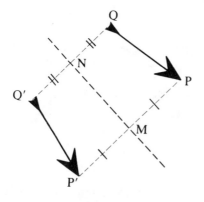

Fig. 25.4

The mirror line bisects PP′ and QQ′. So we join P to P′ and mark M the mid-point. We also find N, the mid-point of QQ′. Then MN is the mirror line.

Rotations

In Fig. 25.5, T′ is the image of T after an anticlockwise *rotation* of 90° (or a quarter turn) about point K. Each point of T moves along an arc of a circle, centre K.

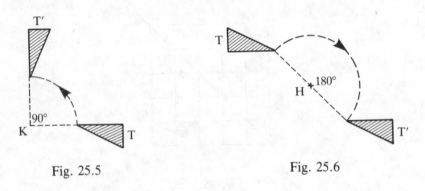

Fig. 25.5   Fig. 25.6

Figure 25.6 shows a clockwise rotation of 180° (or a half turn) about the point H.

Exercise 25.1

1. Copy Fig. 25.7 onto squared paper.

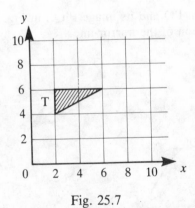

Fig. 25.7

(i) Draw the image of triangle T under a translation of 4 units in the positive *x* direction, and label it A.

(ii) Draw the image of T under a translation of 3 units in the positive *y* direction and label it B.

2. Using axes numbered from 0 to 8, mark the points (1, 1), (3, 1) and (3, 2) and join them to form a triangle. Label the triangle S.

(i) Draw the image of S under the translation $\begin{pmatrix} 5 \\ 2 \end{pmatrix}$ and label it D.

(ii) Draw the image of D under the translation $\begin{pmatrix} -3 \\ 1 \end{pmatrix}$. Label it E.

(iii) State the translation for which S is the image of E.

3. Copy Fig. 25.8 where P is the point (2, 4). On your copy, mark the following points and state their coordinates:

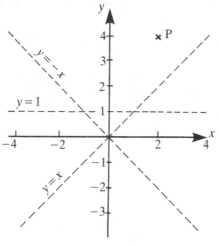

Fig. 25.8

(i) Q, the reflection of P in the *y*-axis
(ii) R, the reflection of P in the line $y = 1$
(iii) S, the reflection of P in the line $y = x$
(iv) T, the reflection of P in the line $y = -x$.

4. Draw *x* and *y* axes and number them from −5 to +5.

(i) Mark P at (3, 3) and join OP. Show OP′, the image of OP under an anticlockwise rotation of 90° about O. State the coordinates of P′.

(ii) Mark Q at (4, 2) and join OQ. Show OQ′, the image of OQ under a clockwise rotation of 90° about O. State the coordinates of Q′.

(iii) Mark R at (3, 5) and join RP. Show R′P, the image of RP under a clockwise rotation of 90° about P. State the coordinates of R′.

5. Using axes numbered from 0 to 8, draw triangle T with vertices at (1, 3), (3, 3) and (1, 7).

   (i) Show the image of T under a clockwise rotation of 90° about (3, 3), and label it V.

   (ii) Show the image of T under a clockwise rotation of 90° about (1, 3), and label it W.

   (iii) State the translation for which W is the image of V.

6. Use an x-axis numbered from 0 to 4 and a y-axis numbered from -4 to $+4$. Join the points (1, 2), (1, 4) and (2, 4) to form a triangle, and label it T. Draw R, the reflection of T in the line $y = x$. Draw S, the reflection of R in the x-axis.

   For what single transformation is S the image of T?

7. With the aid of a sketch, find the coordinates of the image of (2, 3) under a reflection in:

   (i) the y-axis   (ii) the x-axis   (iii) $y = 4$ (iv) $x = -1$ (v) $y = x$
   (vi) $y = -x$.

8. With the aid of a sketch, find the coordinates of the image of (3, 1) under an anticlockwise rotation of 90° about:

   (i) the origin   (ii) the point (3, 3)   (iii) the point $(-1, -1)$.

9. Copy Fig. 25.9.

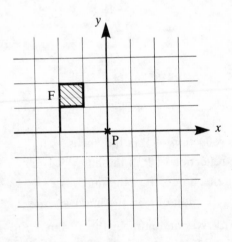

Fig. 25.9

   (i) Show the image of F after reflection in the y-axis, and label it F_1.

   (ii) Show the image of F after a rotation through 180° about P, and label it F_2.

   (iii) Under what transformation is F_2 the image of F_1?

10. Using axes numbered from 0 to 8, mark the following points: P(3, 6), P'(1, 4), Q(5, 6), Q'(1, 2), R(6, 4), R'(3, 1). P', Q' and R' are the images of P, Q and R under reflection in a certain mirror line. Draw this mirror line and state its equation.

Under reflection in the same line, the image of (a, b) is $(k - b, k - a)$. Find the value of k.

11. Copy Fig. 25.10.

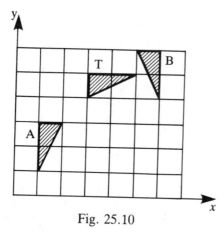

Fig. 25.10

(i) Triangle A is the image of triangle T under reflection in a certain mirror line. Draw this mirror line in your diagram, and label it m.

(ii) Triangle B is the image of T under a clockwise rotation of 90°. Mark P, the centre of rotation, and state its coordinates.

(iii) Draw triangle C, the image of triangle A under reflection in the line $x = 3$. State in the form $\begin{pmatrix} p \\ q \end{pmatrix}$ the translation for which C is the image of B.

12. On squared paper, draw triangle ABC, where A is (1, 3), B is (1,5) and C is (2, 5). Use a scale of 2 cm for one unit on both axes and allow space on your paper for x to vary from −3 to 6 and y to vary from −6 to 6.

P is the reflection in the line $x = 3$ and **Q** is the reflection in the x-axis. **R** is the clockwise rotation through 90° about the point (5, 2). ABC is mapped onto $A_1B_1C_1$ by **P**, $A_1B_1C_1$ is mapped onto $A_2B_2C_2$ by **Q** and $A_2B_2C_2$ is mapped onto $A_3B_3C_3$ by **R**. Draw these triangles in your diagram.

(i) Describe the transformation **H** where **H** = **QP**.

(ii) Describe the transformation **K** which maps ABC onto $A_3B_3C_3$.

(W)

(N.B. **QP** means carry out the transformation **P** first, then transformation **Q**. Therefore **H** is the transformation which maps ABC onto $A_1B_2C_2$).

13. (i) Copy Fig. 25.11. On your diagram draw and label $M_1(F)$, which
is the image of F under reflection in m. Also draw and label
$M_2M_1(F)$, which is the image of $M_1(F)$ under reflection in m_2.

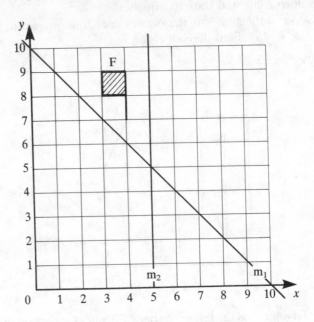

Fig. 25.11

R is a clockwise rotation through 90° about a certain point P
such that **R**(F) is the same as $M_2M_1(F)$. State the coordinates
of P.

(ii) On a separate diagram draw $M_2(F)$ and $M_1M_2(F)$.
Is $M_1M_2 = M_2M_1$?

Enlargements

In Fig. 25.12, triangle A'B'C' is an *enlargement* of triangle ABC, with
P as the *centre of enlargement* and a *scale factor* of $1\frac{1}{2}$. To form the
image A'B'C', PA was extended to A' so that $PA' = 1\frac{1}{2}PA$, PB was
extended to B' with $PB' = 1\frac{1}{2}PB$ and PC was extended to C' with
$PC' = 1\frac{1}{2}PC$. Every point on the image is $1\frac{1}{2}$ times as far from P as the
corresponding point on the object. Also, any length on the image is

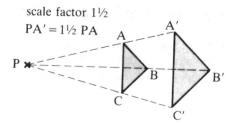

Fig. 25.12

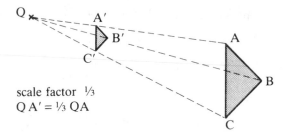

Fig. 25.13

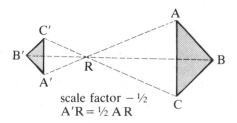

Fig. 25.14

$1\frac{1}{2}$ times the corresponding length on the object. For example $A'B' = 1\frac{1}{2}AB$. Notice that the image and object have the same shape but different sizes. They are similar triangles.

Figure 25.13 shows an enlargement of scale factor $\frac{1}{3}$. $QA' = \frac{1}{3}QA$, and so on. Although the image is smaller than the object, we still call the transformation an enlargement.

Figure 25.14 shows an enlargement of scale factor $-\frac{1}{2}$. When the scale factor is negative, the image and object are on opposite sides of the centre of enlargement, and the image is inverted.

To find the centre of enlargement by construction

If the centre of enlargement is not known, it can be found by construction. We draw straight lines through corresponding object and image points. These intersect at the centre of enlargement. For example in Fig. 25.12 the lines AA', BB' and CC' intersect at P. P is the centre of enlargement.

Stretches

In a *stretch*, distances are increased (or decreased) in one direction only.

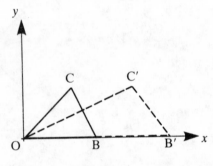

Fig. 25.15

Figure 25.15 shows a stretch of magnitude 2 applied to triangle OBC, in the direction of the *x*-axis. All distances from the *y*-axis are doubled. OB' = 2 OB and OC' = 2 OC. Distances from the *x*-axis are not changed.

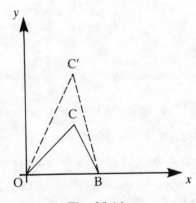

Fig. 25.16

Figure 25.16 shows a stretch of magnitude 2 applied to triangle OBC, in the direction of the *y*-axis.

In a stretch, the size, shape and area are all changed.
To specify a stretch we need to state:

(i) the direction of the stretch

(ii) the base line from which distances are measured

(iii) the magnitude of the stretch.

Exercise 25.2

1. For this question you need a space on squared paper, 12 units wide
   and 6 units high. Copy Fig. 25.17, placing O at the centre of your
   space.

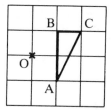

Fig. 25.17

(i) With O as centre of enlargement and scale factor 3, draw the
   image of triangle ABC and label it $A_1B_1C_1$.

(ii) With O as centre of enlargement and scale factor -2, draw the
   image of triangle ABC and label it $A_2B_2C_2$.

2. On squared paper draw a square PQRS with sides of 9 units. Mark
   C inside the square so that it is 3 units from both PQ and PS.

(i) With centre C and scale factor $\frac{1}{3}$, draw the image of PQRS and
   label it $P_1Q_1R_1S_1$.

(ii) With centre C and scale factor 2, draw the image of $P_1Q_1R_1S_1$
   and label it $P_2Q_2R_2S_2$.

(iii) What single enlargement changes PQRS to $P_2Q_2R_2S_2$?

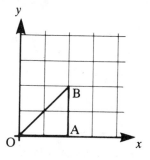

Fig. 25.18

3. Use three different colours in this question. Use one colour to copy Fig. 25.18. Use a second colour to show OA_1B_1, which is the image of OAB under a stretch of magnitude 2 in the x-direction, with the y-axis as base line.

   Use the third colour to show OA_2B_2, the image of OAB under a stretch of magnitude 2 in the y-direction, with the x-axis as base line.

4. Using axes numbered from 0 to 12, draw rectangle ABCD with A (4, 8), B (10, 8), C (10, 12) and D (4, 12). With O as centre of enlargement and scale factor $-\frac{1}{2}$, draw the image of ABCD and label it $A_1B_1C_1D_1$.

   Draw rectangle $A_2B_2C_2D_2$ as the image of $A_1B_1C_1D_1$ under a translation of 6 units in the positive x direction.

   ABCD is the image of $A_2B_2C_2D_2$ under an enlargement with centre P. Show P in your diagram and state its coordinates. Also state the scale factor of this enlargement.

5. Using an x-axis numbered from 0 to 6 and a y-axis numbered from 0 to 4, draw a rectangle ABCD with A(0, 0), B(3, 0), C(3, 2) and D(0, 2).

   (i) With A as centre of enlargement and scale factor 2, show the image of ABCD and label it A'B'C'D'.

   (ii) With C' as centre of enlargement and scale factor $\frac{1}{2}$, show the image of A'B'C'D' and label it A"B"C"D".

   (iii) State the transformation for which A"B"C"D" is the image of ABCD.

6. In Fig. 25.19, triangle ADE is the image of triangle ABC, under an enlargement with centre A and scale factor $\frac{5}{3}$.

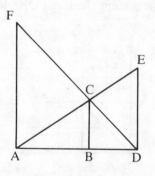

Fig. 25.19

   (i) If AB = 6 cm, state the lengths of AD and BD.

  (ii) Triangle AFD is an enlargement of triangle BCD. Find the centre of enlargement and the scale factor.

 (iii) If BC = 4.2 cm, calculate ED and AF.

 (iv) Triangle AFC is the image of triangle EDC under an enlargement with centre C. Find the scale factor.

26 Matrices and transformations

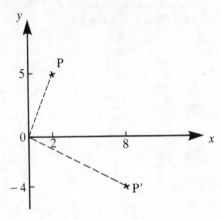

Fig. 26.1

The point P (2, 5) has position vector $\overrightarrow{OP} = \begin{pmatrix} 2 \\ 5 \end{pmatrix}$.

Multiplying this position vector by the matrix $\begin{pmatrix} -1 & 2 \\ 3 & -2 \end{pmatrix}$ we have

$\begin{pmatrix} -1 & 2 \\ 3 & -2 \end{pmatrix}\begin{pmatrix} 2 \\ 5 \end{pmatrix} = \begin{pmatrix} 8 \\ -4 \end{pmatrix}$ which is the position vector of the point

P'(8, −4). Thus the matrix $\begin{pmatrix} -1 & 2 \\ 3 & -2 \end{pmatrix}$ transforms P(2, 5) to P'(8, −4).

We can write this change in position as $(2, 5) \to (8, -4)$.

To apply a matrix to several points, we place the corresponding position vectors in one matrix as in the following example.

Example 1 Apply the matrix $\begin{pmatrix} -1 & 2 \\ 3 & -2 \end{pmatrix}$ to the points $(1, 2)$, $(-2, 3)$
and $(5, 0)$.

$$\begin{pmatrix} -1 & 2 \\ 3 & -2 \end{pmatrix}\begin{pmatrix} 1 & -2 & 5 \\ 2 & 3 & 0 \end{pmatrix} = \begin{pmatrix} 3 & 8 & -5 \\ -1 & -12 & 15 \end{pmatrix}$$

and so $(1, 2) \to (3, -1)$, $(-2, 3) \to (8, -12)$ and
$(5, 0) \to (-5, 15)$.

Example 2 Apply $\begin{pmatrix} -1 & 0 \\ 0 & 1 \end{pmatrix}$ to the rectangle joining the points $(0, 0)$,
$(2, 0)$, $(2, 4)$ and $(0, 4)$. Show the original rectangle and its

image in a diagram. Describe the resulting geometric transformation.

$$\begin{pmatrix} -1 & 0 \\ 0 & 1 \end{pmatrix} \begin{pmatrix} 0 & 2 & 2 & 0 \\ 0 & 0 & 4 & 4 \end{pmatrix} = \begin{pmatrix} 0 & -2 & -2 & 0 \\ 0 & 0 & 4 & 4 \end{pmatrix}$$

The original rectangle and
its image are shown in
Fig. 26.2.

The geometric transformation
is a reflection in the y-axis.

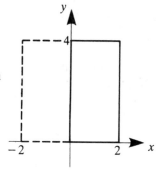

Fig. 26.2

The table below shows the geometric transformations produced by
certain simple matrices.

Matrix	Result	Geometric transformation produced
$\begin{pmatrix} 1 & 0 \\ 0 & -1 \end{pmatrix}$	$(x, y) \rightarrow (x, -y)$	Reflection in x-axis
$\begin{pmatrix} -1 & 0 \\ 0 & 1 \end{pmatrix}$	$(x, y) \rightarrow (-x, y)$	Reflection in y-axis
$\begin{pmatrix} 0 & 1 \\ 1 & 0 \end{pmatrix}$	$(x, y) \rightarrow (y, x)$	Reflection in line $y = x$
$\begin{pmatrix} -1 & 0 \\ 0 & -1 \end{pmatrix}$	$(x, y) \rightarrow (-x, -y)$	Rotation through 180° about O
$\begin{pmatrix} 0 & -1 \\ 1 & 0 \end{pmatrix}$	$(x, y) \rightarrow (-y, x)$	Anticlockwise rotation through 90° about O
$\begin{pmatrix} 0 & 1 \\ -1 & 0 \end{pmatrix}$	$(x, y) \rightarrow (y, -x)$	Clockwise rotation through 90° about O
$\begin{pmatrix} k & 0 \\ 0 & k \end{pmatrix}$	$(x, y) \rightarrow (kx, ky)$	Enlargement, scale factor k, centre O
$\begin{pmatrix} k & 0 \\ 0 & 1 \end{pmatrix}$	$(x, y) \rightarrow (kx, y)$	Stretch in x direction, magnitude k

Exercise 26.1

1. Apply each of the matrices $\begin{pmatrix} 1 & 0 \\ 0 & -1 \end{pmatrix}$, $\begin{pmatrix} -1 & 0 \\ 0 & -1 \end{pmatrix}$, $\begin{pmatrix} 0 & 1 \\ -1 & 0 \end{pmatrix}$, $\begin{pmatrix} 0 & 1 \\ 1 & 0 \end{pmatrix}$ to the point $(5, 2)$. In each case show $(5, 2)$ and its image in a diagram and verify that the transformation given in the above table has taken place.

2. Apply $\begin{pmatrix} 0 & -1 \\ 1 & 0 \end{pmatrix}$ to the vertices of triangle ABC where A is $(2, 4)$, B is $(5, 1)$ and C is $(4, 6)$. On squared paper show triangle ABC and its image triangle A'B'C'. Describe the transformation geometrically.

3. Apply $\begin{pmatrix} 0 & 1 \\ 1 & 0 \end{pmatrix}$ to the vertices of triangle PQR where P is $(3, 2)$, Q is $(5, 4)$ and R is $(6, 2)$. On squared paper show triangle PQR and its image triangle P'Q'R', and check that P'Q'R' is the reflection of PQR in the line $y = x$.

4. Apply $\begin{pmatrix} 0 & -1 \\ -1 & 0 \end{pmatrix}$ to the vertices of triangle STV where S is $(1, 1)$, T is $(3, 1)$ and V is $(3, 2)$. On squared paper show triangle STV and its image triangle S'T'V', and describe the transformation.

5. Apply the matrix $\begin{pmatrix} 1\frac{1}{2} & 0 \\ 0 & 1 \end{pmatrix}$ to the square with vertices at $(0, 0)$, $(2, 0)$, $(2, 2)$ and $(0, 2)$. Show the square and its image in a diagram and describe the transformation.

   State the matrix which represents a stretch of magnitude 3 in the x direction.

6. Apply the matrix $\begin{pmatrix} -0.6 & 0.8 \\ 0.8 & 0.6 \end{pmatrix}$ to the vertices of triangle ABC where A is $(5, 10)$, B is $(-5, 10)$ and C is $(-5, 5)$. On squared paper show triangle ABC and its image triangle A'B'C'. A'B'C' is the reflection of ABC in a certain line. Show the line in your diagram and state its equation.

7. (i) Apply the matrix $\begin{pmatrix} 3 & 0 \\ 0 & 3 \end{pmatrix}$ to the rectangle with vertices at $(1, 1)$, $(3, 1)$, $(3, 2)$ and $(1, 2)$. Sketch the original rectangle and its image and describe the transformation.

(ii) What transformations are represented by the matrices $\begin{pmatrix} 5 & 0 \\ 0 & 5 \end{pmatrix}$ and $\begin{pmatrix} \frac{1}{2} & 0 \\ 0 & \frac{1}{2} \end{pmatrix}$?

8. Find the images of the points $(4, 5)$, $(3, 2)$ and $(1, -2)$ under the transformation with matrix $\begin{pmatrix} -2 & 3 \\ -1 & 2 \end{pmatrix}$. Now apply $\begin{pmatrix} -2 & 3 \\ -1 & 2 \end{pmatrix}$ to the image points and comment on the results.

9. On squared paper draw triangle ABC where A is $(1, 0)$, B is $(4, 0)$ and C is $(3, 2)$, using a scale of 2 cm for 1 unit on both axes and allowing space on your paper for x to vary from 0 to 8 and y to vary from 0 to 4.

Triangle ABC is mapped onto $A_1B_1C_1$ by a transformation $\mathbf{M}$ which is determined by the matrix $\begin{pmatrix} 2 & 0 \\ 0 & 2 \end{pmatrix}$. Draw triangle $A_1B_1C_1$ in your diagram and describe the transformation $\mathbf{M}$.

Triangle ABC is mapped onto triangle $A_2B_2C_2$ by transformation $\mathbf{K}$, where A_2 is $(0, 1)$, B_2 is $(0, 4)$ and C_2 is $(2, 3)$. Draw triangle $A_2B_2C_2$ in your diagram and find the matrix which determines $\mathbf{K}$. (W)

Combining transformations

Let $\mathbf{A} = \begin{pmatrix} 0 & 1 \\ 1 & 0 \end{pmatrix}$ and $\mathbf{B} = \begin{pmatrix} 1 & 0 \\ 0 & -1 \end{pmatrix}$.

Applying $\mathbf{A}$ to the point P $(2, 5)$ we obtain P_1 $(5, 2)$, since $\begin{pmatrix} 0 & 1 \\ 1 & 0 \end{pmatrix}\begin{pmatrix} 2 \\ 5 \end{pmatrix} = \begin{pmatrix} 5 \\ 2 \end{pmatrix}$. P_1 is the reflection of P in the line $y = x$.

Applying $\mathbf{B}$ to P_1 we obtain P_2 $(5, -2)$, since $\begin{pmatrix} 1 & 0 \\ 0 & -1 \end{pmatrix}\begin{pmatrix} 5 \\ 2 \end{pmatrix} = \begin{pmatrix} 5 \\ -2 \end{pmatrix}$.

P_2 is the reflection of P_1 in the x-axis.

Now $\mathbf{BA} = \begin{pmatrix} 1 & 0 \\ 0 & -1 \end{pmatrix}\begin{pmatrix} 0 & 1 \\ 1 & 0 \end{pmatrix} = \begin{pmatrix} 0 & 1 \\ -1 & 0 \end{pmatrix}$. Call this matrix $\mathbf{C}$.

Applying $\mathbf{C}$ to the original point P, we have $\begin{pmatrix} 0 & 1 \\ -1 & 0 \end{pmatrix}\begin{pmatrix} 2 \\ 5 \end{pmatrix} = \begin{pmatrix} 5 \\ -2 \end{pmatrix}$,

so that P becomes P_2.

Thus the matrix $\mathbf{C}$ (which is $\mathbf{BA}$) transforms point P to P_2 in one operation. $\mathbf{C}$ is the matrix for a rotation of $90°$ clockwise about O.

Since $\mathbf{C} = \mathbf{BA}$, then a reflection in the line $y = x$ followed by a reflection in the x-axis is equivalent to a rotation of $90°$ clockwise about O. In general, if matrix $\mathbf{A}$ transforms P to P_1 and matrix $\mathbf{B}$ transforms P_1

to P_2, then matrix $\mathbf{C} = \mathbf{BA}$ transforms P to P_2.

The matrices must be in the correct order. Remember that $\mathbf{AB}$ is not usually the same as $\mathbf{BA}$.

Exercise 26.2

1. (i) $\mathbf{G} = \begin{pmatrix} 2 & -1 \\ 0 & 1 \end{pmatrix}$ and $\mathbf{H} = \begin{pmatrix} -1 & 1 \\ 2 & 1 \end{pmatrix}$. Apply $\mathbf{G}$ to the point P $(1, 3)$ to give P_1 and then apply $\mathbf{H}$ to P_1 to give P_2.

   Form the single matrix $\mathbf{K} = \mathbf{HG}$ and show that when $\mathbf{K}$ is applied to P the point P_2 is obtained.

   (ii) Form the single matrix $\mathbf{M} = \mathbf{GH}$ and apply it to P to give P_3. Is P_3 the same as P_2?

2. Repeat question 1 using $\begin{pmatrix} -3 & 4 \\ 2 & -2 \end{pmatrix}$ for $\mathbf{G}$ and $\begin{pmatrix} 2 & 0 \\ -1 & 3 \end{pmatrix}$ for $\mathbf{H}$.

3. On squared paper draw triangle ABC where A is $(1, 3)$, B is $(1, 1)$ and C is $(2, 1)$. Triangle ABC is mapped onto triangle $A_1 B_1 C_1$ by a transformation given by the matrix $\mathbf{K} = \begin{pmatrix} 0 & -1 \\ -1 & 0 \end{pmatrix}$. Draw triangle $A_1 B_1 C_1$ in your diagram and describe the transformation.

   Triangle $A_1 B_1 C_1$ is mapped onto triangle $A_2 B_2 C_2$ by the matrix $\mathbf{M} = \begin{pmatrix} -1 & 0 \\ 0 & 1 \end{pmatrix}$. Draw triangle $A_2 B_2 C_2$ in your diagram and describe the transformation given by matrix $\mathbf{M}$.

   Describe the transformation which maps triangle ABC onto triangle $A_2 B_2 C_2$, and determine its matrix.

4. (i) Let $\mathbf{N} = \begin{pmatrix} 0 & -1 \\ 1 & 0 \end{pmatrix}$. Find $\mathbf{N}^2$ and $\mathbf{N}^3$. What transformations are represented by $\mathbf{N}$, $\mathbf{N}^2$ and $\mathbf{N}^3$?

   (ii) Let $\mathbf{H} = \begin{pmatrix} 3 & 0 \\ 0 & 3 \end{pmatrix}$. Find $\mathbf{H}^2$ and $\mathbf{H}^3$. What transformations are represented by $\mathbf{H}$, $\mathbf{H}^2$ and $\mathbf{H}^3$?

5. Apply the matrix $\begin{pmatrix} 2 & 0 \\ 0 & 1 \end{pmatrix}$ to the triangle OPQ where O is $(0, 0)$, P is $(2, 0)$ and Q is $(2, 2)$. Show triangle OPQ and its image triangle $OP_1 Q_1$ in a diagram and describe the transformation.

   Apply the matrix $\begin{pmatrix} 0 & 1 \\ 1 & 0 \end{pmatrix}$ to triangle $OP_1 Q_1$ and show the image triangle $OP_2 Q_2$ in your diagram.

   What single matrix maps OPQ onto $OP_2 Q_2$?

Inverse mappings

The matrix $\begin{pmatrix} 5 & -2 \\ 3 & -1 \end{pmatrix}$ maps $(4, 3)$ onto $(14, 9)$, since $\begin{pmatrix} 5 & -2 \\ 3 & -1 \end{pmatrix}\begin{pmatrix} 4 \\ 3 \end{pmatrix} = \begin{pmatrix} 14 \\ 9 \end{pmatrix}$. The inverse of $\begin{pmatrix} 5 & -2 \\ 3 & -1 \end{pmatrix}$ is $\begin{pmatrix} -1 & 2 \\ -3 & 5 \end{pmatrix}$.

$\begin{pmatrix} -1 & 2 \\ -3 & 5 \end{pmatrix}\begin{pmatrix} 14 \\ 9 \end{pmatrix} = \begin{pmatrix} 4 \\ 3 \end{pmatrix}$ so $\begin{pmatrix} -1 & 2 \\ -3 & 5 \end{pmatrix}$ maps $(14, 9)$ back onto $(4, 3)$.

In general, if matrix $\mathbf{M}$ maps point P onto P', then $\mathbf{M}^{-1}$ maps P' back onto P.

Exercise 26.3

1. Apply the matrix $\begin{pmatrix} 3 & 5 \\ 1 & 2 \end{pmatrix}$ to the points $(4, -1)$ and $(-2, 7)$. State the inverse of the matrix and show that it maps the new points back onto the original points.

2. Apply $\begin{pmatrix} 5 & -3 \\ 4 & -2 \end{pmatrix}$ to the points $(2, 1)$ and $(-3, -2)$. State the inverse of the matrix and show that it maps the new points back onto the original points.

3. $\begin{pmatrix} 8 & 5 \\ 3 & 2 \end{pmatrix}$ maps A $(3, -4)$ onto B. Find the coordinates of B.

State the matrix which maps B onto A and check that it does so.

4. $\begin{pmatrix} 4 & -3 \\ 2 & -1 \end{pmatrix}$ maps C $(7, 5)$ onto D. Find the coordinates of D.

State the matrix which maps D onto C and check that it does so.

5. $\begin{pmatrix} -1 & -2 \\ 1 & 1 \end{pmatrix}$ maps the point (x, y) onto $(1, -3)$. Find (x, y).

6. Apply the matrix $\begin{pmatrix} 4 & 2 \\ 2 & 1 \end{pmatrix}$ to the points $(0, 0)$, $(3, 0)$, $(3, 2)$ and $(0, 2)$. Mark the image points on a diagram and show that they all lie on the straight line $y = \frac{1}{2}x$. State the determinant of the matrix. What kind of matrix is it?

7. Apply the singular matrix $\begin{pmatrix} 4 & -2 \\ 6 & -3 \end{pmatrix}$ to the points $(4, 3)$, $(2, 5)$ and $(-1, -4)$. Show that the image points all lie on the straight line $y = \frac{3}{2}x$.

8. Apply $\begin{pmatrix} 1 & 2 \\ 0 & 3 \end{pmatrix}$ to the rectangle joining the points $(1, 0)$, $(1, 1)$, $(-1, 1)$ and $(-1, 0)$. Show the rectangle and the image in a diagram. From your diagram find the area of the image. State the ratio of the image area to the object area. Check that this ratio is equal to the determinant of the matrix.

9. The matrix $A = \begin{pmatrix} 2 & -3 \\ 4 & -4 \end{pmatrix}$ represents the transformation

$T: \begin{pmatrix} x \\ y \end{pmatrix} \rightarrow \begin{pmatrix} 2 & -3 \\ 4 & -4 \end{pmatrix} \begin{pmatrix} x \\ y \end{pmatrix}$.

(i) Find the inverse matrix A^{-1}.

(ii) Find the coordinates of the point which is mapped to $(9, 16)$ under the transformation T.

(iii) Find the matrix A^2. Show that A^2 can be expressed in the form kA^{-1}, and find the value of the number k.

(iv) Find the matrix A^3. Give a geometrical description of the transformation whose matrix is A^3.   (JMB)

10. Using an x-axis numbered from -4 to 6 and a y-axis numbered from 0 to 10, draw and label the rectangle with vertices O $(0, 0)$, A $(4, 0)$, B $(4, 2)$ and C $(0, 2)$,

The points O, A, B and C are transformed by the matrix $\begin{pmatrix} 1 & -2 \\ 2 & 1 \end{pmatrix}$

into O, A', B' and C'. Draw the rectangle OA'B'C' on your diagram. The transformation can be described as an enlargement followed by a rotation. Measure the angle of rotation.

From your diagram, find the ratio of the area of OA'B'C' to the area of OABC and check that this ratio is equal to the determinant of the matrix.

State the scale factor of the enlargement.

Find the matrix that will transform OA'B'C' into OABC.

11. On graph paper, draw axes (with the same scale on both axes) so that x can range from at least -8 to $+8$ and y from -2 to $+14$.

(a) Given that O is $(0, 0)$, A is $(5, 0)$, B is $(5, 5)$ and C is $(0, 5)$, plot the points O, A, B, C on your axes and join them to form square S.

(b) Calculate the matrix product $T\begin{pmatrix} 0 & 5 & 5 & 0 \\ 0 & 0 & 5 & 5 \end{pmatrix}$, where T is the matrix $\begin{pmatrix} 1.6 & -1.2 \\ 1.2 & 1.6 \end{pmatrix}$.

(c) Use the result of (b) to draw the image OA'B'C' of S under the transformation whose matrix is **T**.

(d) Find the length of OA' and hence the area of OA'B'C'.

(e) How could you have obtained the area of OA'B'C' from the original square S and the matrix **T** without drawing an accurate diagram?

(f) Find the inverse of the matrix $\begin{pmatrix} 1.6 & -1.2 \\ 1.2 & 1.6 \end{pmatrix}$. Hence find the coordinates of the point which is mapped to $(4, -2)$ under the transformation whose matrix is **T**. (L)

(Hint: For part (e) refer to question 8 above.)

27 Angles

Types of angles

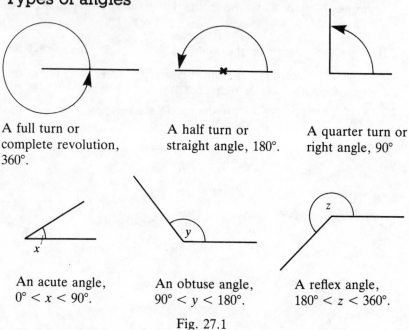

A full turn or complete revolution, 360°.

A half turn or straight angle, 180°.

A quarter turn or right angle, 90°

An acute angle, $0° < x < 90°$.

An obtuse angle, $90° < y < 180°$.

A reflex angle, $180° < z < 360°$.

Fig. 27.1

Supplementary angles are angles that add up to 180° (for example 125° and 55°).

Complementary angles are angles that add up to 90° (for example 28° and 62°).

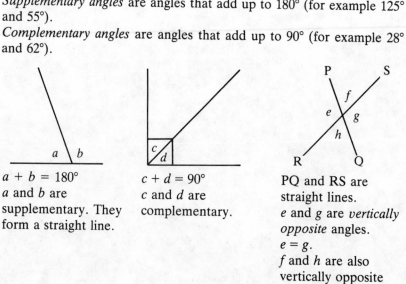

$a + b = 180°$
a and b are supplementary. They form a straight line.

$c + d = 90°$
c and d are complementary.

PQ and RS are straight lines.
e and g are *vertically opposite* angles.
$e = g$.
f and h are also vertically opposite angles. $f = h$.

Fig. 27.2

Example 1 The three angles in Fig. 27.3 form a straight line, so

$$x + 63° + 40° = 180°$$

from which $x = 77°$.

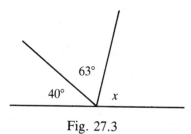

Fig. 27.3

Example 2 The four angles in Fig. 27.4 form a complete turn, so

$$y + 3y + 150° + 74° = 360°$$

from which $y = 34°$.

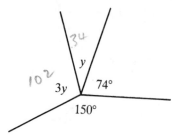

Fig. 27.4

Exercise 27.1

1. Here is a list of angles: 23°, 108°, 84°, 305°, 144°, 230°, 67°, 98°, 195°.

  (i) Which of them are acute?
  (ii) Which are obtuse
  (iii) Which are reflex?

2. (i) If $a = 70°$, find b.
  (ii) If $b = 133°$, find a.
  (iii) If $b = 2a$, find a.

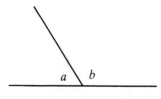

Fig. 27.5

3. (i) If $p = 40°$ and $q = 60°$, find r.

(ii) If $p = 70°$ and $q = r$, find q.

(iii) If $p = q = r$, find p.

(iv) If $p = 2x°$, $q = 3x°$ and $r = 4x°$, find x.

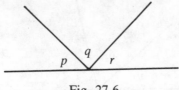

Fig. 27.6

4. What is the supplement of: (i) 50° (ii) 110° (iii) 90° (iv) 168°?

5. What is the complement of: (i) 20° (ii) 80° (iii) 35° (iv) 72°?

6. (i) If $a = 110°$, $b = 80°$ and $c = 70°$, find d.

(ii) If $b = 60°$, $c = 50°$ and $d = 130°$, find a.

(iii) If $b = 70°$, $c = 60°$ and $a = d$, find a.

(iv) If $d = 110°$ and $a = b = 2c$, find c.

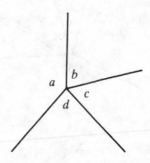

Fig. 27.7

7. (i) If $e = 50°$, find f, g and h.   (ii) If $h = 116°$, find e, f and g.

(iii) If $f = 3g$, find e.   (iv) If $f + h = 260°$, find g.

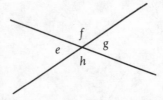

Fig. 27.8

8. Through what angle does the minute hand of a clock turn in:

(i) 15 minutes (ii) 20 minutes (iii) 50 minutes?

9. Through what angle does the hour hand of a clock turn in:

(i) 1 hour (ii) 7 hours (iii) 20 minutes?

10. Through what angle does the Earth turn in:

(i) 1 hour (ii) 3 hours 40 minutes (iii) 2 days?

11. Through what angle does the hour hand of a clock turn, between 16.12 hours and 18.36 hours?

Parallel lines

Parallel lines have the same direction. They are always the same distance apart, so they never meet. We place arrows on lines to show that they are parallel.

A *transversal* is a line cutting two or more parallel lines.

Angle properties of parallel lines

1. Corresponding angles are equal.

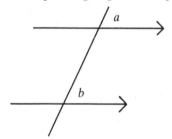 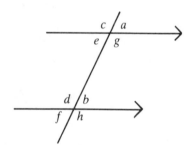

a and *b* are called corresponding angles. $a = b$.

Other pairs of corresponding angles are *c* and *d*, *e* and *f*, *g* and *h*.
$c = d, e = f, g = h.$

Fig. 27.9

2. Alternate angles are equal.

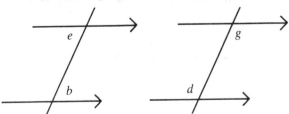

e and *b* are alternate angles, so $e = b$. *d* and *g* are also alternate angles, so $d = g$.

Fig. 27.10

3. Allied angles (or interior angles) are supplementary.

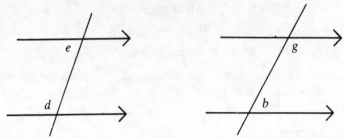

e and d are allied angles. So are b and g. e + d = 180° and
b + g = 180°.

Fig. 27.11

Example In Fig. 27.12, find a, b and c, giving reasons for your
statements.

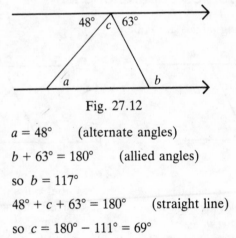

Fig. 27.12

$a = 48°$ (alternate angles)

$b + 63° = 180°$ (allied angles)

so $b = 117°$

$48° + c + 63° = 180°$ (straight line)

so $c = 180° - 111° = 69°$

Exercise 27.2

In questions 1 to 9, find the unknown angles. Give reasons for your
statements.

1.

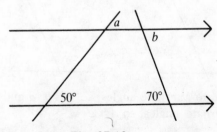

Fig. 27.13

2.

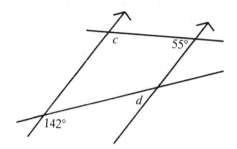

Fig. 27.14

3.

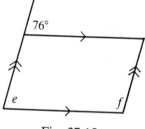

Fig. 27.15

4.

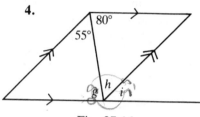

Fig. 27.16

5.

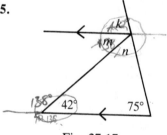

Fig. 27.17

6.

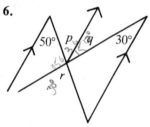

Fig. 27.18

7.

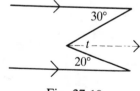

Fig. 27.19

8.

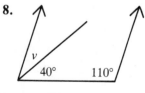

Fig. 27.20

9.

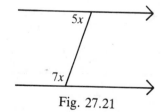

Fig. 27.21

10.

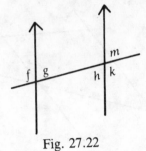

Fig. 27.22

Using Fig. 27.22:
  (i) name a pair of corresponding angles
  (ii) name a pair of alternate angles
  (iii) name a pair of vertically opposite angles
  (iv) find f, if $m = 65°$.

11.

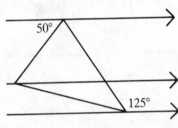

Fig. 27.23

Copy Fig. 27.23 and mark on it the sizes of as many angles as possible.

12.
Find x and y.

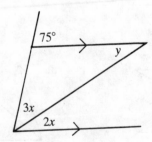

Fig. 27.24

13.
Find a, b and c.

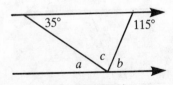

Fig. 27.25

Types of triangles

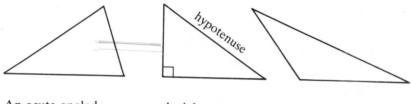

An acute-angled triangle. Each angle is less than 90°.

A right-angled triangle.

An obtuse-angled triangle. One angle is greater than 90°.

Fig. 27.26

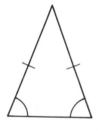
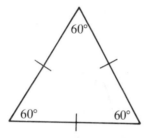
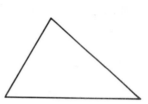

An isosceles triangle. Two sides are equal and two angles are equal.

An equilateral triangle. All three sides are equal. Each angle is 60°.

A scalene triangle. All three sides are different.

Fig. 27.27

Angle properties of triangles

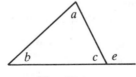

Fig. 27.28

1. The sum of the interior angles is 180°. So, in Fig. 27.28:

$$a + b + c = 180°$$

2. If one side is produced (extended), the exterior angle formed is equal to the sum of the two opposite interior angles. So in Fig. 27.28:

$$e = a + b$$

Exercise 27.3

1. In Fig. 27.28:

    (i) if $a = 75°$ and $b = 50°$, calculate c.

    (ii) if $b = 88°$ and $c = 44°$, calculate a.

    (iii) if $c = 80°$ and $a = b$, calculate a.

2. In Fig. 27.28:

    (i) if $a = 70°$ and $b = 55°$, calculate e.

    (ii) if $e = 130°$ and $a = 48°$, calculate b.

    (iii) if $e = 150°$ and $a = b$, calculate a.

3. Say whether it is possible to have a triangle with angles of:

    (i) 50°, 60°, 80°   (ii) 43°, 64°, 51°

    (iii) 36°, 74°, 70°   (iv) 106°, 37°, 47°.

4. In Fig. 27.29, PQ = PR.

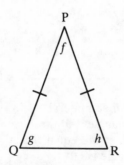

Fig. 27.29

    (i) What type of triangle is PQR?

    (ii) If $g = 52°$, calculate h and f.

    (iii) If $f = 40°$, calculate g.

    (iv) If $f = 96°$, calculate h.

    (v) If $f = 60°$, calculate g. What type of triangle is PQR in this case?

In questions 5 to 13, find the sizes of the angles marked by letters. Give reasons for your statements.

6.

5.

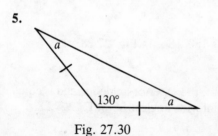

Fig. 27.30   Fig. 27.31

7.

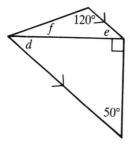

Fig. 27.32

8.

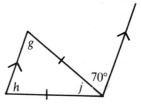

Fig. 27.33

9.

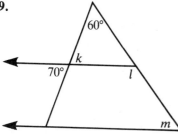

Fig. 27.34

10.

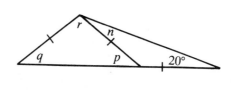

Fig. 27.35

11.

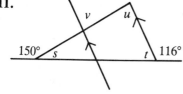

Fig. 27.36

12.

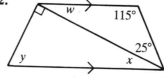

Fig. 27.37

13.

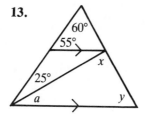

Fig. 27.38

14. The angles of a triangle are $2x°$, $3x°$ and $4x°$. Calculate x.

15. The angles of a triangle are $4y°$, $8y°$ and $3y°$. Calculate the size of the largest angle of the triangle.

16. The angles of a triangle are $(2w + 7)°$, $(6w + 12)°$ and $(w + 26)°$. Calculate w and state the sizes of the three angles.

Polygons

Figure 27.39 shows a 7-sided polygon divided into 5 triangles.

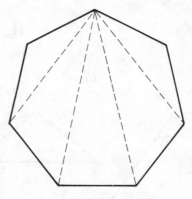

Fig. 27.39

The sum of the 7 interior angles of the polygon = the sum of the angles of the 5 triangles = $5 × 180° = 900°$.

A polygon with n sides can be divided into $(n - 2)$ triangles, so we have the following result:

the sum of the interior angles of a polygon with n sides is $(n - 2) × 180°$.

Example A polygon has 10 sides.

Sum of interior angles is $(10 - 2) × 180° = 8 × 180° = 1440°$

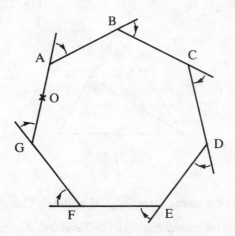

Fig. 27.40

In Fig. 27.40, the sides of a polygon have been produced to form exterior angles. Suppose a large copy of this polygon is drawn on the ground. Starting at O and walking clockwise round the polygon, you would turn through each exterior angle and would finally make a complete revolution. That is, you would turn through 360° altogether. Hence:

the sum of the exterior angles of a polygon is 360°.

This is true for all polygons, no matter how many sides they have.

Regular polygons

A *regular polygon* is one that has all sides equal and all angles equal. To find the size of an interior angle of a regular polygon, do not use the formula for the sum of the interior angles. It is easier to find first an exterior angle and then an interior angle.

Example 1   A regular polygon has 9 sides. Find the size of an interior angle.

The sum of the exterior angles is 360°.

Each exterior angle is therefore 360° ÷ 9 = 40°.

At each corner, the exterior and interior angles add up to 180°.

Each interior angle is therefore 180° − 40° = 140°.

Example 2   Each interior angle of a regular polygon is 160°. How many sides has the polygon?

Each exterior angle is 180° − 160° = 20°.

The number of exterior angles is 360° ÷ 20° = 18.

There are 18 sides.

Quadrilaterals

A *quadrilateral* is a polygon with 4 sides.

The sum of the interior angles is 2 × 180° = 360°.

Exercise 27.4

1. Find the sum of the interior angles of a pentagon (5 sides).

   A pentagon has angles of 84°, 98°, 120° and 132°. Calculate the fifth angle.

2. Find the sum of the interior angles of a hexagon (6 sides).

   A hexagon has angles of 130°, 140°, 100°, 110° and 110°. Calculate the other angle.

3. Calculate the sum of the interior angles of a polygon that has:

   (i) 12 sides   (ii) 17 sides.

4. (i) A regular polygon has 8 sides. As in Example 1 above, calculate the size of an exterior angle and then the size of an interior angle.

   (ii) Do the same for a regular polygon with 5 sides.

   (iii) Do the same for a regular polygon with 12 sides.

5. Calculate an interior angle for a regular polygon with:

   (i) 20 sides   (ii) 120 sides   (iii) 7 sides.

6. Find the number of sides for a regular polygon with an exterior angle of:

   (i) 60°   (ii) 18°   (iii) 10°.

7. Is it possible to have regular polygons with the following sizes of interior angles? For those which are possible, state the number of sides:

   (i) 140°   (ii) 110°   (iii) 155°   (iv) 135°   (v) 170°.

8. Calculate the fourth angle of a quadrilateral if three angles are:

   (i) 70°, 140°, 100°   (ii) 105°, 82°, 71°.

9. A quadrilateral has angles of $2x°$, $3x°$, $4x°$ and $3x°$. Calculate x.

10. A quadrilateral has angles of $(x + 40)°$, $(x + 80)°$, $(2x + 20)°$ and $x°$. Calculate x.

11. Six angles of an octagon (8 sides) are each 130°. The remaining two angles are equal. Find the size of each.

12. Two sides AB and DC of a regular pentagon ABCDE are produced to meet at H. Calculate the size of the angle at H.

13. ABCDE is a regular pentagon. Calculate the sizes of the angles of triangles ABC and ACD.

14. PQRSTV is a regular hexagon. Calculate the angles of triangle PRS.

28 Similarity and congruence

Similar triangles

Similar triangles have the same shape, but they may have different sizes. For two triangles to have the same shape, the three angles of one triangle must be the same as the three angles of the other. Then the corresponding sides are all in the same ratio.

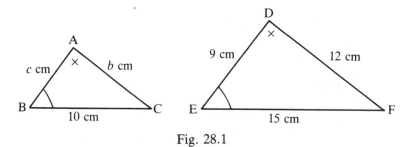

Fig. 28.1

Example 1   In Fig. 28.1, $\hat{A} = \hat{D}$ and $\hat{B} = \hat{E}$. It follows that $\hat{C} = \hat{F}$, since the angles of each triangle add up to 180°. Hence the triangles are similar. Therefore $AB:DE = BC:EF = CA:FD$.

This can be written as $\dfrac{AB}{DE} = \dfrac{BC}{EF} = \dfrac{CA}{FD}$.

Using the given measurements, we have $\dfrac{c}{9} = \dfrac{10}{15} = \dfrac{b}{12}$ from which $c = \dfrac{9 \times 10}{15} = 6$ and $b = \dfrac{10 \times 12}{15} = 8$.

(Notice that triangle DEF is an enlargement of triangle ABC with scale factor $\dfrac{3}{2}$.)

Example 2   In Fig. 28.2 overleaf, $\hat{P} = \hat{T}$ and $\hat{Q} = \hat{S}$. It follows that $\hat{R} = \hat{V}$. Hence the triangles are similar. Therefore $\dfrac{PQ}{TS} = \dfrac{QR}{SV} = \dfrac{RP}{VT}$. (Notice that corresponding sides are opposite corresponding angles. Thus QR and SV are corresponding sides because they are opposite the angles marked ×.)

Using the given measurements, we have $\dfrac{r}{16} = \dfrac{p}{12} = \dfrac{6}{10}$ from

which $p = \dfrac{6 \times 12}{10} = 7.2$ and $r = \dfrac{6 \times 16}{10} = 9.6$.

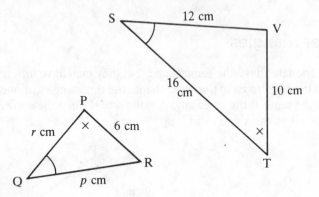

Fig. 28.2

Exercise 28.1

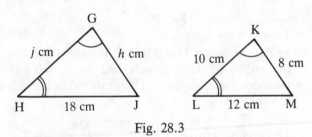

Fig. 28.3

1. Copy and complete the statement $\dfrac{GH}{\quad} = \dfrac{\quad}{LM} = \dfrac{JG}{\quad}$. Hence calculate

j and h.

For questions 2, 3 and 4 write down the statement for the ratios of the corresponding sides, as in the above examples. Then calculate the unknown sides.

2.

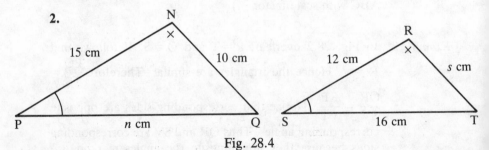

Fig. 28.4

3.

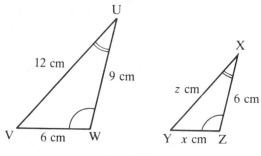

Fig. 28.5

4.

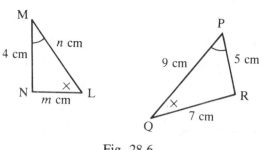

Fig. 28.6

5. DE is parallel to FG. DHG and EHF are straight lines.

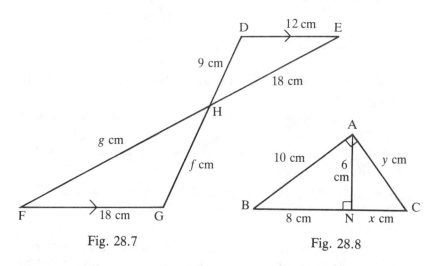

Fig. 28.7   Fig. 28.8

6. In Fig. 28.8, BAC, ANB and ANC are right angles. Show that CÂN = NB̂A. Why are triangles ANC and ABN similar? Calculate *x* and *y*.

7. In Fig. 28.9, ST is parallel to QR. Draw triangles PST and PQR
   separately and mark the equal angles. Calculate PR, TR and QR.

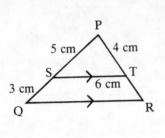

Fig. 28.9   Fig. 28.10

8. Find x in Fig. 28.10.

9. For Fig. 28.11, name the two triangles that are similar to triangle
   ABC.

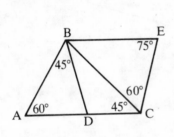

Fig. 28.11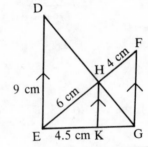

Fig. 28.12

10. In Fig. 28.12, DE, HK and FG are parallel; DHG and EHF are
    straight lines.
    (i) Name a triangle similar to triangle HFG and hence calculate
        FG.
    (ii) Name a triangle similar to triangle EFG and hence calculate
        HK, EG and KG.

11. In triangle PQR, PQ = 9.6 cm and PR = 8 cm. In triangle TVS,
    TS = 12 cm and SV = 9.6 cm. Also $\hat{P} = \hat{T}$ and $\hat{Q} = \hat{V}$. Calculate QR
    and TV.

12. A pole of height 2 m casts a shadow of length 3 m on horizontal
    ground. At the same moment a building casts a shadow of length
    34.5 m. Calculate the height of the building.

13. The two sides of a step-ladder are 3 m long. They are joined by a
    cord 1.2 m long, which is attached to each side at a point which is

1 m from the bottom end of each side. Calculate the distance between the feet of the ladder, when they are fully apart.

Similar figures: ratio of areas

Similar figures have the same shape. All corresponding angles are equal and the ratios of corresponding sides are the same. One figure is an enlargement of the other.
For the rectangles of Fig. 28.13:

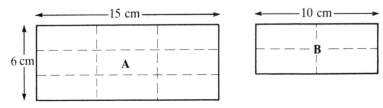

Fig. 28.13

ratio of lengths $= \dfrac{15}{10} = \dfrac{3}{2}$ and ratio of heights $= \dfrac{6}{4} = \dfrac{3}{2}$. The corresponding angles are equal because all are 90°. Hence the rectangles are similar.

A is an enlargement of B with a scale factor of $\dfrac{3}{2}$.

The ratio of the areas is $\dfrac{15 \times 6}{10 \times 4} = \dfrac{90}{40} = \dfrac{9}{4} = \left(\dfrac{3}{2}\right)^2$.

Notice that the ratio of the areas = (ratio of lengths)².

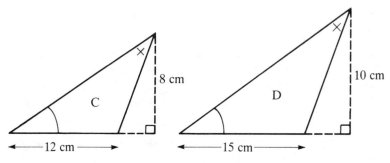

Fig. 28.14

In Fig. 28.14, the triangles are similar since the angles are the same.
Area of C $= \frac{1}{2} \times 12 \times 8 = 48$
Area of D $= \frac{1}{2} \times 15 \times 10 = 75$

Ratio of bases $= \dfrac{12}{15} = \dfrac{4}{5}$ and ratio of heights $= \dfrac{8}{10} = \dfrac{4}{5}$

Ratio of areas $= \dfrac{48}{75} = \dfrac{16}{25} = \left(\dfrac{4}{5}\right)^2$

Again, the ratio of the areas = (ratio of corresponding sides)^2.
In general, if two figures are similar:

ratio of areas = (ratio of corresponding sides)^2.

If the scale factor of the enlargement is k, then the ratio of the areas is k^2.

Exercise 28.2

1. Rectangle E has sides of 3 cm and 2 cm. Rectangle F has sides of 12 cm and 8 cm. State: (i) the ratio of the lengths (ii) the ratio of the widths (iii) the ratio of the perimeters (iv) the area of E (v) the area of F (vi) the ratio of the areas.
   Verify that ratio of areas = (ratio of lengths)^2.

2. A carpet has an area of 8 m^2. Another carpet has the same shape but three times the length. What is its area?

3. A circle has an area of 20 cm^2. What is the area of a circle of:
   (i) twice the radius (ii) half the radius? (All circles are similar.)

4. On a photograph, a building is 3 cm high and its area is 7 cm^2. On an enlargement of the photograph, the building is 12 cm high. What is the scale factor of the enlargement? What is the area of the building on the enlargement?

5. Two triangles are similar. The base of one is 16 cm and of the other is 12 cm. The area of the first is 144 cm^2. What is the area of the second?

6. Two similar triangles have areas of 8 cm^2 and 18 cm^2. The base of the smaller one is 6 cm. What is the base of the larger?

7. A rectangular lawn measures 14 m by 12 m. A second lawn measures 21 m by 18 m. State the ratio of: (i) their lengths (ii) their widths. Are the lawns similar? State the ratio of their areas.
   20 kg of fertiliser are needed for the first lawn. How much is needed for the second lawn?

8. On a map, the area representing a lake measures 126 cm^2. On a second map, the same lake is represented by 350 cm^2.
   (i) State the ratio of the areas in its simplest form. Hence state the ratio of the lengths of the lakes on the maps.
   (ii) The scale of the first map is 1 cm to 300 m. What is the scale of the second map?

Similar solids: ratio of volumes

Fig. 28.15 shows two cuboids.

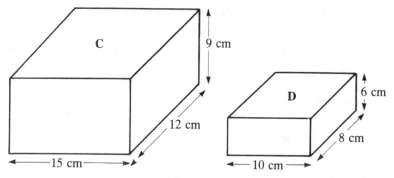

Fig. 28.15

$$\frac{\text{Length of C}}{\text{Length of D}} = \frac{15}{10} = \frac{3}{2}$$

$$\frac{\text{Width of C}}{\text{Width of D}} = \frac{12}{8} = \frac{3}{2}$$

$$\frac{\text{Height of C}}{\text{Height of D}} = \frac{9}{6} = \frac{3}{2}$$

Since all three ratios are the same, the cuboids must be similar.

$$\frac{\text{Volume of C}}{\text{Volume of D}} = \frac{15 \times 12 \times 9}{10 \times 8 \times 6} = \frac{15}{10} \times \frac{12}{8} \times \frac{9}{6} = \frac{3}{2} \times \frac{3}{2} \times \frac{3}{2} = \left(\frac{3}{2}\right)^3.$$

This illustrates the general result that, if two figures are similar:

ratio of volumes = (ratio of corresponding sides)³.

If the scale factor of the enlargement is k, then the ratio of the volumes is k^3.

Exercise 28.3

1. Two cuboids are similar. The ratio of their lengths is $4:3$. The volume of the smaller one is 54 cm^3. What is the volume of the larger one?

2. Two cuboids are similar. The ratio of their heights is $5:2$. The volume of the larger one is 500 cm^3. What is the volume of the smaller one?

3. The volume of a cube is 128 cm^3. What is the volume of a cube with an edge of half the size?

4. The volume of a certain sphere is 20 cm^3. What is the volume of a sphere of double the radius?

5. Cube H has an edge of 6 cm and cube K has an edge of 4 cm.

(i) Calculate the ratio of the edges, in its simplest form.

(ii) Calculate the total surface area of each cube and hence the ratio of the surface areas, in its simplest form.

(iii) Calculate the volume of each cube and hence the ratio of the volumes, in its simplest form.

Verify that the answer to (ii) is the square of the answer to (i), and that the answer to (iii) is the cube of the answer to (i).

6. One sphere has a diameter of 10 cm and another has a diameter of 15 cm. State, in its simplest form, the ratio of:

(i) the diameters

(ii) the surface areas

(iii) the volumes.

(Do not calculate the surface area and volume for each sphere. Use a shorter way to find the answers.)

7. The ratio of the surface areas of two cubes is 16:25. Calculate the ratio of: (i) the lengths of the edges (ii) the volumes.

8. Two jugs are the same shape. One is 40 cm high and the other 32 cm high. Find, in its simplest form, the ratio of: (i) the heights (ii) the volumes. The smaller jug holds 800 ml. How much does the larger hold?

9. A packet of washing powder measures 24 cm by 16 cm by 6 cm, and contains 810 g of powder. Another packet is the same shape, but 32 cm high instead of 24 cm. State the ratio of: (i) the heights (ii) the volumes (iii) the areas of cardboard needed for the packets. How much powder is the larger packet likely to contain?

Congruent triangles

Congruent triangles have both the same shape and the same size. Two triangles are congruent if any one of the following sets of conditions is satisfied:

(i) the three sides of one triangle are equal to the three sides of the other.

(ii) two sides and the included angle of one triangle are equal to two sides and the included angle of the other.

(iii) a side of one triangle is equal to a side of the other and two pairs of corresponding angles are equal.

(iv) they are both right-angled triangles, and the hypotenuse and a second side of one are equal to the hypotenuse and a second side of the other.

The diagrams below illustrate these conditions.

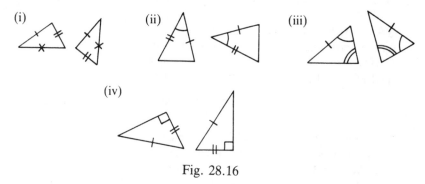

Fig. 28.16

Exercise 28.4

1. Which triangles are congruent to triangle ABC? Name the sides equal to BC.

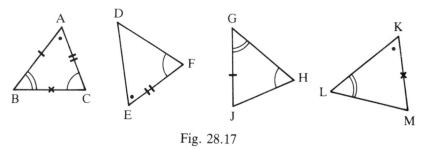

Fig. 28.17

2. Which triangles are congruent to triangle NPQ?
Name the angles equal to N̂.

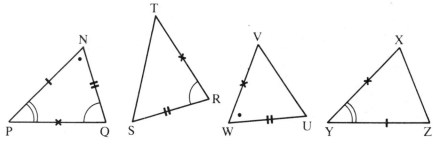

Fig. 28.18

3. By making sketches, find which two of these triangles are congruent:

△ABC with Ĉ = 90°, AB = 8 cm, BC = 5 cm
△DEF with Ê = 90°, DE = 8 cm, EF = 5 cm
△GHJ with Ĥ = 90°, JG = 8 cm, GH = 5 cm.

4. By making sketches, find which three of these triangles are congruent:

△ABC with AB = 7 cm, AC = 9 cm, Â = 40°
△DEF with DE = 7 cm, EF = 9 cm, Ê = 40°
△GHJ with HJ = 7 cm, GH = 9 cm, Ĝ = 40°
△KLM with LM = 7 cm, KM = 9 cm, M̂ = 40°.

5. By making sketches, find which three of these triangles are congruent:

△NPQ with P̂ = 55°, Q̂ = 78°, PQ = 6 cm
△RST with Ŝ = 47°, R̂ = 78°, RT = 6 cm
△UVW with Û = 78°, Ŵ = 55°, UV = 6 cm
△XYZ with Ŷ = 55°, Ẑ = 47°, XY = 6 cm.

29 Symmetry

Bilateral (or line) symmetry

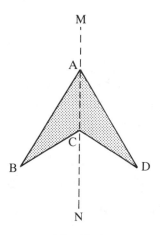

Fig. 29.1

The quadrilateral ABCD in Fig. 29.1 is symmetrical. MN is the *axis of symmetry*. If the quadrilateral is folded about MN, triangle ABC fits exactly onto triangle ADC. If MN is regarded as a mirror line, then D is the image of B, line AD is the image of line AB and line CD is the image of line CB.

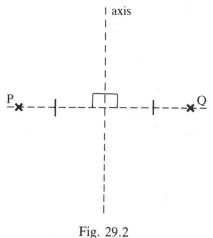

Fig. 29.2

In a symmetrical shape, for every point P on one side of the axis, there is a corresponding point Q on the other side. PQ is perpendicular to the axis and P and Q are equal distances from it (Fig. 29.2).

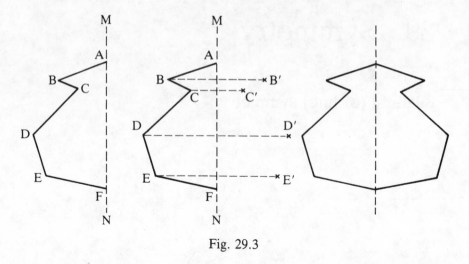

Fig. 29.3

Look at the first shape in Fig. 29.3. To complete it so that it is symmetrical about MN, we draw lines from B, C, D and E perpendicular to MN, and extend them the same distances on the other side of MN to obtain points B′, C′, D′ and E′. Then these points are joined.

Point symmetry

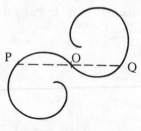

Fig. 29.4

The curve in Fig. 29.4 has symmetry about the point O. For every point P on the curve there is a corresponding point Q, such that POQ is a straight line and QO = PO.

Rotational symmetry

The shape in Fig. 29.5 has rotational symmetry of order 3. If rotated about the centre through 120° or 240° in a clockwise direction, it will still look exactly the same. There are three positions in which it looks the same, so we say the symmetry is of order 3. Notice that 360° ÷ 120° = 3.

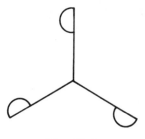

Fig. 29.5

A shape which has rotational symmetry of order *n*, where *n* is even, also has point symmetry.

Exercise 29.1

1. Draw the following, and use dotted lines to show all axes of symmetry: (i) an isosceles triangle (ii) an equilateral triangle (iii) a rectangle (iv) a square.

2. Repeat question 1 for: (i) a diamond (ii) a heart as they appear on playing cards.

3. Repeat question 1 for: (i) a regular pentagon (5 sides) (ii) a regular hexagon (6 sides).

4. Make larger copies of the following block capital letters and show their axes of symmetry: A D H W X.

5. On squared paper, copy each shape in Fig. 29.6, and complete it so that it has symmetry about MN.

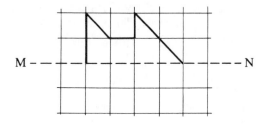

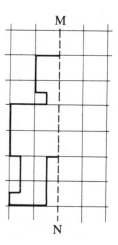

Fig. 29.6

6. Copy and complete the shapes in Fig. 29.7 so that each has two axes of symmetry.

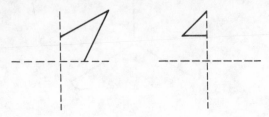

Fig. 29.7

7. Using axes numbered from -4 to $+4$, mark the points A$(1, 1)$, B$(2, 2)$, and C$(4, 1)$. Join them to form a triangle. Complete the shape so that it has symmetry about the x and y axes.

8. The block capital letter 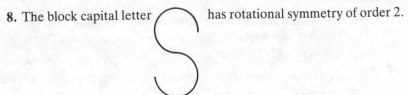 has rotational symmetry of order 2.

Which of the letters below also have such symmetry?

9. The following shapes have rotational symmetry. Draw each shape, mark the centre with a cross, and state the order of symmetry:

   (i) an equilateral triangle   (ii) a rectangle   (iii) a square
   (iv) a regular pentagon.

Fig. 29.8

10. Copy each shape in Fig. 29.8, and complete it so that it has rotational symmetry. State the order of rotational symmetry for each shape. Which shape has no line symmetry?

11. Draw those block capital letters of the alphabet which have point symmetry. On each letter mark the point of symmetry with a cross.

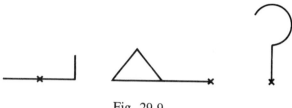

Fig. 29.9

12. Copy and complete each of the shapes in Fig. 29.9, so that it has point symmetry about the cross.

13. On squared paper, draw the triangle with vertices at $(-1, 1)$, $(-3, 1)$ and $(-3, 2)$. Draw another triangle so that the shape made by both triangles together has point symmetry about the origin.

14. Using axes numbered from 0 to 6, mark the points $(1, 4)$, $(1, 5)$, $(2, 5)$ and $(3, 3)$ and join them to form a quadrilateral. Complete the shape so that it has symmetry about each of the lines $x = 3$ and $y = 3$.

On your figure, mark two other axes of symmetry and state their equations.

Special quadrilaterals

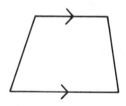

A trapezium. It has 2 parallel sides.

Fig. 29.10

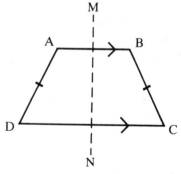

An isosceles trapezium. It also has 2 equal sides. It has one line of symmetry.

Fig. 29.11

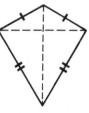

A kite. It has 2 pairs of equal adjacent sides, and one line of symmetry. The diagonals cut at 90°.

Fig. 29.12

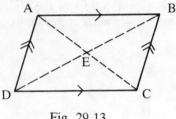

Fig. 29.13

A parallelogram. It has 2 pairs of parallel sides. It has these properties:
1. opposite sides are equal (AB = CD, AD = BC)
2. opposite angles are equal ($\hat{A} = \hat{C}$, $\hat{B} = \hat{D}$)
3. diagonals bisect each other (AE = EC, BE = DE).

A rhombus. It has 4 equal sides, and 2 axes of symmetry (AC and BD). It has these properties:

1. opposite sides are parallel
2. opposite angles are equal
3. diagonals bisect each other
4. diagonals cut at 90°.

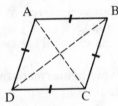

Fig. 29.14

A rectangle. All its angles are equal (90°). It has the same properties as a parallelogram.
   Also its diagonals are equal (AC = BD) and it has two axes of symmetry.

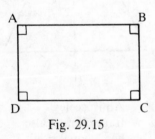

Fig. 29.15

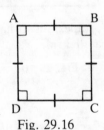

Fig. 29.16

A square. All its sides are equal, and all its angles are equal. It has the same properties as a rectangle and a rhombus.
A rhombus, a rectangle and a square are all special parallelograms.
A rhombus is a special kite.
A square is a special rhombus, and a special rectangle.

Exercise 29.2

1. PQRS is a trapezium, with PQ parallel to SR.
$\hat{P} = 110°$ and $\hat{Q} = 125°$. Calculate $\hat{R}$ and $\hat{S}$.

2. EFGH is a parallelogram. EF = 7 cm, EH =5 cm and E = 68°. Draw
a diagram and mark in the sizes of all sides and angles.

3. PQRS is a kite with PR as the axis of symmetry. PR and QS intersect
at T. Which of the following statements must be true:
   (i) $\hat{P} = \hat{R}$ (ii) $\hat{Q} = \hat{S}$
   (iii) PQ = QR   (iv) PQ = PS
   (v) PR = QS   (vi) PT = TR
(vii) QT = TS?

4. Sketch the following and mark in the axes of symmetry:
   (i) an isosceles trapezium   (ii) a kite   (iii) a rhombus   (iv) a rectangle (v) a square.

For questions 5, 6 and 7, consider each statement in turn. Then say for
which of the shapes in question 4 the statement is true.

5. (i) One pair of opposite sides is equal.

   (ii) Both pairs of opposite sides are equal.

   (iii) All four sides are equal.

6. (i) One pair of opposite angles is equal.

   (ii) Both pairs of opposite angles are equal.

   (iii) All four angles are equal.

7. (i) The diagonals are equal in length.

   (ii) The diagonals are perpendicular.

   (iii) The diagonals do not bisect each other.

8. PQRS is a kite with PR as the axis of symmetry. PR and QS intersect
at T. Triangle QRS is equilateral, and PT = QT = ST. State the
sizes of QT̂P and QR̂S. Calculate the sizes of PQ̂T, PQ̂R and QP̂S.

9. The diagonals AC and BD of square ABCD intersect at E. H is a
point on BC such that BH = BE. Calculate the sizes of the angles
of triangle BEH.

10. The diagonals of rectangle ABCD intersect at E. Draw a diagram
and mark all equal lengths. If AB̂E = 40°, mark the sizes of all
angles on your figure.

11. Draw a square GHJK with its diagonals intersecting at M. Name all
equal lengths. Mark in the sizes of all angles, on your diagram.

30 Pythagoras' theorem

In a right-angled triangle, the square of the hypotenuse is equal to the sum of the squares of the other two sides.

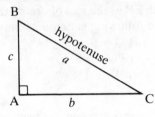

Fig. 30.1

In Fig. 30.1, $BC^2 = AC^2 + AB^2$ or $a^2 = b^2 + c^2$.
(Notice that we use a for the side opposite $\hat{A}$, b for the side opposite $\hat{B}$ and c for the side opposite $\hat{C}$.)

Given the lengths of two sides of a right-angled triangle, we can use Pythagoras' Theorem to calculate the length of the third side.

Example 1
If, in Fig. 30.1, $b = 7\,cm$ and $c = 9\,cm$, calculate a.

$$a^2 = b^2 + c^2$$
$$= 7^2 + 9^2$$
$$= 49 + 81 = 130$$
$$a = \sqrt{130} = 11.4\,cm \text{ to 3 s.f.}$$

Example 2
If, in Fig. 30.1, $a = 8.4\,cm$ and $b = 5.2\,cm$, calculate c.

$$b^2 + c^2 = a^2$$
$$5.2^2 + c^2 = 8.4^2$$
$$27.04 + c^2 = 70.56$$
$$c^2 = 70.56 - 27.04 = 43.52$$
$$c = \sqrt{43.52} = 6.60\,cm \text{ to 3 s.f.}$$

Exercise 30.1

Give the answers to no more than 3 significant figures.

1. If $b = 8\,cm$ and $c = 5\,cm$, calculate a.

2. If $b = 6.4\,cm$ and $c = 9.2\,cm$, calculate a.

3. If $b = 5\,cm$ and $c = 12\,cm$, calculate a.

4. If $b = 3.5\,cm$ and $c = 8.4\,cm$, calculate a.

5. If $a = 9\,cm$ and $b = 6\,cm$, calculate c.

6. If $a = 17\,cm$ and $c = 15\,cm$, calculate b.

7. If $a = 16.8\,cm$ and $c = 10.3\,cm$, calculate b.

8. If $a = 6.5$ cm and $b = 5.2$ cm, calculate c.

9. Calculate the length of the hypotenuse of a right-angled triangle, if the other two sides are:

(i) 6 cm and 8 cm   (ii) 9 cm and 11 cm.

10. Calculate the third side of a triangle, given that the hypotenuse and the second side are:

(i) 15 cm and 9 cm   (ii) 14 cm and 8 cm.

11. A ladder of length 6 m is placed on horizontal ground with its foot 3 m from a vertical wall. How far up the wall does it reach?

12. A ship sails 15 nautical miles due south and then 10 nautical miles due east. How far is it from its starting point?

13. A rectangle has sides of 16 cm and 12 cm. Calculate the length of a diagonal.

14. A square has sides of 10 cm. Calculate the length of a diagonal.

15. On squared paper show points P(9, 2) and Q(3, 10). Calculate the length of PQ.

16. Calculate the distance between the points (2, 7) and (11, 1).

17. Calculate the lengths of the vectors $\begin{pmatrix} 4 \\ 3 \end{pmatrix}$, $\begin{pmatrix} 5 \\ 7 \end{pmatrix}$ and $\begin{pmatrix} 3.6 \\ 2.8 \end{pmatrix}$.

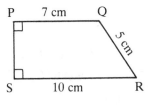

Fig. 30.2

18. Calculate PS in Fig. 30.2. (Divide the trapezium into a rectangle and a triangle.)

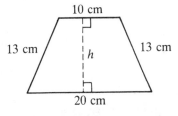

Fig. 30.3

19. Figure 30.3 shows an isosceles trapezium. Calculate h and the area.

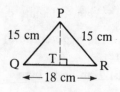

Fig. 30.4

20. In Fig. 30.4, PT is perpendicular to QR, that is, PT is an altitude of △PQR. Since △PQR is isosceles (PQ = PR), it follows that QT = TR. Calculate: (i) PT   (ii) the area of △PQR.

21. In △DEF, DE = DF = 12 cm and EF = 14 cm. Calculate the length of the altitude DG.

22. Figure 30.5 shows a box.

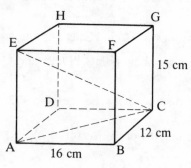

Fig. 30.5

  (i) Using the right-angled triangle ABC, calculate AC.

  (ii) Using the right-angled triangle ACE, calculate CE.

23. Figure 30.6 shows a flagpole held in a vertical position by four ropes of length 3 m. P, Q, R and S are at the corners of a square and A is 2 m above the ground. Calculate:

(i) PB   (ii) PQ.

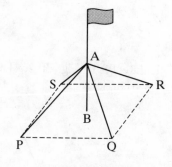

Fig. 30.6

Special right-angled triangles

In some right-angled triangles, all three sides are whole numbers. In Fig. 30.1, for example, if $b = 3$ and $c = 4$ then $a^2 = 3^2 + 4^2 = 25$, so

$a = 5$. We have a 3-4-5 triangle. If $b = 6$ and $c = 8$, we have a triangle twice the size of a 3-4-5 triangle, so $a = 2 \times 5 = 10$. Again, if $b = 5$ and $c = 12$ then $a = 13$. We have a 5-12-13 triangle. It is useful to remember these sets of numbers.

Converse of Pythagoras' theorem

If the square of the longest side of a triangle is equal to the sum of the squares of the other two sides, then the largest angle is a right-angle.

In a triangle PQR, for example, PQ = 53 cm, QR = 45 cm and RP = 28 cm. So $PQ^2 = 53^2 = 2809$ and $PR^2 + QR^2 = 28^2 + 45^2 = 784 + 2025 = 2809$. Since $PQ^2 = PR^2 + QR^2$, there is a right-angle at R.

Suppose the sides of a triangle are 23 cm, 35 cm and 39 cm. $39^2 = 1521$ and $23^2 + 35^2 = 529 + 1225 = 1754$. Since $1521 \neq 1754$, there is no right-angle in the triangle.

The largest angle of a triangle: acute, 90° or obtuse?

The largest angle is opposite the longest side. Suppose the sides are a, b and c with a as the longest side:

if $a^2 < b^2 + c^2$, then $\hat{A} < 90°$ and all angles are acute;

if $a^2 = b^2 + c^2$, then $\hat{A} = 90°$;

if $a^2 > b^2 + c^2$, then $\hat{A} > 90°$, that is, $\hat{A}$ is obtuse.

Exercise 30.2

1. State the length of the hypotenuse, given that the other sides are:
   (i) 3 m and 4 m   (ii) 9 m and 12 m
   (iii) 30 cm and 40 cm   (iv) 15 cm and 20 cm.

2. State the length of the hypotenuse, given that the other sides are:
   (i) 5 cm and 12 cm   (ii) 10 cm and 24 cm
   (iii) 20 m and 48 m   (iv) 500 m and 1200 m.

3. State the length of the third side, given that the hypotenuse and another side are:
   (i) 5 cm and 3 cm   (ii) 5 cm and 4 cm
   (iii) 10 m and 8 m   (iv) 13 m and 5 m.

4. In $\triangle$KLM, KL = 20 cm, LM = 21 cm and MK = 29 cm. Show that $\hat{L} = 90°$.

5. In △STV, ST = 37 cm, TV = 12 cm and VS = 35 cm. Show that the triangle has a right-angle, and name it.

6. In △DEF, DE = 19 mm, EF = 23 mm and FD = 27 mm. Show that this triangle does not have a right-angle.

7. Show that a triangle with sides of 15, 13 and 19 cm does not have a right-angle.

8. In Fig. 30.7, calculate PQ and PR. Hence show that △PQR has a right-angle.

9. In Fig. 30.8, calculate p and show that $\theta = 90°$.

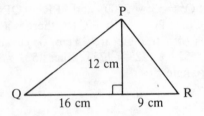

Fig. 30.7

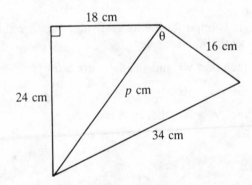

Fig. 30.8

10. For each set of numbers, find whether they give a triangle with three acute angles, a right-angle or an obtuse angle:

   (i) 5, 7, 9   (ii) 6, 8, 10   (iii) 7, 9, 11

   (iv) 8, 13, 16   (v) 13, 16, 20   (vi) 21, 28, 35.

31 Sine, cosine and tangent

The sine, cosine and tangent ratios

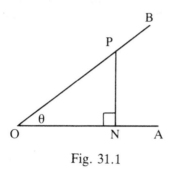

Fig. 31.1

In Fig. 31.1, the angle between the lines OA and OB is θ. For all positions of P on OB, the ratio $\dfrac{NP}{OP}$ remains the same. It is called the *sine* of θ. This is shortened to sin θ. Similarly, the ratio $\dfrac{ON}{OP}$ is called the *cosine* of θ (cos θ) and the ratio $\dfrac{NP}{ON}$ is called the *tangent* of θ (tan θ).

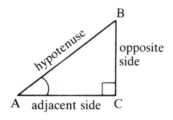

Fig. 31.2

Consider $\hat{A}$ in Fig. 31.2. BC is called the *opposite side* to $\hat{A}$, and AC is called the *adjacent side*. ('Adjacent' means 'next to'.)

$$\sin A = \frac{\text{opposite side}}{\text{hypotenuse}} = \frac{BC}{AB}$$

$$\cos A = \frac{\text{adjacent side}}{\text{hypotenuse}} = \frac{AC}{AB}$$

$$\tan A = \frac{\text{opposite side}}{\text{adjacent side}} = \frac{BC}{AC}$$

Using the first letters of $\sin = \dfrac{\text{opp}}{\text{hyp}}$, $\cos = \dfrac{\text{adj}}{\text{hyp}}$ and $\tan = \dfrac{\text{opp}}{\text{adj}}$, we have the word 'sohcahtoa'. Learn this word and it will help you to remember the three ratios.

For most values of A the ratios are not rational numbers. They can be obtained using a calculator, or from tables, to a certain number of decimal places. For example, a calculator might give 0.342020143 for $\sin 20°$.

The ratios can be used to calculate an unknown side or an unknown angle in a right-angled triangle.

Unless instructed otherwise, results of calculations should be given correct to 3 s.f.

Finding a side

Example 1   Find the length of BC in Fig. 31.3.

BC is opposite the angle of 27°, and we are given the length of the hypotenuse. We therefore use the sine ratio.

$$\sin = \frac{\text{opp}}{\text{hyp}} \quad \text{so} \quad \sin 27° = \frac{\text{BC}}{18}$$

$$\text{BC} = 18 \sin 27° = 18 \times 0.45399\ldots \text{ using a calculator}$$

$$= 8.17182\ldots$$

$$= 8.17\,\text{cm to 3 s.f.}$$

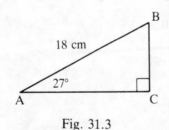

Fig. 31.3

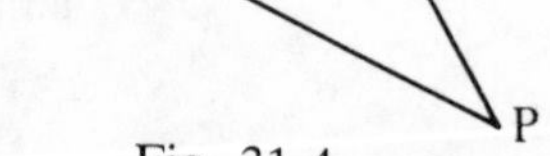

Fig. 31.4

Example 2   Find the length of QP in Fig. 31.4.

QP is opposite to 55° and we are given the length of the adjacent side QR. We therefore use the tangent ratio.

$$\tan = \frac{\text{opp}}{\text{adj}} \quad \text{so} \quad \tan 55° = \frac{\text{QP}}{7.3}$$

$$\text{QP} = 7.3 \times \tan 55° = 7.3 \times 1.4281\ldots$$

$$= 10.425\ldots = 10.4\,\text{cm to 3 s.f.}$$

Exercise 31.1

For questions 1 to 6, use Fig. 31.5:

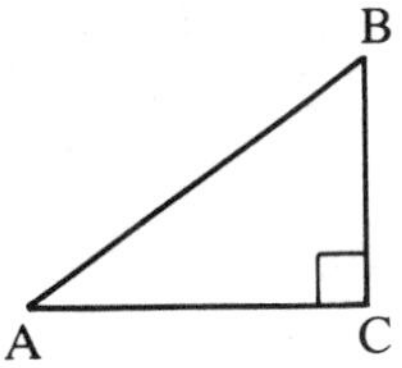

1. If AB = 10 cm and Â = 67°, calculate BC.

2. If AB = 7 cm and Â = 34°, calculate AC.

3. If AC = 14 cm and Â = 26°, calculate BC.

4. If AB = 8 cm and B̂ = 72°, calculate BC.
(Note: BC is adjacent to B̂.)

5. If BC = 6.2 cm and B̂ = 59°, calculate AC.

6. If AB = 9.3 cm and B̂ = 28°, calculate AC.

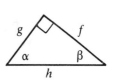

Fig. 31.5

For questions 7 to 12, use Fig. 31.6:

7. $\sin \alpha = \dfrac{f}{h}$. Write down $\cos \alpha$ and $\tan \alpha$ in the same way.

8. Write down fractions equal to $\sin \beta$, $\cos \beta$ and $\tan \beta$.

9. If $\alpha = 22°$ and $g = 5.7$ cm, calculate f.

10. If $\beta = 36°$ and $h = 8.5$ cm, calculate g.

11. If $\beta = 53.7°$ and $f = 28.4$ cm, calculate g.

12. If $\alpha = 65.2°$ and $h = 36.3$ cm, calculate g.

Fig. 31.6

For questions 13 to 16, calculate the side marked x:

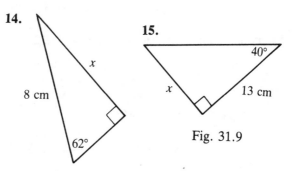

13.

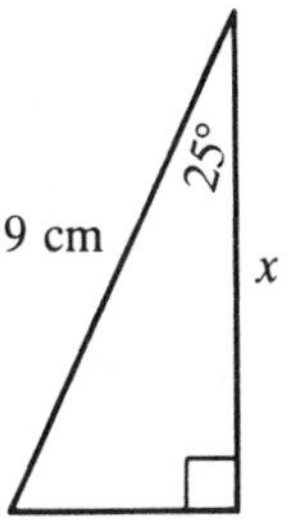

9 cm   25° x

Fig. 31.7

14.

x 8 cm   62°

Fig. 31.8

15.

40° x 13 cm

Fig. 31.9

16.

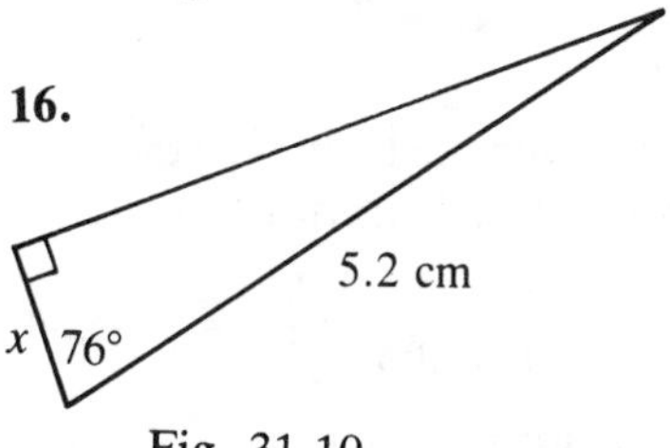

x 76°   5.2 cm

Fig. 31.10

Finding an angle

Example 1 Find the size of $\hat{P}$ in Fig. 31.11.

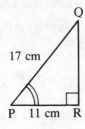

Fig. 31.11

We are given the lengths of the side adjacent to $\hat{P}$ and the hypotenuse, so we use the cosine ratio.

$$\cos P = \frac{\text{adj}}{\text{hyp}} = \frac{PR}{PQ} = \frac{11}{17} = 0.64705\ldots$$

So $\hat{P}$ is the angle with a cosine of 0.64705...
We write this as: $\hat{P} = \cos^{-1} 0.64705\ldots$
 or $\hat{P} = \text{arc cos } 0.64705\ldots$
 or $\hat{P} = \text{inv cos } 0.64705\ldots$

A calculator gives $\hat{P}$ as 49.7° to 3 s.f.

(On your calculator, you will probably have to press either '0.64705 arc cos' or '0.64705 inv cos'.)

Example 2 Find the size of $\hat{V}$ in Fig. 31.12.

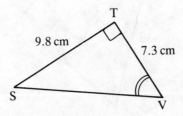

Fig. 31.12

We are given the lengths of the opposite and adjacent sides, so we use the tangent ratio.

$$\tan V = \frac{\text{opp}}{\text{adj}} = \frac{ST}{TV} = \frac{9.8}{7.3} = 1.34246\ldots$$

$$\hat{V} = 53.3° \text{ to 3 s.f.}$$

Exercise 31.2

For questions 1 to 6, use Fig. 31.13.

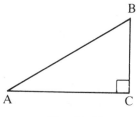

Fig. 31.13

1. If AC = 4 cm and AB = 7 cm, calculate Â.

2. If BC = 9 cm and AC = 12 cm, calculate Â.

3. If BC = 13 cm and AB = 26 cm, calculate Â.

4. If AC = 8 cm and BC = 5 cm, calculate B̂.

5. If AC = 23 cm and AB = 36 cm, calculate B̂.

6. If AB = 18 cm and BC = 12 cm, calculate B̂.

For questions 7 to 10, use Fig. 31.14:

7. If $r = 5$ and $t = 9$, calculate φ.

8. If $t = 7$ and $p = 8$, calculate θ.

9. If $t = 8$ and $p = 10$, calculate φ.

10. If $t = 6.6$ and $r = 5.3$, calculate θ.

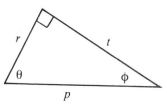

Fig. 31.14

For questions 11 to 14, calculate the angle θ.

11.

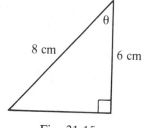

Fig. 31.15

12.

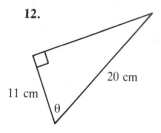

Fig. 31.16

13.

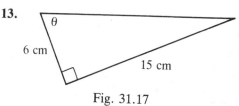

Fig. 31.17

14.

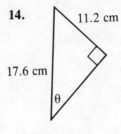

Fig. 31.18

Using an isosceles triangle

In Fig. 31.19, DN divides the isosceles triangle DEF into two congruent triangles, so EN = NF = 3 cm.

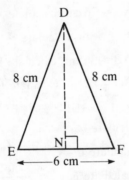

Fig. 31.19

From $\triangle$DEN, $\cos E = \dfrac{EN}{ED} = \dfrac{3}{8} = 0.375$

$$\hat{E} = 68.0° \text{ to 3 s.f.}$$

Calculating a hypotenuse

In Fig. 31.20, $\dfrac{6}{h} = \cos 32°$

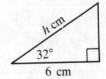

Fig. 31.20

Inverting: $\dfrac{h}{6} = \dfrac{1}{\cos 32°}$

Multiplying by 6: $h = \dfrac{6}{\cos 32°} = \dfrac{6}{0.8480...} = 7.0750...$

The hypotenuse is 7.08 cm to 3 s.f.

Exercise 31.3

For questions 1 to 5, use Fig. 31.21.

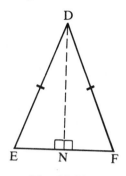

Fig. 31.21

1. If DE = DF = 12 cm and EF = 10 cm, calculate Ê.

2. If DE = DF = 10 cm and Ê = 72°, calculate EN and EF.

3. If DE = DF = 24 cm and EF = 30 cm, calculate Ê and D̂.

4. If DE = DF = 6.8 cm and D̂ = 50°, calculate EF.

5. If DE = DF = 9 cm and Ê = 65°, calculate the altitude DN.

6. An isosceles triangle has sides of 8 cm, 11 cm and 11 cm. Calculate its angles.

7. PQ is a chord of a circle, centre O. If PQ = 12 cm and PO = QO = 10 cm, calculate angle POQ.

8. The five corners of a regular pentagon ABCDE lie on a circle of radius 10 cm and centre O. Calculate: (i) angle AOB (ii) side AB.

For questions 9 to 12, use Fig. 31.22.

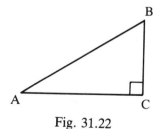

Fig. 31.22

9. If AC = 10 cm and Â = 28°, calculate AB.

10. If BC = 6 cm and B̂ = 55°, calculate AB.

11. If BC = 9 cm and Â = 18°, calculate AB.

12. If AC = 14.6 cm and B̂ = 63.2°, calculate AB.

13. $\triangle$GHK is isosceles, with $\hat{H} = \hat{K} = 70°$ and HK = 16 cm. N is the mid-point of HK, so GN is perpendicular to HK. Using $\triangle$GHN, calculate GH.

14. In $\triangle$STV, $\hat{T} = \hat{V} = 68°$ and TV = 12 cm. Calculate ST.

Finding ratios using Pythagoras' theorem

Example 1   If $\cos \alpha = \dfrac{12}{13}$, find $\sin \alpha$ and $\tan \alpha$.

Sketch a triangle as in Fig. 31.23.

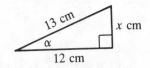

Fig. 31.23

Then $x^2 + 12^2 = 13^2$ so $x = 5$

Hence $\sin \alpha = \dfrac{5}{13}$ and $\tan \alpha = \dfrac{5}{12}$

Example 2   If $\tan \beta = \dfrac{5}{4}$, find $\sin \beta$ and $\cos^2 \beta$.

Sketch a triangle as in Fig. 31.24.

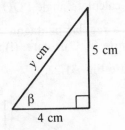

Fig. 31.24

Then $y^2 = 4^2 + 5^2 = 41$ and so $y = \sqrt{41}$

$\sin \beta = \dfrac{5}{y} = \dfrac{5}{\sqrt{41}}$ and $\cos^2 \beta = (\cos \beta)^2 = \left(\dfrac{4}{\sqrt{41}}\right)^2 = \dfrac{16}{41}$

Exercise 31.4

1. If $\sin \theta = \dfrac{3}{5}$, find $\cos \theta$ and $\tan \theta$.

2. If $\tan \varphi = \dfrac{4}{3}$, find $\sin \varphi$ and $\cos \varphi$.

3. If $\cos \alpha = \dfrac{15}{17}$, find $\sin \alpha$ and $\tan \alpha$.

4. If $\tan \beta = \dfrac{2}{3}$, find $\cos \beta$ and $\cos^2 \beta$.

5. If $\sin \gamma = \dfrac{3}{7}$, find $\cos \gamma$ and $\cos^2 \gamma$.

6. If $\cos \delta = \dfrac{2}{5}$, find $\tan \delta$ and $\sin^2 \delta$.

7. If $\sin^2 \theta = \dfrac{1}{5}$, find $\tan \theta$.

8. $\sin 30° = \dfrac{1}{2}$. Use this to find, in fraction form, $\cos 30°$ and $\tan 30°$.

(The fractions will contain the square root of a number, as for $\sin \beta$ in Example 2 above.)

9. Sketch a triangle with two sides of 1 cm and an angle of 90° between them. From the triangle, what is the value of $\tan 45°$?

Use the triangle to find, in fraction form, $\sin 45°$ and $\cos 45°$.

(As in question 8, the fractions will contain the square root of a number.)

Angles of elevation and depression

In Fig. 31.25, the *angle of elevation* of B (the basket of the balloon) from A is the angle AB makes with the horizontal.

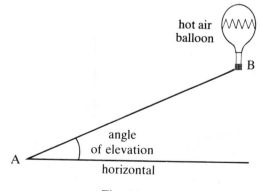

Fig. 31.25

In Fig. 31.26, the *angle of depression* of D (the man in the boat) from C is the angle CD makes with the horizontal.

Example   If in Fig. 31.26 the height of the cliff (CN) is 120 m and the angle of depression is 38°, calculate the distance of D from the base of the cliff.

$D\hat{C}N = 90° - 38° = 52°$

From $\triangle$ CND, $\dfrac{ND}{CN} = \tan D\hat{C}N$ so $\dfrac{ND}{120} = \tan 52°$

$ND = 120 \tan 52° = 120 \times 1.2799... = 153.59...$
The distance is 153 m to 3 s.f.

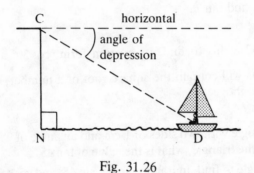

Fig. 31.26

Bearings

Bearings are measured clockwise from the north.

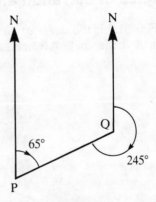

The bearing of Q from P is 065°.
The bearing of P from Q is 245°.   Fig. 31.27

East is 090°; south is 180°.
Note: 3 digits are always used, for example 065° instead of 65°.

Example A ship sails 7 km due east and then 9 km due south. Find its bearing from its original position.

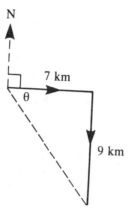

Fig. 31.28

$$\tan \theta = \frac{9}{7} = 1.2857...$$

$\theta = 52.125...° = 52°$ to the nearest degree.

The bearing is $90° + 52° = 142°$ to the nearest degree.

Exercise 31.5

1. The foot of a ladder is placed 4 m from the wall of a house. The ladder makes an angle of 68° with the ground. How far up the wall does it reach?

2. A rectangle has sides of 10 cm and 14 cm. Calculate the angle between a diagonal and a long side.

3. At a distance of 20 m from a church tower, the angle of elevation of the top is 56°. Calculate the height of the tower.

4. From the top of a vertical cliff of height 50 m, the angle of depression of a small boat is 32°. Calculate the distance of the boat from the foot of the cliff.

5. A ship sails 8 km due east and then 13 km due north. Find the bearing of the new position from the old.

6. An aircraft flies for 40 km on a bearing of 282°. How far is it then:
   (i) north or south (ii) east or west
   of its starting point?

7. At a distance of 70 m from the base of a tower, I measure the angle of elevation of the top as 35°. If my eye is 1.50 m above ground level, calculate the height of the tower.

8. Standing on top of a mountain 700 m above sea level, a walker sees another peak which is at a horizontal distance of 2400 m from her position and at an angle of elevation of 15°. Find the height of the second peak above sea level, correct to the nearest 100 m.

9. A helicopter hovers vertically above a boat which is 800 m from the shore. The angle of elevation of the helicopter from the shore is 28°. Calculate the height of the helicopter.

10. After take-off, an aircraft climbs for 15 km at an angle of 22° to the horizontal. Calculate: (i) the height it reaches (ii) the horizontal distance it travels.

11. From the top of a cliff 70 m high, the angles of depression of two boats at sea are 10° and 15°. Calculate the distance of each boat from the cliff, and hence the distance between the boats. (The boats and the base of the cliff are in a straight line.)

12. From a stationary balloon at a height of 120 m, the angle of depression of church A, which is due west of the balloon, is 28°, and the angle of depression of church B, due east of the balloon, is 36°. Calculate the distance between the churches.

13. Figure 31.29 shows a tower of height 25 m and a river PQ. From T the angles of depression of P and Q are 64° and 35° respectively.

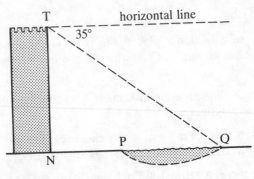

Fig. 31.29

Calculate: (i) NP   (ii) NQ   (iii) the width of the river.

14. A rhombus has four sides of 6 cm and two angles of 70°. Calculate:
    (i) the length of the longer diagonal
    (ii) the length of the shorter diagonal.

15. In Fig. 31.30, on the next page, T is the top and B is the bottom of a vertical cliff of height 100 m. From T the angles of depression of two marker buoys, P and Q, are 35° and 20° respectively. P is due east of B and Q is due south of P.

Draw △TBP and mark in the length of TB and the sizes of TB̂P and TP̂B. From this triangle, calculate BP.

In a similar way, use △TBQ to calculate BQ.

Draw △BPQ and mark in the lengths of BP and BQ. Hence calculate PB̂Q and state the bearing of Q from B. Also calculate PQ, using Pythagoras' Theorem.

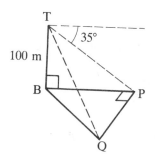

Fig. 31.30

16. Figure 31.31 shows part of a ski slope. AB is a straight horizontal footpath; BC is the track of a ski lift up the line of greatest slope. Franz skis in a straight line from C to A. Calculate:

  (i) the distance he skis (CA).
 (ii) the height of C above B (CN).
(iii) the slope of his track (CÂN).

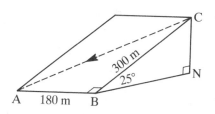

Fig. 31.31

17. From a point A, 12 m south of the base B of a flag pole, the angle of elevation of the top, T, of the pole is 40°. C is a point 18 m east of A. Calculate:

  (i) the height of the flag pole.
 (ii) the distance from C to B.
(iii) the angle of elevation of T from C.

18. A vertical tower AT, of height 50 m, stands at a point A on a horizontal plane. The points A, B and C lie on the same horizontal plane. B is due west of A and C is due south of A. The angles of

elevation of the top, T, of the tower from B and C are 25° and 30° respectively. Calculate:

(i) the distances AB, AC and BC, giving your answers to the nearest metre

(ii) the angle of elevation of T from the mid-point M of AB.   (L)

Ratios of angles of any size

A straight line OP of length 1 unit is rotated anticlockwise about O starting from the x-axis.

In Fig. 31.32, OP has turned through an acute angle θ. If P has coordinates (x, y), then

$$\sin \theta = \frac{PN}{OP} = \frac{y}{1} = y,$$

$$\cos \theta = \frac{ON}{OP} = \frac{x}{1} = x,$$

$$\tan \theta = \frac{PN}{ON} = \frac{y}{x}.$$

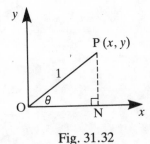

Fig. 31.32

For values of $\theta > 90°$ and for negative values of θ we define $\sin \theta$, $\cos \theta$ and $\tan \theta$ in exactly the same way, as

$$\sin \theta = y, \cos \theta = x, \tan \theta = \frac{y}{x}.$$

x or y or both may be negative.

In Fig. 31.33, θ is obtuse. y is positive but x is negative, so $\sin \theta$ is positive, $\cos \theta$ is negative and $\tan \theta$ is negative.

In Fig. 31.34, $180° < \theta < 270°$. Here $\sin \theta < 0$, $\cos \theta < 0$ and $\tan \theta > 0$.

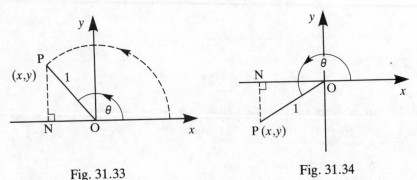

Fig. 31.33   Fig. 31.34

Exercise 31.6

1. Draw a diagram showing the position of OP when $270° < \theta < 360°$. From the diagram state the signs of $\sin \theta$, $\cos \theta$ and $\tan \theta$ in this case. On your calculator, check your answers for $\theta = 310°$.

2. By considering Figs. 31.33 and 31.34 state the values of sin 90°, sin 180°, sin 270°, cos 90°, cos 180° and cos 270°. Check your answers on your calculator.

3. By considering PN in Figs 31.32, 31.33 and 31.34, describe how sin θ changes as θ increases from 0° to 360°.

4. From a calculator, obtain sin 25° and sin 385°. Explain why they are the same.

5. From a calculator, obtain cos (−75°) and cos 285°. Explain why they are the same.

6. In Fig. 31.35, P_{70}, P_{110}, P_{250} and P_{290} are the positions of P when OP has turned through 70°, 110°, 250° and 290°.

    (i) Use the symmetry of the figure to explain why sin 110° = sin 70°, sin 250° = −sin 70° and sin 290° = −sin 70°.

    (ii) Make similar statements for the cosines of 110°, 250° and 290°.

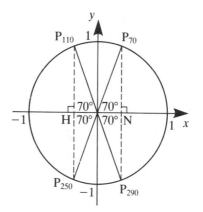

Fig. 31.35

7. From a calculator, find sin 33° and sin 147°. Note that 33 + 147 = 180. For any other acute angle, φ, find sin φ and sin (180° − φ).

8. Repeat question 7 for cosines instead of sines.

9. Repeat question 7 for tangents instead of sines.

10. Using a calculator, draw up a table for the sines of angles from 0° to 360° at intervals of 15°. Give each correct to 3 s.f.

    Use the table to draw the graph of y = sin x for 0° ≤ x ≤ 360°. Suitable scales are 2 mm to 5° on the x-axis and 4 cm to 1 unit on the y-axis.

    On your graph, mark as A and B the points where sin x = 0.5 and state the values of x at those points.

    Also mark, as C and D, the points where sin x = −0.5 and state the values of x at those points.

    Describe fully the symmetry of the part of the graph:

    (i) from 0° to 180°   (ii) from 180° to 360°   (iii) from 0° to 360°.

    Use these symmetries to find acute angles θ and φ such that sin θ = sin 160° and sin φ = −sin 295°.

11. Draw the graph of $y = \cos x$ for values of x from $0°$ to $360°$, plotting points every $15°$.

Mark as E and F the points where $\cos x = 0.5$ and as G and H the points where $\cos x = -0.5$. State the values of x at these points.

Describe the symmetry of the graph. Use the symmetry to find an angle which has the same cosine as: (i) $25°$ (ii) $160°$.

Describe the symmetry of the part of the graph from $0°$ to $180°$. Use this symmetry to find an obtuse angle θ such that $\cos \theta = -\cos 25°$.

12. Draw the graph of $y = \tan x$ for values of x from $0°$ to $70°$, from $110°$ to $250°$ and from $290°$ to $360°$. Suitable scales are 2 mm to $5°$ on the x-axis and 2 cm to 1 unit on the y-axis.

Mark as K and L the points where $\tan x = 1.6$ and state the values of x at these points.

Also mark as P and Q the points where $\tan x = -1.6$ and state the values of x.

From your calculator, state the values of $\tan 85°$, $\tan 88°$ and $\tan 93°$. What happens to the graph as x approaches $90°$?

13. For values of x from $0°$ to $360°$, sketch, on separate axes, the graphs of: (i) $y = 1 + \sin x$ (ii) $y = \frac{1}{2} + \cos x$. Place suitable numbers on each axis. Where does the graph of $y = \frac{1}{2} + \cos x$ cut the x-axis?

Summary for obtuse angles

If θ is obtuse, $\sin \theta = \sin(180° - \theta)$,
                $\cos \theta = -\cos(180° - \theta)$,
                $\tan \theta = -\tan(180° - \theta)$.

Finding obtuse angles

Suppose that θ is obtuse and that $\sin \theta = 0.8480$.

A calculator does not give an obtuse angle as the inverse of $\sin \theta$. It gives $58°$, the acute angle which has a sine of 0.8480. To obtain the obtuse angle you must subtract $58°$ from $180°$. The answer is $122°$.

If $\cos \theta = -0.3090$, a calculator gives the obtuse angle $108°$.

If $\tan \theta = -0.9004$, a calculator gives the negative angle $-42°$. The obtuse value of θ is $180° - 42° = 138°$.

Exercise 31.7

1. ϕ is an obtuse angle. State its value if:

   (i) $\sin \phi = \sin 20°$ (ii) $\cos \phi = -\cos 80°$
   (iii) $\tan \phi = -\tan 50°$ (iv) $\cos \phi = -\cos 32°$

2. tan $105°$ = $-$tan $75°$. In the same way, copy and complete the following:

   (i) cos $110°$ = . . .   (ii) sin $130°$ = . . .   (iii) tan $165°$ = . . .
   (iv) sin $153°$ = . . .   (v) cos $147°$ = . . .   (vi) tan $99°$ = . . .

Questions 3 to 7: give your answers correct to the nearest degree.

3. Find θ if:

   (i) cos θ = 0.4067   (ii) cos θ = -0.4067
   (iii) cos θ = 0.7880   (iv) cos θ = -0.7880

4. Find ϕ if it is obtuse and:

   (i) sin ϕ = 0.7660   (ii) sin ϕ = 0.4226
   (iii) sin ϕ = 0.3746   (iv) sin ϕ = 0.9613

5. State two possible values of θ if:

   (i) sin θ = 0.8829   (ii) sin θ = 0.5446

6. Find ϕ if it is between $0°$ and $180°$ and:

   (i) tan ϕ = 0.5774   (ii) tan ϕ = -0.5774
   (iii) tan ϕ = -1.4826 (iv) tan ϕ = 1.4826

7. Solve the following equations for values of θ between $0°$ and $180°$:

   (i) cos θ = 0.5577   (ii) cos θ = -0.5577 (iii) sin θ = 0.8973
   (iv) tan θ = 0.6494   (v) tan θ = -0.6494

32 Chord, angle and tangent properties of circles

Chords of circles

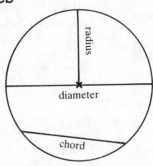

Fig. 32.1

A *chord* is a straight line joining two points on the circumference of a circle. A *diameter* is a chord that passes through the centre of the circle. A *radius* is a straight line joining a point on the circumference to the centre of the circle.

Properties of chords

1. If a straight line is drawn from the centre of a circle to the mid-point of a chord, then it is perpendicular to the chord. In Fig. 32.2, M is the mid-point of AB. The statement tells us that OM̂A = OM̂B = 90°.

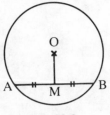

Fig. 32.2

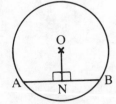

Fig. 32.3

Conversely, if a straight line is drawn from the centre so that it is perpendicular to a chord, then it bisects that chord. In Fig. 32.3, ON is perpendicular to AB. Therefore AN = NB.

2. If two chords are equal, then they are the same distance from the centre of the circle. This means that if AB = CD, in Fig. 32.4, then OP = OQ.

Conversely, if chords are the same distance from the centre, then they are equal in length. This means that if OP = OQ, in Fig. 32.4, then AB = CD.

Example For a circle of radius 15 cm, calculate:

    (i) the distance from the centre, for a chord of length 24 cm

    (ii) the length of a chord which is 11 cm from the centre.

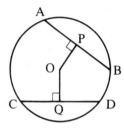

Fig. 32.4

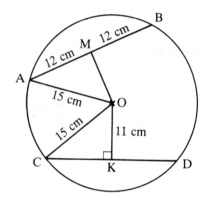

Fig. 32.5

(i) In Fig. 32.5, M is the mid-point of AB so $A\hat{M}O = 90°$ and AM = ½ AB = 12 cm.
Applying Pythagoras' Theorem to △AMO:
$AO^2 = AM^2 + MO^2$ so $15^2 = 12^2 + MO^2$
from which MO = 9 cm.

(ii) In Fig. 32.5, $O\hat{K}C = 90°$ so CK = KD.
Applying Pythagoras' Theorem to △OCK:
$CK^2 + 11^2 = 15^2$
from which $CK = \sqrt{104} = 10.20$ to 4 s.f.
Therefore CD = 20.4 cm to 3 s.f.

Exercise 32.1

Questions 1 to 10 refer to Fig. 32.6.

1. If AB = 8 cm and AO = 5 cm, calculate OM.

2. If OM = 5 cm and AB = 24 cm, calculate AO.

3. If AB = 30 cm and AO = 17 cm, calculate OM.

4. If AO = 20 cm and OM = 16 cm, calculate AB.

5. If AB = 14 cm and OM = 5 cm, calculate AO.

6. If AB = 18 cm and AO = 11 cm, calculate OM.

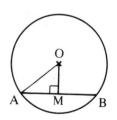

Fig. 32.6

7. If AO = 8 cm and OM = 4 cm, calculate AB.

8. If OM = 6 cm and AB = 12 cm, what is AM? State the sizes of AÔM and AÔB.

9. If AO = 10 cm and AB = 14 cm, state the value of sin AÔM. Hence find AÔM and AÔB.

10. If AÔB = 110° and AO = 8 cm, calculate AB.

11. PQ and RS are parallel chords of a circle of radius 10 cm. Their lengths are 12 cm and 16 cm. Calculate the distance between the chords if they are: (i) on opposite sides of the centre (ii) on the same side of the centre.

12. A circle is drawn through the four corners of a rectangle with sides of 8 cm and 6 cm. Calculate the diameter of the circle.

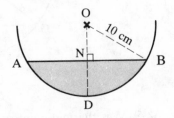

Fig. 32.7

13. Figure 32.7 shows a vertical section of a hemispherical bowl containing water. The radius of the bowl is 10 cm and AB is 12 cm.

(i) Calculate the greatest depth of the water.

(ii) If the maximum depth is increased to 4 cm, calculate the new width of AB.

Intersecting chords

1. If chords AB and CD intersect inside the circle at point P (Fig. 32.8), then PA × PB = PC × PD.

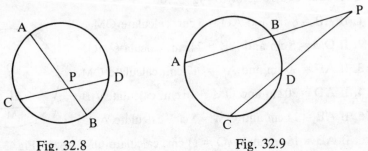

Fig. 32.8   Fig. 32.9

2. If chords AB and CD do not intersect inside the circle, but are produced (continued) to meet outside the circle (Fig. 32.9), then PA × PB = PC × PD. Notice that this is exactly the same as statement 1 above.

3. If points C and D in Fig. 32.9 are brought closer together until they coincide at T (Fig. 32.10), then we have:

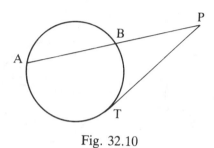

Fig. 32.10

PA × PB = PT × PT   or   PA × PB = PT².

Notice that all distances are measured from P, in all three statements above.

Example 1   If, in Fig. 32.8, PA = 5 cm, PB = 3 cm and PC = 4 cm, find PD.
Since PC × PD = PA × PB

then $4 \times PD = 5 \times 3 = 15$

so $PD = \dfrac{15}{4} = 3\tfrac{3}{4}$ cm.

Example 2   If, in Figs. 32.9 and 32.10, AB = 7 cm, PB = 9 cm and PD = 10 cm, calculate CD and PT.
AP = AB + BP = 7 + 9 = 16
Since   PA × PB = PC × PD
then   16 × 9 = PC × 10
from which PC = 14.4 cm
so   CD = PC − PD = 4.4 cm.
            PT² = PA × PB = 16 × 9 = 144
so   PT = 12 cm

Exercise 32.2

The questions for this exercise are on pages 284 and 285. All lengths are in centimetres. Use Fig. 32.11 for questions 1 to 6.

1. If $a = 3$, $b = 4$ and $c = 6$, calculate d.

2. If $a = 4$, $b = 12$ and $d = 8$, calculate c.

3. If $a = 10$, $b = 9$ and $c = 12$, calculate d.

4. If $b = 5$, $c = 6$ and $d = 7$, calculate a.

5. If AB = 11, PB = 6 and PD = 2, calculate CD.

6. If PC = 4, PD = 9 and PA = PB, calculate AB.

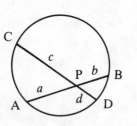
Fig. 32.11

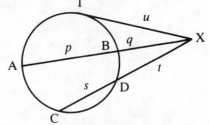
Fig. 32.12

Use Fig. 32.12 for questions 7 to 14.

7. If $p = 2$, $q = 4$ and $t = 3$ calculate s.

8. If $q = 5$, $s = 4$ and $t = 6$ calculate p.

9. If $p = 6$ and $q = 2$ find u.

10. If $s = 15$ and $t = 5$, calculate u.

11. If $u = 6$ and $t = 4$, calculate s.

12. If $u = 8$ and $q = 5$, calculate p.

13. If $p = 3$, $q = 5$ and $s = 6$, show that $t(t + 6) = 40$. By solving this quadratic equation, find t.

14. If $u = 6$ and $p = 9$, form a quadratic equation for q and solve it.

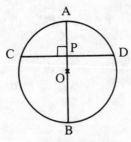
Fig. 32.13

15. In Fig. 32.13, AB is a diameter and CD is perpendicular to AB. What can you say about CP and PD? If AP = 2 and CP = 4, calculate PB, AB and the radius of the circle.

16. In Fig. 32.13, if AP = 4 and the radius is 6.5, calculate AB, PB, CP and CD.

17. In Fig. 32.14, LJT and GHJK are straight lines and GT is a tangent. Calculate *a* and *b*.

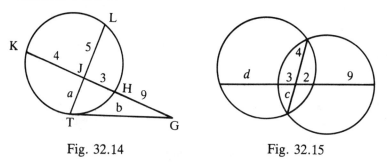

Fig. 32.14   Fig. 32.15

18. In Fig. 32.15, there are two straight lines. Calculate *c* and *d*.

19. In Fig. 32.16, ABP is a straight line; PT and PV are tangents.
  (i) If AB = 5 and BP = 4, calculate PT and PV.
  (ii) If AB = *x* and BP = *y*, express the lengths of PT and PV in terms of *x* and *y*.

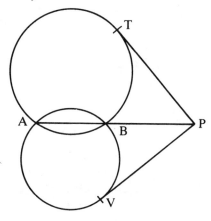

Fig. 32.16

20. In Fig. 32.13, let CD = 6, AB = 10, and AP = *x*. State the length of PB in terms of *x*. Using the property of intersecting chords, form an equation for *x* and solve it.

Angles in a circle

1. The angle at the centre of a circle is twice any angle at the circumference standing on the same arc.

Fig. 32.17

Fig. 32.18

Figure 32.17 shows angles standing on a minor arc (less than half the circumference). The angle at the centre (2θ) is twice the angle at the circumference (θ).

Figure 32.18 also shows angles standing on a minor arc.

Figure 32.19 shows angles standing on a major arc (more than half the circumference).

Figure 32.20 shows the special case where the arc is a semi-circle. Since the angle at the centre is now 180°, the angle at the circumference is 90°. This result is sometimes stated as 'The angle in a semi-circle is a right-angle' (Fig. 32.21).

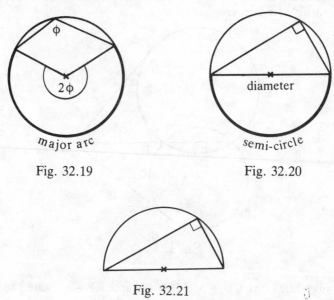

Fig. 32.19

Fig. 32.20

Fig. 32.21

2. Angles at the circumference are equal if they stand on the same arc. In Fig. 32.22, the three angles AP̂B, AQ̂B and AR̂B all stand on the minor arc AB. They are all equal since each is half the angle at the centre of the circle (2θ in Fig. 32.17).

   In Fig. 32.23, AŜB and AT̂B both stand on the major arc AB. Each is half the angle at the centre (2φ in Fig. 32.19) so they are equal.

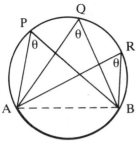

Fig. 32.22

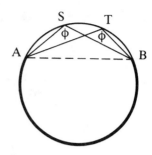

Fig. 32.23

In Fig. 32.24, chord AB divides the circle into two parts called *segments*.

In Fig. 32.25, $A\hat{P}B$, $A\hat{Q}B$ and $A\hat{R}B$ are in the major segment; $A\hat{S}B$ and $A\hat{T}B$ are in the minor segment.

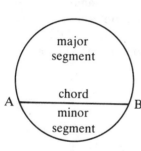

Fig. 32.24

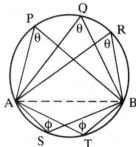

Fig. 32.25

We have $A\hat{P}B = A\hat{Q}B = A\hat{R}B$ and $A\hat{S}B = A\hat{T}B$, so property 2 above is sometimes stated as 'Angles in the same segment are equal'.

Example 1 In Fig. 32.26, if $A\hat{P}B = 63°$, calculate $A\hat{O}B$, reflex $A\hat{O}B$ and $A\hat{Q}B$.

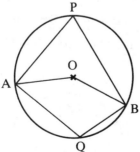

Fig. 32.26

$A\hat{O}B = 2 \times A\hat{P}B = 2 \times 63° = 126°$.
Reflex $A\hat{O}B = 360° - A\hat{O}B = 360° - 126° = 234°$.
$A\hat{Q}B = \frac{1}{2}$ reflex $A\hat{O}B = \frac{1}{2} \times 234° = 117°$.

Example 2   In Fig. 32.27, calculate *a*, *b*, *c*, *d* and *e*.

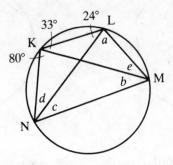

Fig. 32.27

a = 80° since *a* and MK̂N stand on same arc.
b = 24° since *b* and KL̂N stand on same arc.
c = 33° since *c* and LK̂M stand on same arc.
d = 180° − 80° − 33° − 24° = 43°, from △KLN.
e = *d* = 43° since *e* and *d* stand on the same arc.

Exercise 32.3

For questions 1 to 8, use Fig. 32.26.

1. If AP̂B = 70°, calculate AÔB.

2. If AÔB = 108°, calculate AP̂B.

3. If AQ̂B = 115°, calculate reflex AÔB.

4. If reflex AÔB = 310°, calculate AQ̂B.

5. If AP̂B = 50°, calculate AÔB, reflex AÔB and AQ̂B.

6. If AP̂B = 46°, calculate AQ̂B.

7. If AQ̂B = 100°, calculate reflex AÔB, obtuse AÔB and AP̂B.

8. If AP̂B = 60°, calculate AÔB, OÂB and OB̂A.

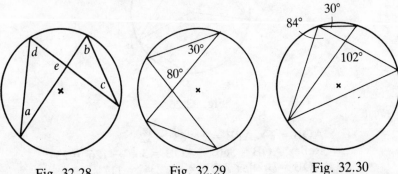

Fig. 32.28   Fig. 32.29   Fig. 32.30

9. In Fig. 32.28, if $d = 70°$ and $e = 80°$, calculate a and state the sizes of b and c.

10. In Fig. 32.28, if $a = 28°$ and $b = 65°$, calculate c, d and e.

11. In Fig. 32.28, if $b = 76°$ and $c = 24°$, calculate a, d and e.

12. Copy Fig. 32.29 and fill in the sizes of all the angles.

13. Copy Fig. 32.30 and fill in the sizes of all the angles.

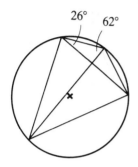

Fig. 32.31

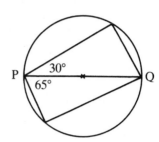

Fig. 32.32

14. Copy Fig. 32.31 and fill in the sizes of as many angles as possible.

15. AB is the diameter of a circle, centre O. P is a point on the circumference. State the size of AP̂B. If PÂB = 38°, calculate AB̂P.

16. Copy Fig. 32.32 in which PQ is a diameter, and fill in the sizes of all the angles.

17. In Fig. 32.33, if PŜQ = 58°, find PR̂Q, PÔQ and PQ̂O.

18. In Fig. 32.33, if PQ̂O = 22°, find PÔQ and PŜQ.

19. In Fig. 32.34, find PQ̂R, obtuse PÔR and x.

20. F, G, H and K are points on a circle. If HĜK = 60° and GĤK = 70°, calculate GK̂H, KF̂H and GF̂H.

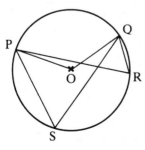

Fig. 32.33

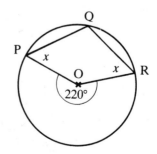

Fig. 32.34

21. In Fig. 32.35, O is the centre of the circle, $P\hat{O}Q = 112°$ and RO is parallel to PQ. Calculate *a*, *b*, *c*, *d* and *e*.

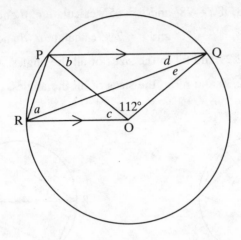

Fig. 32.35

Cyclic quadrilaterals

A quadrilateral that has its four vertices on a circle is called a *cyclic quadrilateral*.

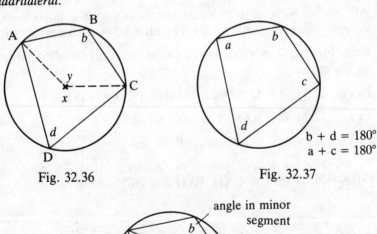

Fig. 32.36

$b + d = 180°$
$a + c = 180°$

Fig. 32.37

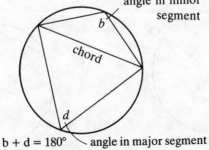

angle in minor segment

chord

$b + d = 180°$ angle in major segment

Fig. 32.38

In Fig. 32.36, $b = \frac{1}{2}x$ and $d = \frac{1}{2}y$, so $b + d = \frac{1}{2}(x + y)$.
But $x + y = 360°$ so $b + d = \frac{1}{2} \times 360° = 180°$.
This result can be stated as:

the sum of two opposite angles of a cyclic quadrilateral is 180°.
(See Fig. 32.37.)

It can also be stated as:

the angles in opposite segments are supplementary (add up to 180°).
(See Fig. 32.38.)

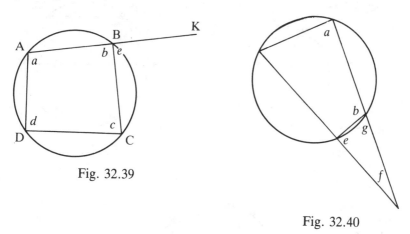

Fig. 32.39

Fig. 32.40

In Fig. 32.39, $b + d = 180°$ and $b + e = 180°$, therefore $d = e$. This result can be stated as follows:

when a side of a cyclic quadrilateral is produced, the exterior angle formed is equal to the opposite interior angle.

Exercise 32.4

1. In Fig. 32.39, if $b = 110°$, find d.

2. In Fig. 32.39, if $a = 78°$, find c.

3. In Fig. 32.39, if $c = 135°$ and $d = 65°$, find a and b.

4. In Fig. 32.39, if $d = 72°$, find e.

5. In Fig. 32.40, if $a = 56°$ and $b = 100°$, find e and f.

6. In Fig. 32.40, if $f = 30°$ and $g = 100°$, find e and a.

7. In Fig. 32.40, if $e = 40°$ and $b = 70°$, copy the figure and mark in the sizes of all the other angles.

8. In cyclic quadrilateral ABCD, AB = BC and AD̂C = 56°. Calculate AB̂C and BÂC.

9. In Fig. 32.41, PAQ and RBS are straight lines. Calculate AB̂S and AQ̂S. What follows for PR and QS?

10. In Fig. 32.42, chord PQ is parallel to diameter AB. State the sizes of u, w, x, y and z, giving reasons.

11. In Fig. 32.43, AB = BD and BC = CD. Calculate a, b and c.

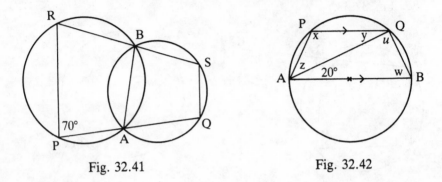

Fig. 32.41   Fig. 32.42

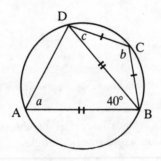

Fig. 32.43

Tangent properties

A straight line which touches a circle is called a *tangent* to the circle. The tangent and the circle have just one common point, which is called the *point of contact*. In Fig. 32.44, T is the point of contact.

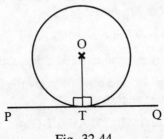

Fig. 32.44

1. A tangent is perpendicular to the radius at the point of contact. In Fig. 32.44, $P\hat{T}O = Q\hat{T}O = 90°$. This follows from the symmetry of the figure.

2. If two tangents are drawn from a point outside the circle then:

   (a) they are equal in length

   (b) the line joining the point to the centre of the circle bisects the angle between the two tangents.

In Fig. 32.45, this means that (a) PA = PB, and (b) $A\hat{P}O = B\hat{P}O$. These results follow either from the fact that the figure is symmetrical about line OP, or by proving that triangles AOP and BOP are congruent.

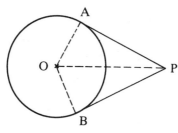

Fig. 32.45

Example   In Fig. 32.45, if OP = 13 cm and the radius is 5 cm, calculate AP and $A\hat{P}B$.

By Pythagoras' theorem, $OA^2 + AP^2 = OP^2$,

so $5^2 + AP^2 = 13^2$

from which AP = 12 cm.

From $\triangle POA$, $\sin O\hat{P}A = \dfrac{OA}{OP} = \dfrac{5}{13} = 0.3846$

$O\hat{P}A = 22.62°$ to 2 d.p.

$A\hat{P}B = 2 \times O\hat{P}A = 45.24°$ to 2 d.p.

$= 45.2°$ to 3 s.f.

Exercise 32.5

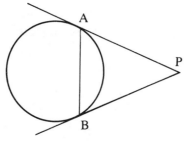

Fig. 32.46

1. In Fig. 32.46, if $A\hat{P}B = 52°$, calculate $B\hat{A}P$.

2. In Fig. 32.46, if $B\hat{A}P = 73°$, calculate $A\hat{P}B$.

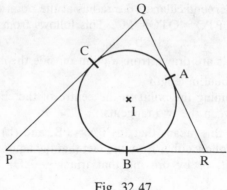

Fig. 32.47

3. In Fig. 32.47, if AR = 7 cm, BP = 6 cm and CQ = 4 cm, calculate the lengths of the sides of △ PQR.

4. In Fig. 32.47, if Q̂ = 58° and R̂ = 66°, calculate:
(i) QÂC (ii) RÂB (iii) BÂC (iv) AB̂C (v) BĈA.

5. In Fig. 32.47, I is the centre of the circle. If P̂ = 30°, calculate BÎC.

6. In Fig. 32.47, if AÎB = 126° and BÎC = 140°, calculate the angles of △ PQR.

For questions 7 to 12, use Fig. 32.45.

7. If OA = 6 cm and OP = 10 cm, calculate AP and OP̂A.

8. If OA = 8 cm and OP = 17 cm, calculate AP and AP̂B.

9. If OA = 7 cm and OP = 11 cm, calculate AP and AP̂B.

10. If OA = 9.2 cm and OP = 12.7 cm, calculate AP and AP̂B.

11. If OA = 10 cm and AÔP = 50°, calculate AP and OP.

12. If OA = 8 cm and AÔB = 128°, calculate AP and OP.

The property of angles in alternate segments

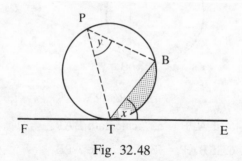

Fig. 32.48

In Fig. 32.48, x is the angle between the chord TB and the tangent TE. Chord TB divides the circle into two segments. The shaded segment contains part of angle x. The other segment (unshaded) is called the *alternate segment*. Angle y is in the alternate segment. It can be shown that $x = y$. This result can be stated as:

the angle between a chord and a tangent is equal to the angle in the alternate segment.

The result is also true for the obtuse angle between a chord and a tangent. In Fig. 32.49, $v = w$. (v covers the shaded segment; w is in the alternate segment.)

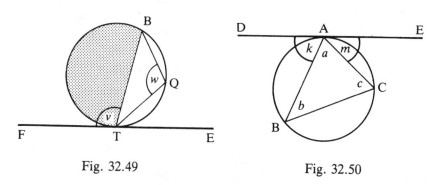

Fig. 32.49   Fig. 32.50

Example   In Fig. 32.50, if $k = 75°$, and $m = 58°$, calculate b and c.
$b = m = 58°$, using chord AC and tangent AE.
$c = k = 75°$, using chord AB and tangent AD.

Exercise 32.6

1. In Fig. 32.50, if $k = 83°$ and $m = 49°$, state the sizes of b and c.

2. In Fig. 32.50, if $b = 35°$ and $c = 68°$, state the sizes of k and m.

3. In Fig. 32.50, if $a = 60°$ and $k = 70°$, calculate b.

4. State the sizes of e and f in Fig. 32.51.

5. State the sizes of g, h and n in Fig. 32.52.

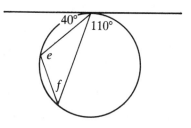

Fig. 32.51

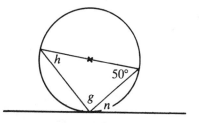

Fig. 32.52

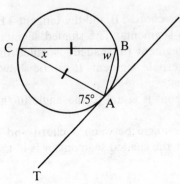

Fig. 32.53

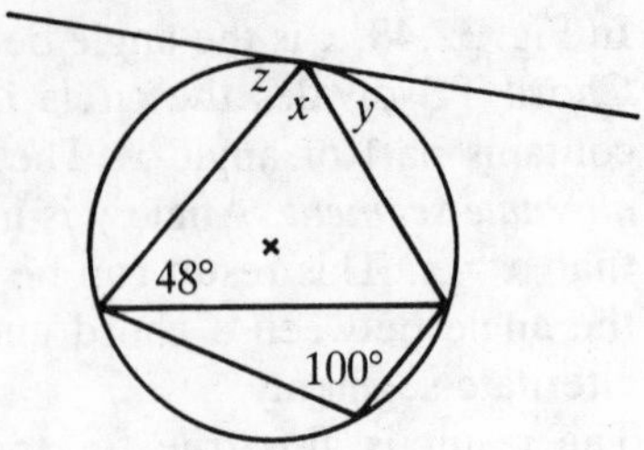

Fig. 32.54

6. In Fig. 32.53, AC = BC and TÂC = 75°. Calculate *x*.

7. State the sizes of *x*, *y* and *z* in Fig. 32.54, giving reasons.

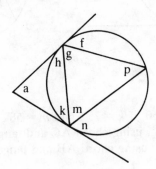

Fig. 32.55

For questions 8 to 12, use Fig. 32.55.

8. If *h* = 65°, find *p*, *k* and *a*.

9. If *p* = 52°, find *h*, *k* and *a*.

10. If *a* = 72°, find *h* and *p*.

11. If *n* = 70° and *p* = 48°, find *f*.

12. If *f* = 64° and *n* = 70°, find *p* and *a*.

33 Constructions and loci

Constructions

1. The bisector of an angle.
   These are the steps for constructing the line which bisects BÂC in Fig. 33.1. With centre A draw arcs cutting AB and AC at P and Q. With centres P and Q draw arcs intersecting at R. Join AR. AR bisects BÂC.

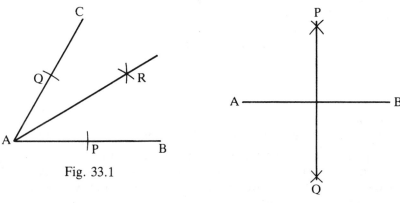

Fig. 33.1

Fig. 33.2

2. The perpendicular bisector (or mediator) of a given line segment.
   These are the steps for constructing the perpendicular bisector of AB, in Fig. 33.2. With A and B as centres and with a suitable radius, draw arcs which intersect at P and Q. Join PQ. PQ bisects AB and is perpendicular to it.

3. An angle of 60°.
   These are the steps for constructing an angle of 60° at point A on the given line AB (Fig. 33.3). With centre A and a suitable radius, draw an arc to cut AB at P. With centre P and the same radius, draw a second arc to cut the first at Q. Join AQ. Then BÂQ = 60°.

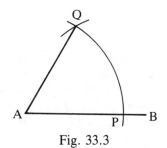

Fig. 33.3

4. An angle of 30°.
   Draw an angle of 60° and then bisect it.

5. A line perpendicular to a given line at a given point on the line.
   In Fig. 33.4, A is the given point on BC. With centre A draw two
   arcs cutting BC at P and Q. With centres P and Q and a radius greater
   than before, draw arcs to intersect at R. Join AR. AR is perpendicular
   to BC.

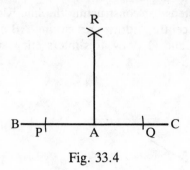

Fig. 33.4

6. An angle of 90°.
   This is the same as for 5 above. RÂC = 90°.

7. A perpendicular to a line from a point outside the line.
   Figure 33.5 shows the construction of a perpendicular from A to BC.
   With centre A draw two arcs to cut BC at P and Q. With centres P
   and Q draw two arcs to intersect at R. Join AR. AR is perpendicular
   to BC.

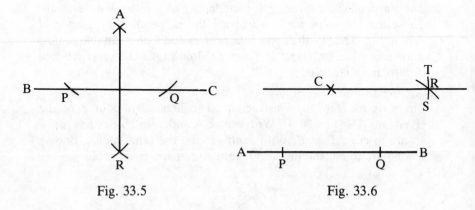

Fig. 33.5   Fig. 33.6

8. A line through a given point, parallel to a given line.
   Figure 33.6 shows the construction of a line through C, parallel to
   AB. Mark any two points P and Q on AB. With centre C and radius
   PQ, draw an arc ST. With centre Q and radius PC, draw an arc to
   cut ST at R. Draw the line through C and R. This line is parallel to
   AB.

Exercise 33.1

1. Practise each of the above constructions.

2. Construct an angle of 90° and bisect it to obtain two angles of 45°.

3. Construct an angle of 60° and bisect it to obtain two angles of 30°.

4. Draw a line segment of length 9 cm. Bisect it, and then bisect each half, so that the line segment is divided into four equal parts. Check by measuring the parts.

5. Draw an obtuse angle and divide it into four equal parts. Check by measuring with a protractor.

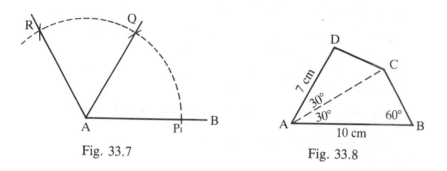

Fig. 33.7   Fig. 33.8

6. Copy Fig. 33.7 to obtain an angle of 120° by drawing two angles of 60°.

7. Copy Fig. 33.7 again and construct AT to bisect QÂR. Then BÂT should be 90°.

8. Draw a circle, centre O, radius 4 cm. Mark a point A on the circumference. At A draw a line perpendicular to OA. This should be a tangent to the circle.

9. Draw a circle, centre O, radius 5 cm. Mark point A on the circumference. With centre A and radius 8 cm, draw an arc to cut the circle at B. Join AB. Construct the perpendicular bisector of AB. It should pass through O.

10. Draw a line segment AB of length 9 cm. With centre A draw an arc of radius 8 cm, and with centre B draw an arc of radius 7 cm, so that the two arcs intersect at C. Join AC and BC. You now have a triangle with sides of 9, 8 and 7 cm. Construct the bisectors of the three angles of the triangle. They should all pass through a certain point D. Measure AD.

11. Draw quadrilateral ABCD having the lengths and angles shown in Fig. 33.8. Measure BC, CD, angle BCD and angle ADC. Check the sum of the four angles of your quadrilateral.

12. Draw quadrilateral PQRS with PQ = 8 cm, angle QPS = 60°, angle PQR = 90°, PS = 5 cm and QR = 6 cm. Measure PR, RS, angle PSR and angle SRQ.

13. Draw triangle PQR with PQ = 8 cm, QR = 7 cm and Q = 60°. Measure PR. Draw the bisector of P. Let it meet QR at T. Measure QT.

14. Draw triangle DEF with DE = 6.9 cm, EF = 9.2 cm and FD = 8.3 cm. From D draw the line perpendicular to EF. Let it meet EF at H. Measure EH.

15. Two buoys, B and C, mark the entrance to a harbour. C is 120 m east of B. A boat is moored in the harbour at D which is 100 m north of B. The boat leaves D sailing on a course of 150°.

    Using ruler and compasses only

    (i) draw a diagram to show B, C, D and the course of the boat, using a scale of 1 cm to represent 10 m

    (ii) construct the point E at which the boat is closest to B and the point F at which it is closest to C.

    By measurement, find the distances BE and CF, correct to the nearest metre.
    State the bearing of B from E.

Loci

A *locus* is a set of points which satisfy a given condition. The plural of locus is *loci*. It often helps to think of a locus as the path traced out by a point moving in such a way that it satisfies a given condition. For example, the locus of a point P which moves so that it is always 5 cm from a fixed point A is a circle, with centre A and radius 5 cm. We might describe this locus in set language as {points P : PA = 5 cm}.

The shape of a locus can often be found by marking several points that satisfy the condition. For example, suppose that we wish to find the locus of a point which is the same distance from AC as from AB (Fig. 33.9). Several points that satisfy this condition are marked with crosses. They suggest that the locus is a straight line bisecting Â.

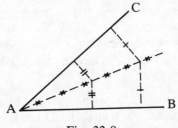

Fig. 33.9

Some standard loci

Some standard loci are shown by broken lines, in the diagrams below.

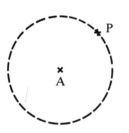

P is a fixed distance from A. The locus is a circle, centre A.

Fig. 33.10

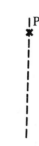

P is equidistant from A and B. The locus is the mediator of AB.

Fig. 33.11

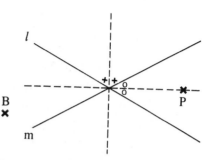

P is equidistant from lines *l* and *m*. The locus consists of the two bisectors of the angles between *l* and *m*.

Fig. 33.12

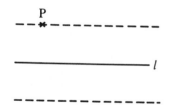

P is a fixed distance from line *l*. The locus is two lines parallel to *l*.

Fig. 33.13

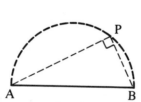

P is above AB and $A\hat{P}B = 90°$. The locus is a semicircle with AB as diameter. (See page 286.)

Fig. 33.14

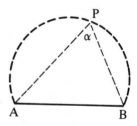

P is above AB and $A\hat{P}B$ is a fixed size, α. The locus is a major arc of a circle. (See page 286.)

Fig. 33.15

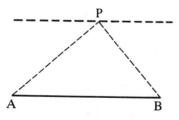

P is above AB and the area of $\triangle APB$ is constant. Since the height of the triangle is constant, the locus is a line parallel to AB.

Fig. 33.16

Exercise 33.2

1. For conditions (a) to (d) below:
   (i) make a drawing and mark several points that satisfy the condition
   (ii) show the locus with a suitable straight line or curve
   (iii) describe the locus.
   (a) Points 3 cm from a given point.
   (b) Points 1 cm from a given straight line, assumed to continue indefinitely in each direction.
   (c) Points equidistant from two fixed points C and D where CD = 4 cm. (For any point P on the locus, PC = PD.)
   (d) Points equidistant from two fixed straight lines EF and EG.

2. On graph paper, draw axes Ox and Oy, each marked from −5 to +5.
   (i) Draw the locus of points which are 5 units from O.
   (ii) Draw the locus of points which are 3 units from the x-axis.
   (iii) State the coordinates of the four points which are on both loci.

3. On squared paper, draw △KLM so that $\hat{K} = 90°$, KL = 8 cm and KM = 6 cm. Draw:
   (i) the locus of points equidistant from K and L, and
   (ii) the locus of points equidistant from K and M.
   Comment on the point which is common to both loci.

4. On squared paper draw lines AB and AC of length 5 cm so that $\hat{A} = 90°$. Construct:
   (i) the locus of points equidistant from AB and AC
   (ii) the locus of points 4 cm from B.
   Label E and F, the two points which belong to both loci, and measure EF.

5. Mark two points, C and D, 5 cm apart. Draw the following loci:
   (i) {P : PC = PD}   (ii) {Q : QD = 3.5 cm}.
   Label G and H, the points which are in both sets.

6. Across the page, draw a line segment AB of length 5 cm. Draw the following loci:
   (i) {P : A$\hat{P}$B = 90° and P is below AB}
   (ii) {Q : Q is below AB and 2 cm from it}.
   Label K and L, the points which are in both sets. Measure KL.

7. (i) A wheel of radius 10 cm rolls in a straight line on level ground. Describe the locus of its centre.

(ii) A wheel of radius 10 cm rolls round the outside of a fixed wheel of radius 20 cm, centre C. Describe the locus of the centre of the moving wheel.

(iii) A plank is placed flat against a vertical wall with its base on horizontal ground. The top of the plank is pushed away from the wall and falls to the ground. During the motion the bottom end remains fixed. Describe the locus of the top end.

8. Draw a line segment AB of length 8 cm. Construct:

(i) the locus of P so that $A\hat{P}B = 90°$ and P is above AB

(ii) the locus of Q so that the area of $\triangle AQB$ is 12 cm^2 and Q is above AB.

Label as R and S the points which are on both loci, and measure RS.

9. The fixed points X and Y are 8 cm apart in a given plane. For each of the following cases state the complete locus of the point Z which moves in the given plane:

(a) so that ZX = ZY;

(b) so that the area of $\triangle XYZ = 20$ cm^2;

(c) so that $X\hat{Z}Y = 40°$.   (L)

10. Construct the parallelogram ABCD in which AB = 9 cm, AD = 5 cm and $B\hat{A}D = 60°$. Measure, and write down, the length of AC. On the same diagram, construct:

(i) the locus of a point X which moves so that it is equidistant from A and C

(ii) the locus of a point Y which moves so that $B\hat{Y}D = 90°$.

The position of a point P, which lies inside the parallelogram, is such that $AP \geqslant PC$ and $B\hat{P}D \leqslant 90°$. Indicate clearly, by shading, the region in which the point P must lie.   (C)

34 The sine and cosine rules

In Chapter 31, we calculated the sizes of sides and angles in right-angled triangles. For a triangle which does not have a right-angle (and is not isosceles—see page 268) we use the sine rule or the cosine rule.

The sine rule

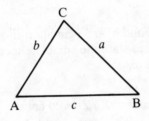

Fig. 34.1

$$\frac{a}{\sin A} = \frac{b}{\sin B} = \frac{c}{\sin C}$$

Notice that each fraction is made up of a side and the sine of the angle opposite that side. Notice also that we use a for the side opposite $\hat{A}$, b for the side opposite $\hat{B}$ and c for the side opposite $\hat{C}$.

Example 1 In $\triangle ABC$, $\hat{A} = 42°$, $\hat{B} = 74°$ and $b = 9.6$ cm. Find a.

Substituting in $\dfrac{a}{\sin A} = \dfrac{b}{\sin B}$ we have $\dfrac{a}{\sin 42°} = \dfrac{9.6}{\sin 74°}$

$a = \dfrac{9.6 \times \sin 42°}{\sin 74°} = 6.68$ cm to 3 s.f.

Example 2 In $\triangle PQR$, $q = 7.2$ cm, $r = 5.8$ cm and $\hat{Q} = 26°$. Find $\hat{P}$.

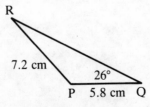

Fig. 34.2

We must first find R using $\dfrac{q}{\sin Q} = \dfrac{r}{\sin R}$

Inverting these fractions:

$\dfrac{\sin R}{r} = \dfrac{\sin Q}{q}$, so $\dfrac{\sin R}{5.8} = \dfrac{\sin 26°}{7.2}$

$\sin R = \dfrac{5.8 \times \sin 26°}{7.2} = 0.3513$

$\hat{R} = 20.7°$ to 3 s.f.

$\hat{P} = 180° - 26° - 20.7° = 133.3°$

Exercise 34.1

For each question, draw a diagram, insert the sizes of known sides and angles and mark the side or angle to be calculated. Then write down a statement about two sides and their opposite angles.
Questions 1 to 6 refer to $\triangle ABC$ (Fig. 34.1).

1. $\hat{A} = 58°$, $\hat{B} = 42°$ and $b = 6.6$ cm. Calculate a.

2. $\hat{B} = 36°$, $\hat{C} = 70°$ and $c = 7.9$ cm. Calculate b.

3. $\hat{A} = 118°$, $\hat{C} = 34°$ and $a = 8.5$ cm. Calculate c.

4. $\hat{B} = 52°$, $\hat{C} = 122°$ and $b = 15.7$ cm. Calculate c.

5. $\hat{A} = 43°$, $\hat{B} = 72°$ and $c = 9.5$ cm. Calculate $\hat{C}$ and a.

6. $\hat{B} = 55°$, $\hat{C} = 78°$ and $a = 25.7$ cm. Calculate $\hat{A}$ and b.

7. In $\triangle DEF$, $\hat{D} = 55°$, $\hat{E} = 30°$ and $EF = 14$ cm. Calculate DF.

8. In $\triangle PQR$, $QR = 6.3$ cm, $\hat{P} = 42°$ and $\hat{Q} = 63°$. Calculate PQ.

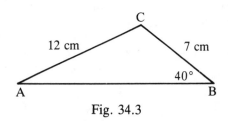

Fig. 34.3

9. In Fig. 34.3, $\dfrac{\sin A}{7} = \dfrac{\sin 40°}{12}$.

Find the value of sin A and hence find angle A.

10. In $\triangle DEF$, $e = 15$ cm, $f = 22$ cm and $\hat{F} = 38°$. Calculate $\hat{E}$.

11. In $\triangle PQR$, $p = 7$ cm, $q = 15$ cm and $\hat{Q} = 124°$. Calculate $\hat{P}$.

12. In $\triangle KLM$, $LM = 5.9$ cm, $MK = 4.7$ cm and $\hat{K} = 106°$. Calculate $\hat{L}$ and $\hat{M}$.

The cosine rule

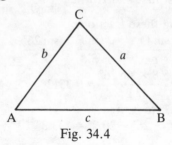

Fig. 34.4

The cosine rule is used when we are given:
either (i) 2 sides and the angle between them (for example $b, c, \hat{A}$)
or   (ii) all 3 sides.
   For (i) it is used in the form

$$a^2 = b^2 + c^2 - 2bc \cos A$$

For (ii) it can be used in the same form or in the alternative form:

$$\cos A = \frac{b^2 + c^2 - a^2}{2bc}$$

Example 1 If $b = 7$ cm, $c = 9$ cm and $\hat{A} = 52°$, calculate a.
   $a^2 = b^2 + c^2 - 2bc \cos A$
      $= 7^2 + 9^2 - 2 \times 7 \times 9 \times \cos 52°$
      $= 49 + 81 - 126 \cos 52°$
      $= 130 - 126 \times 0.6156...$
      $= 130 - 77.573... = 52.426...$
   $a = \sqrt{52.426...} = 7.24$ cm to 3 s.f.

   [A common error here is to subtract 126 from 130 (in the
   third line from the end) which gives 4, and then multiply
   4 by 0.6156, which gives 2.4624.]

Example 2 If $a = 6.3$ cm, $b = 9.2$ cm and $\hat{C} = 135°$, calculate c.
   For side c the cosine formula is:
   $c^2 = a^2 + b^2 - 2ab \cos C$
   $c^2 = 6.3^2 + 9.2^2 - 2 \times 6.3 \times 9.2 \times \cos 135°$
      $= 39.69 + 84.64 - 115.92 \times (-0.7071...)$
      $= 124.33 + 81.97$
      $= 206.30$
   $c = \sqrt{206.30} = 14.4$ cm to 3 s.f.

   (The sign before 81.97 is + because the cosine of 135° is
   negative.)

Example 3 If $a = 5$ cm, $b = 11$ cm and $c = 8$ cm, calculate the largest
   angle of the triangle.
   The largest angle is opposite the largest side, so it is $\hat{B}$.

$$\cos B = \frac{a^2 + c^2 - b^2}{2ac} = \frac{25 + 64 - 121}{80} = -\frac{32}{80} = -0.4$$

Since cos B is negative, $\hat{B}$ is obtuse.
$\hat{B} = 113.6°$ to 1 d.p.

Right-angled triangles

If $\hat{A} = 90°$, cos A $= 0$ so $a^2 = b^2 + c^2 - 2bc$ cos A becomes $a^2 = b^2 + c^2$, which is Pythagoras' theorem.

Exercise 34.2

Questions 1–8 and 11–18 are for a triangle ABC.

1. If $\hat{A} = 49°$, $b = 11$ cm and $c = 8$ cm, calculate a.

2. If $\hat{A} = 114°$, $b = 7$ cm and $c = 9$ cm, calculate a.

3. If $\hat{B} = 73°$, $a = 13$ cm and $c = 10$ cm, calculate b.

4. If $\hat{C} = 33°$, $a = 5$ cm and $b = 7$ cm, calculate c.

5. If $\hat{B} = 120°$, $a = 8$ cm and $c = 6$ cm, calculate b.

6. If $\hat{C} = 144°$, $a = 7$ cm and $b = 11$ cm, calculate c.

7. If $\hat{A} = 35.8°$, $b = 9.6$ cm and $c = 7.8$ cm, calculate a.

8. If $\hat{B} = 76.5°$, $a = 13.4$ cm and $c = 19.2$ cm, calculate b.

9. For $\triangle$PQR, express p^2 in terms of q, r and $\hat{P}$. Hence calculate p given that $\hat{P} = 72°$, $q = 9$ cm and $r = 6$ cm.

10. For $\triangle$PQR express q^2 in terms of p, r and $\hat{Q}$. Calculate q given that $\hat{Q} = 134°$, $p = 5.2$ cm and $r = 4.3$ cm.

11. If $a = 9$ cm, $b = 8$ cm and $c = 7$ cm, calculate $\hat{A}$.

12. If $a = 5$ cm, $b = 6$ cm and $c = 4$ cm, calculate $\hat{A}$.

13. If $a = 8$ cm, $b = 10$ cm and $c = 11$ cm, calculate $\hat{B}$.

14. If $a = 12$ cm, $b = 11$ cm and $c = 13$ cm, calculate $\hat{C}$.

15. If $a = 7.3$ cm, $b = 9.2$ cm and $c = 6.5$ cm, calculate $\hat{C}$.

16. If $a = 15.6$ cm, $b = 13.7$ cm and $c = 9.8$ cm, calculate $\hat{B}$.

17. If $a = 7$ cm, $b = 5$ cm and $c = 4$ cm show that cos A $= -0.2$ and calculate $\hat{A}$.

18. If $a = 5$ cm, $b = 9$ cm and $c = 6$ cm, show that cos B $= -\frac{1}{3}$ and calculate $\hat{B}$.

19. In $\triangle$PQR, $p = 9$ cm, $q = 4$ cm and $r = 6$ cm. Calculate $\hat{P}$.

20. In $\triangle ABC$, $a = 5.5$ cm, $b = 8.3$ cm and $c = 4.8$ cm. Calculate $\hat{B}$.

21. In $\triangle ABC$, $a = 4.9$ cm, $b = 5.8$ cm and $c = 3.8$ cm. Calculate the smallest angle.

22. In $\triangle PQR$, $p = 9.3$ cm, $q = 11.6$ cm and $r = 8.4$ cm. Calculate the largest angle.

Problems using the sine and cosine rules

Exercise 34.3

1. Figure 34.5 shows two trees, D and E, on a bank of a river. F is a tree on the other bank. Calculate $D\hat{F}E$, DF and the width of the river, FN.

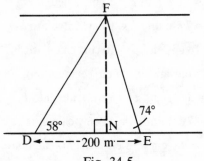

Fig. 34.5

2. A and B are two coastguard stations, B being 8 km east of A. From A the bearing of ship S is 040° and from B the bearing of S is 340°. Calculate $A\hat{S}B$ and then the distances AS and BS.

3. Town B is 35 km due north of town A. Town C is 28 km from A in the direction 070°. Calculate the distance of B from C.

4. Two ships leave port at the same time. One sails on a course of 060° at 14 km/h, and the other on a course of 110° at 12 km/h. How far apart are they after half an hour?

5. A golfer is playing an 80 m hole. He hits the ball 70 m at an angle of 25° to the correct line. How far is it from the hole?

6. Town Q is 18 km due north of town P. Town R is 13 km from P and 15 km from Q. If R is east of the line PQ, calculate the bearing of R from P and from Q.

7. In parallelogram ABCD, AB = 9 cm, AD = 6 cm and BD = 8 cm. Calculate $B\hat{A}D$ and $A\hat{B}C$.

8. A road is straight from A to B. At B its direction changes by 35°, then it is straight from B to C. AB = 800 m and BC = 500 m. There is a straight footpath from A to C. Find its length to the nearest 10 metres.

9. Village E is 9 km due east of village D. Village F is 12 km from D and 6 km from E. Calculate DÊF. If F is south of the line DE, state the bearing of F from E.

10. (a) Two ships, D and F, leave port at the same time. D sails at 16 knots on a bearing of 035° (N 35°E) and F sails at 12 knots on a bearing of 317° (N 43°W). Calculate their distance apart after 1½ hours.

   (b) A stake B is 17.8 metres due north of a stake A. The bearing of a flagpole from A is 037° and from B is 078°. Calculate the distance of the flagpole from A.   (W)

11. A ship was sailing at 12 knots in a straight line on a bearing of 065°. At 14.00 h a lighthouse was observed in the direction 140°. At 14.30 h the lighthouse was in the direction 210°. Calculate

   (i) the distance of the ship from the lighthouse at 14.00 h

   (ii) the shortest distance between the ship and the lighthouse

   (iii) when the ship was at this shortest distance.

Area of a triangle as $\frac{1}{2}bc \sin A$

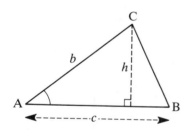

In Fig. 34.6, the area of
$\triangle ABC = \frac{1}{2}hc$ and $h = b \sin A$ so
the area of $\triangle ABC = \frac{1}{2}bc \sin A$.

Fig. 34.6

Exercise 34.4

1. Calculate the area of $\triangle ABC$ if $\hat{A} = 65°$, $b = 9$ cm and $c = 11$ cm.

2. Calculate the area of $\triangle PQR$ if $\hat{P} = 126°$, $q = 12$ cm and $r = 8$ cm.

3. A circle, centre O, radius 6 cm, passes through the vertices of a regular pentagon ABCDE. Calculate: (i) AÔB (ii) the area of $\triangle AOB$ (iii) the area of the pentagon.

4. Use the method of question 3 to calculate the area of a regular hexagon having its vertices on a circle of radius 10 cm.

5. In a parallelogram PQRS, PQ = 8 cm, PS = 5 cm and $\hat{P} = 70°$. Calculate the area of: (i) $\triangle PQS$ (ii) the parallelogram.
   Write down a formula for the area of a parallelogram having sides of x cm and y cm and an angle of $\theta°$.

35 Bar charts and pie charts

This table compares the numbers of people booking holidays in different countries, through a new travel agency:

Country	France	Austria	Spain	Switzerland	Greece	Others
No. of people	72	120	220	80	94	134

Such information is much more effective when shown as a diagram. The two sorts of diagram most commonly used are *bar charts* and *pie charts*.

In a bar chart each number is represented by the height of a bar, all the bars having the same width. Figure 35.1 shows a bar chart for the above data. To draw such a chart we might use 1 mm to represent 4 people, so that the column for 220 people would have a height of 55 mm.

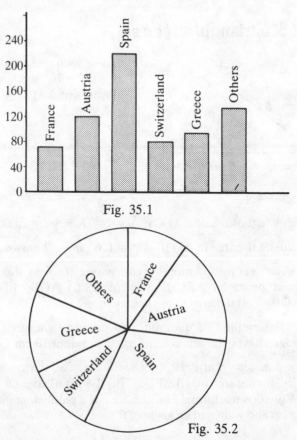

Fig. 35.1

Fig. 35.2

In a pie chart, each number is represented by the area of a sector of a circle and thus by the angle of the sector. Figure 35.2 shows a pie chart for the data above. 720 people are represented by 360°, so 2 people are represented by 1°. For the 72 people booking for France the angle is therefore $72 \div 2 = 36°$.

Exercise 35.1

1. During a peak period, the numbers of different types of vehicle using a certain road were:

Type of vehicle	Bus	Lorry	Car	Van
No. of vehicles	24	60	116	38

Draw a bar chart to show this information. Use 1 mm to represent 2 vehicles.

2. A firm called Shaw Calculators sold the following numbers of electronic adding machines over five years:

Year	1978	1979	1980	1981	1982
No. sold	20	60	100	124	160

Draw a bar chart to show this information, using 1 mm to represent 2 machines.

3. The numbers of students at a certain college entered for various GCSE subjects were:

Subject	Mathematics	Physics	Chemistry	Biology	General science	Geology
No. of students	88	63	54	42	30	20

Show this information as a bar chart, choosing a suitable scale.

4. The methods by which students travelled to a certain college were:

Method	Walk	Train	Bus	Cycle	Motorbike	Car
No. of students	55	42	78	93	69	11

Choosing a suitable scale, draw a bar chart to show this information.

5. This pie chart (Fig. 35.3) represents 40 people.

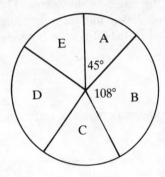

Fig. 35.3

(i) What angle represents 1 person?
(ii) How many people are represented by sector A and how many by sector B?
(iii) If sector C represents 7 people and sector D represents 10 people, what angles do they need?
(iv) What size is the angle of sector E, and how many people does it represent?

6. Each week a certain class has the following numbers of lessons in various subjects: English 4, mathematics 5, science 4, geography 3, history 3, others 11. Draw a pie chart to show this information, and state the angle used for each sector.

7. In a games period, 90 pupils were divided for various activities, like this: athletics 18, cricket 22, tennis 16, swimming 34. Draw a pie chart to show this information, and state the angle used for each sector.

8. Of 120 pupils who left school one term, 8 started work in shops, 17 in offices, 32 in factories and 23 went to universities or colleges. It is not known what happened to the others. Show this information as a pie chart, and state the angle used for each sector.

9. The numbers of various types of books issued by a college library in one day were: literature 12, geography 15, history 12, science 33, hobbies 27, sport 21, others 60. Show this information as a pie chart, and state the angle used for each sector.

36 Mean, median and mode

In eleven tests a student obtained the following marks:

$$8, 4, 5, 2, 10, 4, 4, 10, 7, 4, 8$$

The *mean* of the marks is $\dfrac{\text{sum of marks}}{\text{number of tests}} = \dfrac{66}{11} = 6$.

Arranging the marks in order of size, we have:

$$2, 4, 4, 4, 4, \text{⑤}, 7, 8, 8, 10, 10.$$

The *median* is the middle mark (ringed) in this ordered list. It is 5.
The *mode* is the most frequent mark. It is 4.

For the median of an even number of tests, we take the middle two marks and find their mean. For example, the median of:

$$13, 14, 17, 20, 24, 27$$

is $\frac{1}{2}(17 + 20) = \frac{1}{2} \times 37 = 18\frac{1}{2}$.

If we know the mean and the number of quantities, the sum of the quantities can be found by multiplying the mean by the number of quantities.

Example   The mean mass of a boat crew of eight men is 69.3 kg.
A man of mass 67.9 kg is replaced by one of mass 72.7 kg.
Find the new mean mass.
The total mass of the original crew is 69.3 kg $\times$ 8 = 554.4 kg.
The mass when one man is removed is
554.4 kg $-$ 67.9 kg = 486.5 kg.
The mass when the new man is then added is
486.5 kg + 72.7 kg = 559.2 kg.
The new mean is $\dfrac{559.2}{8} = 69.9$ kg.

Use of a working origin to calculate a mean

If the quantities do not differ much, the working can be simplified by using a *working origin*, as in the next example.

Example   The heights of ten students are:
157, 152, 160, 145, 162, 148, 159, 143, 163, 151 cm.
Find the mean height.
The heights are scattered about 150 cm. They exceed 150 cm
by 7, 2, 10, -5, 12, -2, 9, -7, 13, 1 cm.

(Notice that for those below 150 cm, the excesses are negative.)

The mean of these excesses is $\dfrac{\text{total}}{10} = \dfrac{54-14}{10} = \dfrac{40}{10} = 4$ cm.

The mean height is therefore $150 + 4 = 154$ cm.

Here we have used a working origin of 150 cm.

When choosing a working origin, there are two alternatives.

1. We can choose a number which is a rough estimate of the mean. For example, we might look at a list of students' ages and choose a working origin of 17 years.

2. We can choose a number which allows differences to be calculated easily. For example, for 1035, 1027, 996, 1003, ... we would choose 1000.

Exercise 36.1

Questions 1–12 are on means.

1. Find the means of these sets of numbers:

    (i) 3, 7, 5, 7, 8   (ii) 6, 7, 14, 10, 19, 16

    (iii) 1, 0, 4, 0, 6, 7   (iv) 0.9, 1.3, 1.6, 1.2, 1.5, 1.7, 1.6.

2. Twelve women and six men took a typing course. At one stage in the course the number of words typed per minute were

    women: 12, 19, 13, 17, 20, 10, 12, 22, 13, 21, 14, 19
    men: 9, 13, 14, 16, 10, 16.

    Calculate the mean score for: (i) the women (ii) the men (iii) the full group of 18 students.

3. The ages of the members of a family are 38, 35, 14, 12, 10 and 5. Find the mean age: (i) now (ii) 3 years ago (iii) in 5 years' time.

4. In a novel of 237 pages, the mean number of words per page is 230. Find the number of words in the novel, to the nearest thousand.

5. On a match box is printed 'Average contents 95'. How many matches would you expect to find in six such boxes?

6. The mean of four numbers is 19. What is the sum of the four numbers? Three of them are 21, 25 and 16. What is the fourth?

7. Find x if the mean of 8, 10, x, 12 and 9 is 11.

8. The mean weekly wage of 21 office clerks was £98 and that of 29 apprentices was £84. Calculate the mean wage for these 50 people.

9. The mean age of ten members of a football team is 24.3 years. (The goalkeeper has been left out.) Players aged 31 and 28 are replaced by players aged 19 and 18. Find the new mean age for the ten players.

10. The mean mass of 28 pupils in a class was 47.5 kg. Twins joined the class. Their masses were 45.8 kg and 46.2 kg. What was the new mean mass of the class?

11. The mean height of 10 men is 179 cm and the mean height of 20 women is 173 cm. Calculate the mean height for the 30 people.

12. The mean of five numbers is 7. A sixth number is included, and this changes the mean to 8. Find the sixth number.

Questions 13–18 are on medians.

13. State the median for each of these sets of scores:
    (i) 2, 5, 6, 8, 8   (ii) 15, 16, 18, 19, 24, 24, 25
    (iii) 0, 5, 10, 10, 20, 20, 25, 30   (iv) 0.3, 0.3, 0.4, 0.6, 0.7, 0.7.

14. Arrange the following in order of size and state the median:
    (i) 7, 4, 8, 3, 6   (ii) 8, 13, 11, 15, 16
    (iii) 7, 1, 2, 3, 0, 6   (iv) 0.4, 0.5, 0.3, 0.4, 0.3, 0.4.

15. Find the median mass of seven students whose masses are 61, 69, 60, 63, 66, 72 and 65 kg.

16. The heights of six people are 152, 154, 160, 149, 158 and 157 cm. Find the median height.

17. On a cycling tour, a girl cycled the following distances on five days: 102, 82, 115, 48, 93 km. (i) Find the median distance. (ii) Find the mean distance.

18. The scores of eight people playing darts were 25, 10, 160, 27, 42, 12, 6 and 38. Find: (i) the median score (ii) the mean score.

Questions 19–23 are on modes.

19. State the mode for each of the following sets of marks:
    (i) 4, 6, 7, 9, 9, 9, 10   (ii) 3, 3, 4, 5, 5, 5, 6, 6.

20. State the mode for each of the following sets of marks:
    (i) 7, 3, 6, 3, 7, 3   (ii) 5, 0, 4, 3, 2, 7, 0, 3, 3, 5.

21. State the mode for each of the following sets of scores:
    (i) 3, 7, 9, 2, 5, 3, 7, 5, 7, 1, 7, 5, 7
    (ii) 4, 8, 10, 6, 8, 4, 6, 7, 4, 4, 3, 5, 4.

22. The shoe sizes of ten children were 7, 6, 9, 5, 8, 6, 8, 8, 5 and 8. Find the mode and calculate the mean.

23. A football team scored the following numbers of goals in ten matches: 2, 1, 5, 2, 2, 1, 0, 2, 3, 1. State the mode and calculate the mean.

The remaining questions are on the use of working origins.

24. The ages at which five men retired were 66, 63, 61, 61 and 64. Make a list of the numbers of years by which these ages exceed 60. Find the mean of these excesses, and add it on to 60 to obtain the mean of the original ages.

25. Find the means of the following sets of numbers:

    (i) 79, 72, 75, 74, 80, using a working origin of 70

    (ii) 267, 261, 262, 264, 263, 265, 269, 261, using a working origin of 260.

26. The numbers of cassettes sold at a shop in each of six months were 202, 196, 208, 204, 190 and 212. Make a list of the amounts by which these numbers exceed 200. (Note that 196 exceeds 200 by -4.) Calculate the mean of these excesses and so find the mean of the original numbers.

27. Find the means of the following sets of numbers:

    (i) 27, 33, 29, 35, 30, 32, using a working origin of 30.

    (ii) 98, 102, 104, 95, 90, 88, 107, 96, using a working origin of 100.

28. Over a period of 10 years the numbers of trees planted in a forest were 1710, 1980, 2120, 2170, 1830, 2000, 2130, 1940, 2250 and 1670. Choose a working origin and calculate the mean.

37 Frequency distributions

Frequency

The grades obtained by 40 students in an examination were:

3 3 4 5 4 5 3 4 4 2 4 3 2 4 1 6 4 4 3 5
5 3 2 3 1 2 6 6 5 5 2 4 5 3 4 5 3 5 4 4.

From this list we can form the following combined tally chart and frequency table:

Grade	Tallies	Frequency
1	11	2
2	~~1111~~	5
3	~~1111~~ 1111	9
4	~~1111~~ ~~1111~~ 11	12
5	~~1111~~ 1111	9
6	111	3

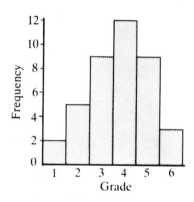

Fig. 37.1

The table shows that 2 students obtained grade 1, 5 obtained grade 2, and so on. The numbers 2, 5, 9, 12, 9 and 3 are called *frequencies*. Grade 1 occurs with a frequency of 2, grade 2 with a frequency of 5, and so on.

Figure 37.1 shows a frequency chart representing this information. The height of each column represents the frequency.

Grade 4 has the greatest frequency, so the mode is 4.

Grouped data

When a variable has a large number of different values, we often place the values in groups or *classes*. Here are three examples
  (i) The ages of workers in a factory might be placed in the classes 16 to 20 years, 21 to 30 years, 31 to 40 years and so on. A frequency table would show the number of workers in each age group. Such a table is shown below.

Age in years (x)	16–20	21–30	31–40	41–50	51–60	61–64
Frequency (f)	35	100	140	180	90	24

 (ii) Examination marks might be placed in the classes 1 to 10, 11 to 20, 21 to 30 and so on.
(iii) The heights of some students might be measured to the nearest centimetre and then placed in the classes 146 to 149 cm, 150 to 153 cm and so on.

Class boundaries and class intervals

When a height is given as 146 cm to the nearest centimetre, it lies between 145.5 cm and 146.5 cm. Therefore a height in the 146 to 149 cm class lies between 145.5 cm and 149.5 cm. These are the *boundaries* of the 146 to 149 cm class. The difference between the two boundaries is 4 cm as shown in Fig. 37.2. We say that the *class interval* is 4 cm.

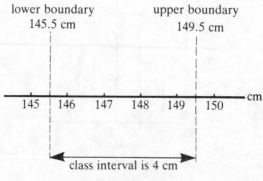

Fig. 37.2

Care must be taken with grouped ages. The 16 to 20 years class contains all workers who have passed their 16th birthdays but have not yet reached their 21st birthdays. The boundaries of this class are 16 years and 21 years and the class interval is 5 years as shown in Fig. 37.3. Similarly, for the 21 to 30 years class the boundaries are 21 years and 31 years and the class interval is 10 years.

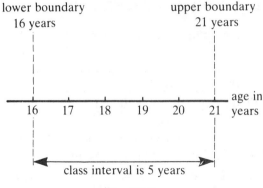

Fig. 37.3

Histograms

Figure 37.4 shows a *histogram* representing the information about the workers' ages, from the frequency table above. A histogram looks like a bar chart, but in a histogram each frequency is represented by an *area*. In Fig. 37.4, one small square represents one person. The height of each column was found by dividing the area needed by the base width.

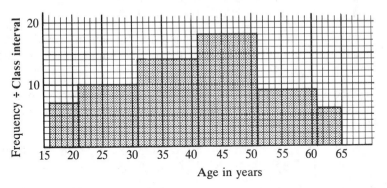

Fig. 37.4

For example, the first group in the table is represented by a column 5 units wide (ages from 16 to 20). This group contains 35 people so the column must have an area of 35 square units. Therefore its height must be $35 \div 5 = 7$ units. The second group is represented by a column 10 units wide (ages from 21 to 30). This group contains 100 people so the column must have an area of 100 square units. Therefore its height must be $100 \div 10 = 10$ units.

For the distribution above we have a *modal class* instead of a mode. The modal class is the class 41–50 years, since this has the greatest frequency.

Median for a frequency distribution

Consider the following distribution of marks obtained by 40 students in a test:

Mark	1	2	3	4	5	6	Total
Frequency	4	8	10	13	4	1	40

The median mark is the average of the 20th. and 21st. marks, when the 40 marks are arranged in order of size like this:

$$\underbrace{1,1,1,1}_{4},\underbrace{2,2,2,2,2,2,2,2}_{8},\underbrace{3,3,3,\ldots\ldots 3,3}_{10},4,4,4,\ldots\ldots$$

The total number of students obtaining 1 or 2 marks is 12 and the total obtaining 1, 2 or 3 marks is 22. Hence the 20th. and the 21st. marks are both 3. The median mark is thus 3.

Exercise 37.1

1. The following is a list of the numbers of goals scored by 30 football teams one Saturday:

6 1 2 3 2 1 4 6 1 1 0 1 2 0 3
1 4 3 2 0 2 1 2 0 3 1 4 2 2 1.

Prepare a tally chart and frequency table for this information, and then draw a frequency chart. State the mode. How many teams scored more than 3 goals?

2. A shoe firm questioned 40 people about the numbers of pairs of shoes their families had bought over the previous twelve months. Here are the replies:

5 9 6 6 7 5 7 6 9 4 8 6 5 8 4 7 6 8 4 9
7 6 6 5 7 5 10 5 6 7 6 8 5 6 8 5 6 4 7 8.

Prepare a tally chart and frequency table for this information, and then draw a frequency chart. State the mode.

3. Find the median and mode for each of the following distributions:
   (i)

Grade	1	2	3	4	5	6
Frequency	1	3	4	6	9	2

   (ii)

Age	14	15	16	17	18
Frequency	4	13	10	8	5

4. This table shows the ages of the buses belonging to a certain bus
   company:

Age in years	0–1	1–3	3–5	5–9
Frequency	12	36	20	16

The class 1–3 is for buses aged over 1 year but under 3 years. Thus
the class interval is 2 years. State the intervals for the other classes.
Draw a histogram. What is the modal class?

5. This table shows the masses of parcels posted in one day by a certain
   firm:

Mass in kilograms	Less than 1	1–2	2–4	4–8	8–16
Frequency	20	36	12	8	4

The class interval 4–8 contains masses of 4 kg or more but less than
8 kg, so the interval is 4 kg. State the modal class and draw a histogram
for the information.

6. The heights of some plants were measured and recorded correct to
   the nearest centimetre. The results were grouped to give this table:

Height of plant (to nearest cm)	15–20	21–23	24–26	27–32	33–38
Frequency	8	9	11	11	4

The class 15–20 contains plants from $14\frac{1}{2}$ to $20\frac{1}{2}$ cm, and the class
interval is 6 cm. State the intervals for the other classes. Draw a
histogram for the information.

Means of frequency distributions

Fifty people were asked to put a coin in a 'silver' collection box. Six
people gave 5p, twelve gave 10p, seventeen gave 20p and fifteen gave
50p. The total sum in the box was:

$6 \times 5p + 12 \times 10p + 17 \times 20p + 15 \times 50p$
$$= 30p + 120p + 340p + 750p = 1240p$$

The mean amount given $= \dfrac{\text{total sum}}{\text{no. of people}} = \dfrac{1240p}{50} = 24.8p$

It is better to set out the working as follows:

Value of coin in pence (x)	Frequency (f)	Coin value × frequency (xf)
5	6	30
10	12	120
20	17	340
50	15	750
Total	50	1240

$$\text{Mean} = \frac{1240\text{p}}{50} = 24.8\text{p}$$

Notice that:

$$\text{mean} = \frac{\text{sum of (coin value} \times \text{frequency) column}}{\text{sum of frequency column}} = \frac{\text{sum of } xf}{\text{sum of } f}.$$

This can be written as: mean $= \dfrac{\Sigma xf}{\Sigma f}$ where Σ stands for 'sum of'.

The mean of any frequency distribution can be found in this way. For grouped data, we use the centre value of each class as the value of x.

Example   Find an estimate of the mean of the ages of the factory workers on page 318.

From Fig. 37.3 we see that the centre of the 16–20 class is $18\frac{1}{2}$. We take this as the value of x.

Age in years	Centre of interval (x)	Frequency (f)	xf
16–20	$18\frac{1}{2}$	35	647.5
21–30	26	100	2600
31–40	36	140	5040
41–50	46	180	8280
51–60	56	90	5040
61–64	63	24	1512
Total		569	23119.5

$$\text{Mean} = \frac{\Sigma xf}{\Sigma f} = \frac{23119.5}{569} = 40.6 \text{ years to 3 s.f.}$$

Note: since the actual ages have not been used, this is just an estimate of the mean.

Exercise 37.2

1. This table shows the coins in a flag-day collection box:

Coin value	5p	10p	20p	50p
Frequency	12	16	10	2

Calculate: (i) the total value of the coins
(ii) the mean value per coin.

2. The marks obtained by 20 students in a test were:

Mark	1	2	3	4	5
Frequency	2	4	7	6	1

Calculate: (i) the total marks obtained by the students
(ii) the mean mark per student.

3. One Saturday the goals scored by 30 football teams were as follows: 6 scored 0, 9 scored 1, 6 scored 2, 5 scored 3, 2 scored 4 and 2 scored 5. State the mode and calculate the mean number of goals per team.

4. The number of people in each of 50 cars stopping at a road junction was recorded. The results are shown in this table.

Number of people per car	1	2	3	4	5
Frequency	17	14	10	7	2

(i) State the modal number of people per car.
(ii) Find the median number of people per car.
(iii) Calculate the mean number of people per car.

5. The number of flowers on each of 50 plants of the same variety were counted and the results were:

Number of flowers per plant	3	4	5	6	7	8	9
Frequency	2	4	9	12	15	6	2

(i) State the modal number of flowers per plant.
(ii) Find the median number of flowers per plant.
(iii) Calculate the mean number of flowers per plant.

6. The times taken by 20 students to run a cross-country course were noted to the nearest minute and the following table was obtained:

Time (minutes)	12–14	15–17	18–20	21–23
Frequency (no. of students)	3	5	8	4

Using the centres of the time classes, calculate an estimate of the mean time. (The centre of the 12–14 class is 13.)

7. 30 parcels were weighed and gave the following table:

Mass in kilograms	Under 2	2–4	4–6	6–8
Frequency	6	11	8	5

Using the centres of the mass classes, calculate an estimate of the mean mass. (The centre of the 2–4 class is 3.)

8. 40 pupils were asked how far they lived from their school. The answers are shown in this table:

Distance in kilometres	Less than 1	1–3	3–5	5–9
Frequency	4	13	17	6

Calculate an estimate of the mean distance the pupils lived from school.

9. A survey was taken of the number of cars passing a road junction during 35 equal intervals of time.

The results were recorded in the following table. For example 4 cars passed the junction during each of 6 intervals.

Number of cars	0	1	2	3	4	5	6
Frequency	8	7	4	4	6	3	3

Find, for this distribution:

(i) the mode,   (ii) the median,   (iii) the mean.   (C)

38 Dispersion

Range

Here are the marks obtained by Mary and Pete in some tests.

   Mary   5, 5, 5, 5, 6, 6, 7, 7   Pete   1, 3, 3, 5, 6, 6, 8, 9

Both have the same median mark of $5\frac{1}{2}$, but Pete's marks are widely spread whereas Mary's are closely packed about the median.

   One way of measuring the dispersion or spread is to use the range. This is the difference between the lowest and highest marks. Mary's range is 2 and Pete's is 8.

Inter-quartile range

Ann's marks in the same tests were 1, 5, 5, 5, 6, 6, 7, 7. Her range is 6, although her marks are the same as Mary's except for the 1. This shows that a freak high or low value makes a lot of difference to the range. To avoid this, we can use the inter-quartile range.

   The median divides an ordered list of numbers into two equal parts. The quartiles divide a list into four quarters — four equal parts. (The middle quartile is the median.) For Pete's marks we have

1		3	↑	3		5	↑	6		6	↑	8		9
			lower				median				upper			
			quartile				$5\frac{1}{2}$				quartile			
			3								7			

The inter-quartile range is (upper quartile) − (lower quartile). For Pete's marks it is $7 - 3 = 4$. For Mary's marks it is $6\frac{1}{2} - 5 = 1\frac{1}{2}$. It is the same for Ann's marks.

Exercise 38.1

1. For each of the following sets of numbers, state:
   - (i) the range (highest value − lowest value)   (ii) the median
   - (iii) the upper quartile, the lower quartile and the inter-quartile range.
     - (a) 4, 7, 9, 11, 11, 12, 14, 16
     - (b) 2, 3, 5, 5, 7, 8, 8, 8, 9, 9, 9, 10
     - (c) 3, 5, 8, 11, 17, 22, 25, 30

2. Arrange the following sets of numbers in order of size. Then state the median, the lower quartile, the upper quartile and the inter-quartile range.
     - (a) 7, 10, 2, 5, 9, 12, 14, 9
     - (b) 6, 9, 4, 4, 8, 6, 4, 8
     - (c) 5, 10, 8, 8, 2, 9, 5, 3, 8, 9, 9, 7

3. Compare the hours of sunshine for eight days at two resorts
   (i) by the range between the highest and lowest
   (ii) by the inter-quartile range

   Sunville   8.6, 6.4, 12.3, 10.7, 9.5, 7.2, 8.8, 8.2
   Rainton   2.4, 0, 0, 5.6, 3.2, 9.8, 1.6, 0.4

4. Compare the average monthly temperatures of two towns by finding
   the inter-quartile ranges

   Exton   4, 9, 11, 15, 18, 20, 26, 23, 20, 18, 10, 6
               degrees Celsius

   Wyebridge   12, 12, 14, 16, 17, 18, 19, 17, 16, 14, 13, 12
               degrees Celsius

Cumulative frequency

The masses of 160 men were measured and recorded correct to the
nearest kilogram. The results were grouped to give the table on the left.

Mass (kg)	Frequency
55–58	2
59–62	11
63–66	19
67–70	36
71–74	42
75–78	31
79–82	13
83–86	6

Mass (kg)	Cumulative frequency
Less than $58\frac{1}{2}$	2
Less than $62\frac{1}{2}$	13
Less than $66\frac{1}{2}$	32
Less than $70\frac{1}{2}$	68
Less than $74\frac{1}{2}$	110
Less than $78\frac{1}{2}$	141
Less than $82\frac{1}{2}$	154
Less than $86\frac{1}{2}$	160

From the table, we find that 2 men had masses less than $58\frac{1}{2}$ kg,
$2 + 11 = 13$ had masses less than $62\frac{1}{2}$ kg, $13 + 19 = 32$ had masses less
than $66\frac{1}{2}$ kg, and so on. These *running totals* or *cumulative frequencies*
are shown in the table on the right. They were used to plot the points for
the *cumulative frequency curve* or *ogive* in Fig. 38.1. This curve can be
used to estimate the median mass and the interquartile range of the
distribution.

The quartiles divide the 160 masses into four groups of 40. They are the
masses corresponding to 40, 80 and 120 on the cumulative frequency
axis and are $67\frac{1}{2}$ kg, $71\frac{1}{2}$ kg and $75\frac{1}{2}$ kg. So the median mass (the middle
quartile) is $71\frac{1}{2}$ kg and the inter-quartile range is $75\frac{1}{2} - 67\frac{1}{2} = 8$ kg.

Suppose that we wish to know how many men had masses greater
than 80 kg. 80 kg on the mass axis corresponds to 148 on the cumulative

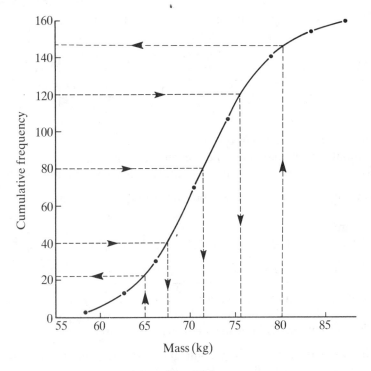

Fig. 38.1

frequency axis. 148 men have masses less than 80 kg and so 12 have masses more than 80 kg.

Note. The results obtained from the graph are only estimates. For accurate answers we need the complete list of 160 masses.

Deciles and percentiles

The deciles divide an ordered list of numbers into 10 equal parts; the percentiles divide it into 100 equal parts. The lower quartile is the 25th percentile.

Exercise 38.2

1. The table shows the distribution of the marks of 132 candidates in an examination.

Marks	0–19	20–39	40–59	60–79	80–99
Frequency	12	34	58	20	8

Draw up a cumulative frequency table and draw the curve.
Estimate the median mark and the inter-quartile range.
How many scored less than 30? How many scored 70 or more?

2. The masses of packets of washing powder gave this frequency distribution.

Mass (grams)	460–470	470–480	480–490	490–500	500–510	510–520	520–530	530–540	540–550
Frequency	2	4	23	40	122	82	17	7	3

Draw up a cumulative frequency table and draw the curve.
Estimate the median mass and the inter-quartile range.
Also estimate the percentage of packets with a mass less than 495 g.

3. The lengths of the journeys of 600 motorists gave this table.

Length (km)	0–5	5–10	10–15	15–20	20–25	25–30	30–35	35–40
Frequency	42	60	125	130	108	74	49	12

Draw up a cumulative frequency table and draw the curve.
Estimate the median journey and the inter-quartile range.
Also estimate the 3rd decile and the 65th percentile.

4. The times of arrival of a certain inter-city train on 200 days were recorded and gave this frequency distribution.

Time	11.17–11.20	11.20–11.23	11.23–11.26	11.26–11.29	11.29–11.32	11.32–11.35	11.35–11.38
Frequency	4	14	49	69	43	17	4

Draw up a cumulative frequency table and draw the curve.
Estimate: (i) the median time   (ii) the number of occasions the train was early, given that its scheduled time of arrival was 11.25   (iii) the number of occasions it was more than 5 minutes late.

5. Eighty women and eighty men estimated the mass of a cake.

Mass (grams)	500–600	600–700	700–800	800–900	900–1000	1000–1100	1100–1200
No. of women	0	6	15	22	24	11	2
No. of men	4	8	11	17	20	15	5

Draw the two cumulative frequency curves in one diagram. State the median estimate for the women and for the men. Use the inter-quartile range to compare their estimates.
The cake had a mass of 930 g. What percentage of women and what percentage of men overestimated the mass?

39   Probability

If a pack of 52 playing cards is well shuffled and one is drawn out, each of the 52 cards has an equal chance of being drawn. The probability of

drawing a heart is $\dfrac{\text{number of hearts in pack}}{\text{number of cards in pack}} = \dfrac{13}{52} = \dfrac{1}{4}$.

The drawing of a card is an example of an 'experiment' or *trial*.
The result is called an *outcome*. There are 52 possible outcomes and each is *equiprobable* (or equally likely).
Drawing a heart is called an *event*. This event occurs in 13 of the possible outcomes. In general:

the probability of an event

$$= \frac{\text{no. of different outcomes in which the event occurs}}{\text{total no. of possible outcomes}}$$

p(A) is used for the probability of event A.

When spinning a coin, $p(\text{head}) = \dfrac{1}{2}$.

When throwing a die, $p(\text{six}) = \dfrac{1}{6}$.

When drawing a card, $p(\text{king}) = \dfrac{4}{52} = \dfrac{1}{13}$.

If an event is impossible, its probability is 0.
If an event is certain to occur, its probability is 1.
If x is the probability of an event happening, then the probability of it not happening is $1 - x$. For, example, when a die is thrown, p (six) =

$\dfrac{1}{6}$ so p (not six) $= 1 - \dfrac{1}{6} = \dfrac{5}{6}$.

If x is the probability of an event happening, then in n trials, we should expect the event to happen nx times. For example, in 300 throws of a

die, the expected number of sixes is $300 \times \dfrac{1}{6} = 50$. (This is a theoretical

number. If a die is actually thrown 300 times, the number of sixes will be close to 50, for example 48 or 53.)

Use of sets

It sometimes helps to use sets, to make the working clear in probability problems.

Example The nine numbers 11 to 19 are written on pieces of paper and placed in a hat. One number is drawn out at random. Find: (i) the probability that it is a prime number (ii) the probability that it is a multiple of 3.

Let the set of possible outcomes be ξ.

Then $\xi = \{11, 12, \ldots, 19\}$ and $n(\xi) = 9$.

(i) Let the set of prime numbers be A.

Then $A = \{11, 13, 17, 19\}$ so $n(A) = 4$.

$$p(\text{prime number}) = \frac{n(A)}{n(\xi)} = \frac{4}{9}$$

(ii) Let the set of multiples of 3 be B.

Then $B = \{12, 15, 18\}$ so $n(B) = 3$.

$$p(\text{multiple of 3}) = \frac{n(B)}{n(\xi)} = \frac{3}{9} = \frac{1}{3}$$

Exercise 39.1

1. A die is thrown. State the probability that the outcome is:

   (i) 1 (ii) an odd number (iii) a number greater than 4 (iv) 7.

2. Nine cards numbered 1 to 9 are placed in a box, and one is drawn out at random. State the probability that the drawn card has:

   (i) an odd number (ii) an even number (iii) 1, 2 or 3 (iv) a number less than 5.

3. A box contains 10 black pens, 9 red pens and 6 pencils. One is taken out at random. What is the probability that it is: (i) a black pen (ii) a red pen (iii) a pencil?

4. A letter is chosen at random from the word PROBABILITY. Find:

   (i) p(vowel) (ii) p(consonant) (iii) p(A or B) (iv) p(Z).

5. In a box of 40 tickets, 20 are red, 15 are white and 5 are blue. A ticket is taken at random. Find:

   (i) p(red) (ii) p(white) (iii) p(not white) (iv) p(blue) (v) p(red or white) (vi) p(yellow).

6. From a well-shuffled pack of 52 playing cards, one card is drawn. Find the probability that the drawn card is:

   (i) a spade (ii) not a spade (iii) an ace (iv) not an ace (v) the king of hearts (vi) a red card (vii) a 9 or 10 (viii) a king, queen or jack.

7. A die is thrown 120 times. How many times would you expect:

   (i) a 6 (ii) an odd number (iii) 1 or 2?

8. If you took a card 60 times from a shuffled pack (returning the card and reshuffling the pack each time), how many times would you expect not to get a heart?

9. Assuming that the chance of being born in January is $\frac{1}{12}$, how many pupils in a school of 600 are likely to have birthdays in January?

10. At a snack bar, the estimated probabilities of a customer asking for tea, coffee or a soft drink are: $p(tea) = \frac{5}{12}$, $p(coffee) = \frac{1}{3}$, $p(soft\ drink) = \frac{1}{4}$. Estimate the likely number of each drink ordered by 600 customers.

11. A box contains 20 beads, some red and the rest green. If one bead is taken out, the probability of it being red is 0.3. How many of each colour are there?

12. A spinner has the four numbers 1, 2, 3 and 4. It is spun twice and the scores are added together. Copy and complete the table showing the possible sums of the two scores. There are 16 possible outcomes. The event 'sum is 6' occurs in three of these outcomes. Hence $p(6) = \frac{3}{16}$. State: (i) p(3) (ii) p(5) (iii) p(8) (iv) p(9).

2nd. spin

+	1	2	3	4
1	2	3	4	5
2	3	4	6	7
3	4	5	6	7
4	5	6	7	8

1st. spin

13. The spinner of question 12 is spun twice and the scores are multiplied together. Draw up a table showing the possible outcomes. State: (i) p(4) (ii) p(5) (iii) p(1, 2 or 3) (iv) p(odd number).

14. Two coins are tossed. Copy and complete the table of possible outcomes. Find: (i) p(two heads) (ii) p(one head) (iii) p(no heads).

2nd. coin

	H	T
H	HH	
T		

1st. coin

15. A die is thrown twice and the scores are added. Prepare a table showing the 36 possible outcomes.
Find: (i) p(7) (ii) p(10) (iii) p(not 5) (iv) p(at most 6).

16. Three dice are thrown together. Explain why there are 216 possible outcomes. Write down all the different ways of obtaining a total of 17 (for example $6 + 6 + 5$). What is the probability of this total?

Mutually exclusive events

Events A and B are *mutually exclusive* if they cannot both happen at the same time. For example, in drawing one card from a pack, the events 'a king is drawn' and 'an even number is drawn' are mutually exclusive. However, the events 'a king is drawn' and 'a heart is drawn' are not mutually exclusive, because the king of hearts might be drawn.

There are 4 kings in a pack, so $p(\text{king}) = \dfrac{4}{52}$.

There are 5 even numbered cards of each suit, so $p(\text{even}) = \dfrac{20}{52}$.

For p(king or even number) there are 24 suitable cards, so $p(\text{king or even number}) = \dfrac{24}{52}$.

Notice that p(king or even number) = p(king) + p(even number).

There are 13 hearts in a pack, so $p(\text{heart}) = \dfrac{13}{52}$.

For p(heart or king) there are 16 suitable cards (13 hearts and 3 non-heart kings), so $p(\text{heart or king}) = \dfrac{16}{52}$.

Notice that p(heart or king) $\neq$ p(heart) + p(king).
In general, when events A and B are mutually exclusive we use the *sum rule*:

$$p(A \text{ or } B) = p(A) + p(B)$$

Independent events

Events C and D are *independent* if one can happen whether or not the other has happened. For example, on spinning a coin and throwing a die, the events 'head' and '6' are independent.

To find the probability that both C and D occur we use the *product rule*:

$$p(C \text{ and } D) = p(C) \times p(D)$$

For a coin and a die, $p(\text{head}) = \dfrac{1}{2}$ and $p(\text{multiple of 3}) = \dfrac{2}{6}$

so p(head and multiple of 3) $= \frac{1}{2} \times \frac{2}{6} = \frac{1}{6}$.

The table shows the 12 possible outcomes and the two which give a head and a multiple of 3.

	1	2	3	4	5	6
H			√			√
T						

Using both rules

Example   A red box contains 8 milk and 4 plain chocolates; a blue box contains 4 milk and 5 plain chocolates. Joyce chooses a box and then takes a chocolate from it, without looking. The probability that she chooses the red box is $\frac{2}{3}$. Find the probability that she takes a milk chocolate.

Let R and B denote the red box and the blue box respectively.

Let MR denote a milk chocolate from the red box and MB a milk chocolate from the blue.

$$p(R \text{ and } MR) = p(R) \times p(MR) = \frac{2}{3} \times \frac{8}{12} = \frac{4}{9}$$

$$p(B \text{ and } MB) = p(B) \times p(MB) = \frac{1}{3} \times \frac{4}{9} = \frac{4}{27}$$

$$p(\text{milk chocolate}) = p(R \text{ and } MR \text{ or } B \text{ and } MB)$$

$$= \frac{4}{9} + \frac{4}{27} = \frac{16}{27}.$$

Exercise 39.2

1. A die is thrown. State: (i) p(odd number)
(ii) p(6)
(iii) p(odd number or 6).

2. In a 100 m race, the probability that Ann wins is $\frac{1}{3}$ and the probability that Betty wins is $\frac{1}{5}$. Find the probability that Ann or Betty wins.

3. A card is drawn from a pack and a coin is spun. State the probability of obtaining: (i) a heart (ii) a head (iii) a heart and a head.

4. A card is drawn from a pack. State the probability of drawing: (i) a queen (ii) a number greater than 8 (iii) a queen or a number greater than 8.

5. A die is thrown and a card is drawn. Find the probability of obtaining a 6 and a spade.

6. A card is drawn from a pack. State the probability of obtaining: (i) a diamond (ii) an ace (iii) a diamond or an ace. Explain why p(diamond) + p(ace) ≠ p(diamond or ace).

7. A box contains 40 red, 30 green and 30 blue beads. One is drawn out. Express as a decimal: (i) p(red) (ii) p(blue) (iii) p(red or blue).

8. A die is thrown twice. Find p(6, 6).

9. A number is chosen from the set {1, 2, 3, 4, ... , 9}. State: (i) p(even number) (ii) p(square number). Explain why p(even or square number) ≠ p(even number) + p(square number). By listing the suitable numbers, find p(even or square number).

10. A green die and a blue die are thrown together. Find the probability of obtaining: (i) 6 on the green die and 6 on the blue (ii) 6 on the green and not 6 on the blue (iii) not 6 on the green and 6 on the blue (iv) one 6 only. (Note that (ii) and (iii) are alternative ways for this.)

11. One card is drawn from each of two packs. Find the probability of drawing: (i) a heart from the first pack and a diamond from the second (ii) a diamond from the first and a heart from the second (iii) a heart and a diamond in either order.

12. A die is thrown twice. Find: (i) p(6 on both throws) (ii) p(5 on both throws) (iii) p(same number on both throws).

13. If a person is chosen at random, his birthday this year is equally likely to fall on any day of the week.
    (i)   State the probability that it falls on a Sunday.
    (ii) For two people, state the probability that both have birthdays on a Sunday.
    (iii) For two people, state the probability that both have birthdays on the same day of the week.

14. A 'lucky' wheel has 3 blue, 5 green and 2 red sectors. For a win, it must stop at a red sector. A player spins the wheel twice. Find the probability that
    (i)   he wins both times
    (ii) he wins exactly once
    (iii) he wins at least once.

15. In an examination room, 68 candidates were taking economics, 40 were taking physics and 36 were taking music. Previous years' results indicated that 75% would pass economics, 65% would pass physics and 50% would pass music.

(i) If a candidate had been picked at random, find the probability that: (a) he took economics, (b) he took music and passed, (c) he passed his examination whatever subject he took, (d) he took physics and failed.

(ii) Find how many candidates were expected to fail their examination.   (L)

Tree diagrams

Tree diagrams are helpful for more complicated questions.

Example   A box contains 6 red and 4 green counters. One counter is drawn out and replaced. A second is then drawn out. Fig. 39.1 shows the results of the two trials. R indicates a red counter and G a green counter. The probability of each event is marked on the branch. At each draw, $p(R) = \dfrac{6}{10} = \dfrac{3}{5}$ and $p(G) = \dfrac{4}{10} = \dfrac{2}{5}$.

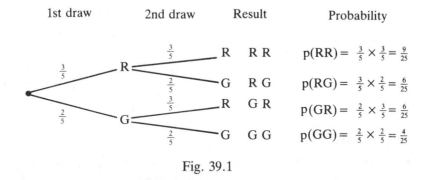

Fig. 39.1

For the probability of R and R we use the product rule:

$$p(RR) = \frac{3}{5} \times \frac{3}{5} = \frac{9}{25}.$$

Similarly for R and then G, $p(RG) = \dfrac{3}{5} \times \dfrac{2}{5} = \dfrac{6}{25}$

and for G and then R, $p(GR) = \dfrac{2}{5} \times \dfrac{3}{5} = \dfrac{6}{25}$

For one of each colour, we can have RG or GR. We use the sum rule.

$$p(RG \text{ or } GR) = p(RG) + p(GR) = \frac{6}{25} + \frac{6}{25} = \frac{12}{25}.$$

As a check we can use the fact that the probabilities of all the possible results should add up to 1. For the tree diagram above:

$$p(RR) + p(RG) + p(GR) + p(GG) = \frac{9}{25} + \frac{6}{25} + \frac{6}{25} + \frac{4}{25} = \frac{25}{25} = 1.$$

Dependent events

Sometimes the probability of an event D depends on what has happened in another event C. Suppose that in the example above the first counter drawn is not replaced. If that counter is red, there remain 5 red counters and 4 green. So for the second draw $p(R) = \frac{5}{9}$. But if a green counter is drawn first, there remain 6 red counters and 3 green. So for the second draw $p(R) = \frac{6}{9}$. The tree is now as below:

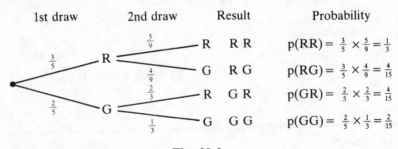

Fig. 39.2

Exercise 39.3

1. For the throw of a die, let A denote '6' and B denote 'not 6'. Copy and complete the tree diagram below for two throws.

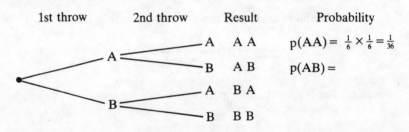

Fig. 39.3

State: (i) p(6, 6)   (ii) p(only one 6)   (iii) p(no 6).

2. A card is drawn from a pack and then replaced. After shuffling, a second card is drawn. Use a probability tree (tree diagram) to find:
   (i) p(two hearts)   (ii) p(one heart)   (iii) p(no hearts).

3. If the chance of a fine day is $\frac{1}{5}$, find, by using a probability tree, the probability of: (i) two fine days in succession (ii) one fine day out of two days (iii) two consecutive days that are not fine.

4. A box contains 5 red and 3 white counters. One is taken out and replaced and then a second draw is made. Prepare a probability tree for the results RR, RW, WR and WW. Find the probability of obtaining: (i) two red counters   (ii) one of each colour   (iii) two white counters.

5. Find the answers to question 4, if the first counter drawn is not replaced.

6. A bag contains 7 red fruit drops and 3 yellow ones. One fruit drop is drawn out and eaten, and then another is taken out. Draw a tree diagram showing the possible results. State the probability of obtaining: (i) two red fruit drops   (ii) one of each colour   (iii) two yellow fruit drops.

7. A soccer player rates his chances of scoring a goal in a match as $\frac{2}{5}$. Find the probability that in his next two matches:
   (i) he scores a goal in each   (ii) he scores a goal in only one
   (iii) he does not score in either match.

8. Six cards are placed in a box. Two have crosses on them and the others are plain. One card is drawn out and not replaced. Then a second is drawn out. Find the probability of drawing:
   (i) two crosses   (ii) one cross only   (iii) no crosses.

9. A card is drawn from a pack and replaced. This is done three times and each time we note whether or not the card is a diamond. Draw a tree diagram using D for 'a diamond' and N for 'not a diamond'. There should be eight branch endings. From the tree find the probability of drawing:
   (i) three diamonds   (ii) two diamonds   (iii) one diamond   (iv) no diamonds.
   Check that the probabilities add up to 1.

10. Each member of a class of 30 boys supports one and only one of three football teams; 13 boys support City, 10 support Rovers and 7 support United.

(a) If a boy is to be chosen at random, what is the probability that he will support City?

(b) If two boys are to be chosen at random, what is the probability that they will both support City?

Draw a tree diagram to show all the possible outcomes of choosing two boys at random, showing the probability of each outcome. Hence find:

(c) the probability that the two boys chosen will support the same team,

(d) the probability that the two boys chosen will support different teams.   (L)

Answers

Exercise 1.1 (page 2)

1. (i) First four letters of the alphabet (ii) suits of playing cards
(iii) eating implements (iv) odd numbers up to 9
2. (i) v, w, x, y, z (ii) January, June, July (iii) 12, 14, 16, 18
4. (i) European countries (ii) birds (iii) planets
5. (i) 7, 12, 5 (ii) true, false, false, true
7. 9, 10, 7, 11; it is an infinite set

Exercise 1.2 (page 4)

2. $V \subset F$ **3.** $\{x, y\}$, $\{x\}$, $\{y\}$, ϕ
4. $\{p, q, r\}$, $\{p, q\}$, $\{q, r\}$, $\{p, r\}$, $\{p\}$, $\{q\}$, $\{r\}$, ϕ
6. 3, 7 **7.** k, n, p **8.** 1, 3, 5, 7, 9, 11, 15
9. k, m, n, p, r, t **11.** (i) True (ii) false (iii) true (iv) false
12. (i) k, m, s, t, x (i) t, v, x, y (iii) x, t (iv) k, m, s, t, x, y, v
13. (i) True (ii) false (iii) false (iv) true (v) false
14. (i) p (ii) m, p, r, k, t (iii) m, p, r; $B \subset A$, $C \subset A$
15. ϕ **16.** 5, 7; 7, 13; 7, 9, 11; 7
19. (i) a, b, c, d, e (ii) a, b, d, e, f, g (iii) a,b, c, f, g
(iv) a, b, c, d, e, f, g
22. 8, 5, 3, 10 **23.** (i) 9 (ii) 8

Exercise 1.3 (page 8)

1. b, d **2.** 1, 2, 3, 5 **3.** (i) {odd numbers} (ii) {females}
(iii) {consonants}
4. (i) I (ii) I′
5. (i) b, c, d, e, f, g, h, k (ii) b, d, e, f (iii) c, g, h, k (iv) c, d, e, k
(v) b, f, g, h
6. (i) p, q, r, t, x, w (ii) q, t, x, w (iii) p, r, (iv) t (v) p, q, r, x, w
10. The fifth diagram
12. (i) a, b, c, d, f, k, m (ii) c, d, f (iii) d, f (iv) a, b, c, e, g, n, p, r
(v) a, b, c (vi) e, n, r
14. (i) b, c, d, g, h, j (ii) c, d, h, i, j, k, l (iii) c, d, h, j
(iv) a, b, e, f, g, i, k, l (v) c, d, h, j (vi) a, b, e, f, g, h, i, k, l
(vii) c, d, j
15. (iii) $(A \cup B)'$ **16.** (i) P' (ii) $P \cap Q$ (iii) $P' \cup Q'$ (iv) $P \cap Q'$
(v) $P \cup Q'$
17. (i) C' (ii) $C \cap D$ (iii) $C \cap D'$ (iv) $C' \cap D'$
18. (i) men not wearing hats
(ii) men carrying newspapers and wearing hats
(iii) men carrying newspapers but not wearing hats
(iv) men not carrying newspapers and not wearing hats
19. (i) {1, 3, 5, 7, 9, 11} (ii) {3, 9} (iii) {2, 3, 4, 6, 8, 9, 10, 12}
(iv) {1, 5, 7, 11} (v) {1, 2, 3, 5, 6, 7, 9, 10, 11} (vi) {2, 6, 10}; $F \subset E$
20. (i) m, t, w, x, y (ii) m, t, x, s, v, z (iii) k, s, w, x, z; m, t, x, s, v, z
(iv) yes

Exercise 1.4 (page 12)

1. (i) 15   (ii) 20   (iii) 6   (iv) 11   (v) 9.
   Those with dogs but not cats; $D \cap C'$
2. (i) 12   (ii) 16   (iii) 19   (iv) 15   (v) 5   (vi) 23   (vii) 31
3. 4   4. 15, 9
5. (i) 31   (ii) 14   (iii) 9   (iv) 2   (v) 7.
   Those with coats but not scarves or gloves; $C \cap (G \cup S)'$
6. (i) 6   (ii) 4   (iii) 14   7. (i) 10   (ii) 6   (iii) 36
8. (i) 7, 8, 12, 27   (ii) 50   (iii) 15   9. (i) 19   (ii) 7; yes   (iii) 15, 3
   (iv) 6
10. (i) 16   (ii) 8   (iii) 6
11. (i) 24   (ii) 6   (iii) 54   (iv) 14
   (v) 16. Pupils who had been to neither the
            pop concert nor the cinema nor the youth club.
12. 35; 0, 1, 2, 3

Exercise 2.1 (page 18)

1. 2, 3, 5, 7, 11, 13, 17, 19, 23, 29, 31, 37   2. 41, 43, 53, 59, 61
3. (i) 1, 2, 3, 6   (ii) 1, 7   (iii) 1, 2, 4, 8   (iv) 1, 3, 9   (v) 1, 2, 5, 10
4. (i) {1, 2, 7, 14}   (ii) {1, 2, 4, 8, 16}   (iii) {1, 2, 4, 5, 10, 20}   (iv) {1, 31}
   (v) {1, 2, 17, 34}
5. (i) 1, 2, 3, 4, 6, 12   (ii) 1, 2, 3, 6, 9, 18   (iii) 1, 2, 3, 6   (iv) 6
6. (i) 1, 2, 4, 8, 16   (ii) 1, 2, 4, 7, 14, 28   (iii) 1, 2, 4   (iv) 4
7. (i) 1, 2, 3, 6, 7, 14, 21, 42   (ii) 1, 3, 7, 9, 21, 63   (iii) 1, 3, 7, 21   (iv) 21
8. (i) 1, 2, 3, 4, 6, 8, 12, 24   (ii) 1, 2, 3, 5, 6, 10, 15, 30   (iii) 1, 2, 3, 6; 6
9. (i) 3   (ii) 8   10. (i) 5   (ii) 9
11. (i) 2×3^2 (ii) $2 \times 3 \times 5$ (iii) $3^2 \times 5$
12. (i) 3×13 (ii) $2 \times 3 \times 7$ (iii) $2^2 \times 3 \times 5$
13. (i) 5, 10, 15, 20, 25, 30, 35, 40, 45
    (ii) 3, 6, 9, 12, 15, 18, 21, 24, 27......, 45   (iii) 15, 30, 45   (iv) 15
14. (i) 4, 8, 12, 16,......, 48   (ii) 6, 12, 18, 24, 30, 36, 42, 48   (iii) 12, 24, 36,
    48   (iv) 12
15. (i) 9, 18, 27, 36,......, 135   (ii) 15, 30, 45,......, 135   (iii) 45, 90, 135
    (iv) 45
16. (i) 6   (ii) 18   (iii) 20   17. Every 60 seconds   18. 4 minutes

Exercise 2.2 (page 19)

1. 20   2. 13   3. 5   4. 2   5. 32   6. 10
7. 6   8. 2   9. 8   10. 2   11. 11   12. 1
13. 11   14. 17   15. 19   16. 23   17. 7   18. 0
19. $11 - (5 - 2)$ 20. $15 - (3 + 7)$ 21. $(8 - 3) \times 2$
22. $40 \div (4 \times 2)$

Exercise 3.1 (page 21)

1. $\dfrac{5}{5}, \dfrac{6}{3}, \dfrac{35}{7}, \dfrac{60}{10}$ 2. $\dfrac{3}{2}, \dfrac{7}{3}, \dfrac{17}{5}, \dfrac{13}{8}, \dfrac{37}{10}$

3. $1\dfrac{1}{4}, 1\dfrac{1}{3}, 2\dfrac{2}{5}, 4\dfrac{3}{4}, 4\dfrac{3}{10}$ 4. $\dfrac{6}{12}, \dfrac{8}{12}, \dfrac{9}{12}, \dfrac{2}{12}, \dfrac{10}{12}$

5. $\dfrac{15}{20}, \dfrac{16}{20}, \dfrac{14}{20}, \dfrac{7}{10}, \dfrac{3}{4}, \dfrac{4}{5}$ 6. $\dfrac{12}{18}, \dfrac{15}{18}, \dfrac{14}{18}, \dfrac{5}{6}, \dfrac{7}{9}, \dfrac{2}{3}$

7. $\dfrac{5}{14}, \dfrac{3}{7}, \dfrac{1}{2}$ **8.** $\dfrac{6}{8}, \dfrac{12}{16}, \dfrac{21}{28}$ **9.** $\dfrac{2}{3}, \dfrac{3}{5}, \dfrac{1}{4}, \dfrac{3}{5}, \dfrac{3}{4}$

10. $\dfrac{3}{4}, \dfrac{8}{9}, \dfrac{9}{11}, \dfrac{1}{4}, \dfrac{4}{7}$ **11.** 30 s   **12.** 2 mm   **13.** 25 p

14. 24 min   **15.** 80 p   **16.** 20 h   **17.** $\dfrac{4}{5}$ **18.** $\dfrac{2}{3}$

19. $\dfrac{3}{5}$ **20.** $\dfrac{7}{12}$ **21.** $\dfrac{7}{10}$ **22.** $\dfrac{2}{5}$ **23.** £6, £24

24. 5 cm, 25 cm   **25.** £256   **26.** 33 l

Exercise 3.2 (page 23)

1. $\dfrac{7}{9}$ **2.** $\dfrac{6}{11}$ **3.** $\dfrac{5}{6}$ **4.** $\dfrac{11}{12}$ **5.** $\dfrac{7}{11}$ **6.** $\dfrac{1}{12}$ **7.** $\dfrac{5}{12}$

8. $\dfrac{1}{2}$ **9.** $\dfrac{3}{4}$ **10.** $1\dfrac{3}{10}$ **11.** $\dfrac{11}{18}$ **12.** $\dfrac{1}{24}$ **13.** $\dfrac{4}{5}$ **14.** $\dfrac{3}{4}$

15. $1\dfrac{1}{2}$ **16.** $1\dfrac{5}{18}$ **17.** $\dfrac{7}{12}$ **18.** 1   **19.** $2\dfrac{7}{12}$ **20.** $2\dfrac{2}{3}$ **21.** $5\dfrac{1}{10}$

22. $7\dfrac{3}{5}$ **23.** $\dfrac{5}{7}$ **24.** $\dfrac{1}{2}$ **25.** $2\dfrac{2}{3}$ **26.** 3   **27.** $2\dfrac{1}{12}$ **28.** $\dfrac{7}{12}$

29. $\dfrac{5}{12}$ **30.** $\dfrac{5}{8}$

Exercise 3.3 (page 24)

1. $\dfrac{6}{35}$ **2.** $\dfrac{3}{8}$ **3.** $\dfrac{3}{14}$ **4.** $\dfrac{1}{2}$ **5.** $1\dfrac{7}{15}$ **6.** $1\dfrac{7}{8}$

7. $5\dfrac{5}{6}$ **8.** $2\dfrac{1}{3}$ **9.** $5\dfrac{1}{10}$ **10.** $6\dfrac{5}{12}$ **11.** $1\dfrac{1}{3}$ **12.** 6

13. $6\dfrac{1}{3}$ **14.** $14\dfrac{2}{3}$ **15.** $\dfrac{3}{8}$ **16.** $\dfrac{20}{21}$ **17.** $\dfrac{4}{5}$ **18.** $\dfrac{2}{3}$

19. $\dfrac{3}{5}$ **20.** $\dfrac{2}{3}$ **21.** $\dfrac{1}{4}$ **22.** $1\dfrac{1}{2}$ **23.** $1\dfrac{1}{3}$ **24.** $2\dfrac{11}{12}$

Exercise 3.4 (page 25)

1. $\dfrac{2}{9}$ **2.** $\dfrac{3}{4}$ **3.** $1\dfrac{1}{4}$ **4.** $2\dfrac{1}{2}$ **5.** $\dfrac{1}{4}$ **6.** $\dfrac{1}{5}$

7. $5\dfrac{1}{2}$ **8.** $19\dfrac{5}{6}$ **9.** $1\dfrac{2}{5}$ **10.** $\dfrac{1}{5}$ **11.** $\dfrac{4}{13}$ **12.** $8\dfrac{3}{4}$

13. $1\dfrac{1}{2}$ **14.** $\dfrac{7}{9}$ **15.** $\dfrac{12}{17}$ **16.** $\dfrac{3}{7}$ **17.** (i) $\dfrac{7}{15}$ (ii) $\dfrac{2}{3}$ (iii) $\dfrac{7}{12}$

Exercise 4.1 (page 27)

1. 0.9, 0.03, 0.007, 0.53, 0.021   **2.** 0.33, 0.209, 0.09, 6.7, 8.03

3. $\dfrac{7}{10}, \dfrac{9}{100}, \dfrac{13}{100}, \dfrac{3}{1000}, \dfrac{207}{1000}$ **4.** $\dfrac{3}{5}, \dfrac{3}{50}, \dfrac{7}{20}, \dfrac{9}{200}, \dfrac{1}{125}$

5. $\dfrac{3}{25}, \dfrac{1}{20}, \dfrac{33}{500}, \dfrac{111}{200}, \dfrac{51}{250}$ **6.** 0.2, 0.5, 0.6, 0.05, 0.35

7. 0.04, 0.52, 0.74, 0.022, 0.852   **8.** (i) 0.38   (ii) 0.85   (iii) 0.057
9. 8.3   **10.** 9.77   **11.** 3.58   **12.** 4.372   **13.** 2.7   **14.** 4.3
15. 3.37   **16.** 0.46   **17.** 4.27   **18.** 23.4   **19.** 4.884   **20.** 6.7

Exercise 4.2 (page 29)

1. 0.3   **2.** 70   **3.** 0.56   **4.** 290   **5.** 2.8   **6.** 720
7. 13.2   **8.** 140   **9.** 0.03   **10.** 0.008   **11.** 0.0053   **12.** 0.0076
13. 0.9   **14.** 0.07   **15.** 0.022   **16.** 0.16   **17.** 0.15   **18.** 0.085
19. 0.09   **20.** 0.008   **21.** 0.0035   **22.** 0.0056   **23.** 0.06   **24.** 0.56
25. 0.024   **26.** 0.008   **27.** 0.12   **28.** 0.85   **29.** 0.02   **30.** 0.003
31. 0.2   **32.** 400   **33.** 20   **34.** 0.6   **35.** 9   **36.** 620
37. 0.2   **38.** 18   **39.** $\dfrac{3}{4}$ **40.** $\dfrac{5}{8}$ **41.** $1\dfrac{1}{5}$ **42.** $\dfrac{1}{20}$
43. 391, 3.91, 0.391, 0.0391   **44.** 1073, 107.3, 1.073, 0.1073
45. 0.06   **46.** 0.69   **47.** 1.2

Exercise 4.3 (page 31)

1. 3 m, 3 m, 8 m, 30 m   **2.** 470 km, 640 km, 60 km, 2310 km
3. 5.4 s, 8.2 s, 0.1 s, 6.1 s   **4.** 15.7, 26.5, 4.0, 47.5
5. 3.74, 7.44, 0.92, 0.07   **6.** 0.616, 0.808, 0.023, 0.005
7. 8700, 540, 0.069, 6.3   **8.** 400, 2000, 0.08, 0.005
9. 7.43, 93 800, 71.1, 0.0350
10. (i) 3.142   (ii) 3.1   (iii) 3.14   (iv) 3.1416
11. (i) 0.008   (ii) 0.00818   (iii) 0.01   (iv) 0.0082   **12.** 2, 3, 3, 2
13. (i) 4250 to 4350   (ii) 171 500 to 172 500   (iii) 0.65 to 0.75
       (iv) 35.5 to 36.5
14. (i) 25.5 to 26.5 km   (ii) 815 to 825 g   (iii) £3150 to £3250   (iv) $2\frac{3}{4}$ to $3\frac{1}{4}$ h
15. (i) 0.335 to 0.345   (ii) 0.065 to 0.075   (iii) 6250 to 6350   (iv) 4.55 to 4.65
16. (i) 8.5 cm, 6.5 cm; 55.25 cm^2 (ii) 7.5 cm, 5.5 cm; 41.25 cm^2
17. 72.25 and 90.25 cm^2

Exercise 4.4 (page 33)

1. 0.778, 0.857, 0.273, 0.385, 0.529   **2.** 0.67, 0.57, 0.56; $\dfrac{5}{9}, \dfrac{4}{7}, \dfrac{2}{3}$

3. $0.\dot{6}, 0.\dot{4}2857\dot{1}, 0.\dot{7}, 0.4\dot{5}, 0.58\dot{3}$ **4.** 0.4815, 0.1935, 0.0465

5. $0.\dot{1}08\dot{9}, 0.1\dot{4}\dot{8}, 0.2\dot{1}, 0.14\dot{5}$ **6.** 20   **7.** 6   **8.** 60
9. 500   **10.** 300   **11.** 0.08   **12.** 0.1   **13.** 3   **14.** 4   **15.** 8
16. 5   **17.** 2   **18.** 30   **19.** 0.02   **20.** 0.6

Exercise 5.1 (page 35)

1. 16, 25, 64, 81   **2.** 1, 100, 10 000, 1 000 000
3. 4, 36, 49, 100   **4.** 900, 1600, 40 000, 250 000
5. ±4, ±6, ±8, ±9   **6.** 5, 7, 10, 1
7. 144, 14 400, 1 440 000   **8.** 961, 96 100, 9 610 000

9. 14   **10.** 35   **11.** 18   **12.** 25   **13.** 22   **14.** 27
15. 28   **16.** 33   **17.** 40, 400, 500   **18.** 90, 200, 3000
19. (i) ±8   (ii) ±5   (iii) ±10   **20.** (i) ±9   (ii) ±5   (iii) ±8
21. 1, 8, 27, 64, 125   **22.** 216, 343, 512, 1000
23. 1, 2, 3, 10   **24.** 4, 5, 6, 7   **25.** 9   **26.** 15   **27.** 12

Exercise 5.2 (page 36)

1. $2\frac{1}{4}$ **2.** $7\frac{1}{9}$ **3.** $3\frac{6}{25}$ **4.** $10\frac{9}{16}$ **5.** $1\frac{1}{2}$ **6.** $2\frac{1}{3}$

7. $1\frac{1}{4}$ **8.** $3\frac{1}{2}$ **9.** 0.16   **10.** 0.04   **11.** 0.0081   **12.** 0.000 144

13. 0.6   **14.** 0.03   **15.** 0.1   **16.** 0.09   **17.** 14, 0.14, 0.014
18. 15, 1.5, 0.15   **19.** 0.008, 0.027, 0.064   **20.** 0.001, 0.125, 0.729

Exercise 5.3 (page 36)

1. 0.5, 0.05, 0.005   **2.** 0.5, 5, 50   **3.** 0.1, 10, 1
4. 0.25, 0.025, 2.5   **5.** 0.4, 0.3, 0.8   **6.** $\frac{1}{3}, \frac{1}{7}, \frac{1}{9}$

7. $2\frac{1}{3}, \frac{3}{5}, \frac{2}{7}$

Exercise 5.4 (page 37)

1. 54.2   **2.** 1150   **3.** 0.0640   **4.** 279 000   **5.** 0.817
6. 18 200 000   **7.** 0.677   **8.** 0.008 59   **9.** 0.0335   **10.** 1790
11. 2.45   **12.** 5.39   **13.** 15.8   **14.** 1.87   **15.** 5.92
16. 0.412   **17.** 0.533   **18.** 0.0825   **19.** 63.2   **20.** 25.2
21. 0.0769   **22.** 0.0175   **23.** 0.106   **24.** 1.28   **25.** 0.001 32
26. 0.206   **27.** 0.0165   **28.** 5.35   **29.** 0.001 92   **30.** 14.6
31. 0.0768   **32.** 3.40   **33.** 0.167   **34.** 129   **35.** 1.47
36. 13.0

Exercise 5.5 (page 38)

1. 7.4×10^2 **2.** 8.6×10^4 **3.** 4×10^3 **4.** 3.09×10^5
5. 4.276×10 **6.** 5.3×10^{-3} **7.** 2.38×10^{-1} **8.** 6×10^{-4}
9. 9.04×10^{-3} **10.** 7×10^{-1} **11.** 8×10^6 **12.** 6×10^{-3}
13. 420   **14.** 58 000   **15.** 6000   **16.** 324.6
17. 0.038   **18.** 0.000 13   **19.** 0.005   **20.** 0.1256
21. 35 > 10, 0.47 < 1, ÷, index not integer

Exercise 5.6 (page 39)

1. 8.4×10^3 **2.** 6.7×10^5 **3.** 1.73×10^4 **4.** 3.74×10^6
5. 6×10^5 **6.** 4.2×10^9 **7.** 2×10^7 **8.** 2.7×10^4
9. 6.4×10^3 **10.** 3×10^4 **11.** 5.3×10^5 **12.** 1.6×10^5
13. 8.7×10^{-4} **14.** 2.2×10^{-4} **15.** 2.2×10^4 **16.** 3.2×10^2
17. 9×10^2 **18.** 4×10^5 **19.** 2.7×10^{-5} **20.** 4×10^{-3}
21. 4×10^{-4} **22.** 4.9×10^{-1} **23.** 6×10^{-3} **24.** 3×10^3
25. (i) 1.4×10^{-4} (ii) 0.000 14
26. 1.1×10^{-5}, 7×10^{-6}, 1.8×10^{-11}, 4.5

27. $2 \times 10^7, 8 \times 10^{-3}, 5.04 \times 10^4$ **28.** (i) 4×10^3 (ii) 6×10^4
29. (i) 5×10^{-3} (ii) 8×10^{-5} **30.** 3×10^2

Exercise 5.7 (page 40)

1. $\sqrt{10}, \sqrt{21}, 4, 6$ **2.** $\sqrt{20}, \sqrt{63}, \sqrt{75}, \sqrt{32}, \sqrt{24}, \sqrt{72}$
3. $3\sqrt{2}, 3\sqrt{3}, 5\sqrt{2}, 3\sqrt{5}, 7\sqrt{2}, 5\sqrt{5}$
4. $\dfrac{\sqrt{3}}{3}, \quad \dfrac{\sqrt{7}}{7}, \quad \dfrac{\sqrt{3}}{6}, \quad \dfrac{\sqrt{2}}{6}, \quad \dfrac{\sqrt{6}}{3}, \quad \dfrac{\sqrt{10}}{2}$
5. $\dfrac{\sqrt{6}}{2}, \quad \dfrac{\sqrt{35}}{7}, \quad \dfrac{1}{2}, \quad \dfrac{1}{3}, \quad \dfrac{\sqrt{2}}{2}, \quad \dfrac{\sqrt{5}}{5}$
6. $3\sqrt{2}, 5\sqrt{2}, 3\sqrt{2}, 7\sqrt{2}$ **7.** $3.4641, 5.1962, 17.3205$

Exercise 6.1 (page 42)

1. $\dfrac{3}{5}$ **2.** $\dfrac{11}{20}$ **3.** $\dfrac{3}{25}$ **4.** $\dfrac{3}{20}$ **5.** $\dfrac{9}{10}$ **6.** $\dfrac{37}{100}$
7. $\dfrac{7}{10}$ **8.** $\dfrac{99}{100}$ **9.** $\dfrac{9}{100}$ **10.** $\dfrac{17}{20}$ **11.** 30%   **12.** 80%
13. 35%   **14.** 54%   **15.** 76%   **16.** 0.31   **17.** 0.68   **18.** 0.8
19. 0.06   **20.** 0.02   **21.** 73%   **22.** 55%   **23.** 30%   **24.** 7%
25. 90%   **26.** $\dfrac{3}{40}$ **27.** $\dfrac{1}{8}$ **28.** $\dfrac{2}{15}$ **29.** $\dfrac{5}{16}$ **30.** $\dfrac{12}{125}$
31. 0.046   **32.** 0.532   **33.** 0.085   **34.** 0.613   **35.** 0.0044   **36.** 82.5%
37. 3.5%   **38.** 5%   **39.** 1.5%   **40.** 62.3%   **41.** $\dfrac{1}{2}, \dfrac{1}{4}, \dfrac{1}{10}, \dfrac{1}{3}$
42. $\dfrac{1}{20}, \dfrac{1}{5}, \dfrac{3}{4}, \dfrac{2}{3}$

Exercise 6.2 (page 43)

1. £6   **2.** £5   **3.** £7   **4.** £6   **5.** £3   **6.** £92
7. 3p   **8.** 12p   **9.** 11p   **10.** 55p   **11.** £14   **12.** £28
13. £6   **14.** 46 g   **15.** 650g   **16.** 18 g   **17.** £9   **18.** £14
19. 9%   **20.** 13%   **21.** 7%   **22.** 55%   **23.** 12%   **24.** 75%
25. 25%   **26.** 70%   **27.** 120%   **28.** 45%   **29.** £72   **30.** £15.30
31. £6.44   **32.** £11.44   **33.** £3.12   **34.** 50p   **35.** £1.56   **36.** £2.31
37. 77p   **38.** £1.17   **39.** £11.95   **40.** 74%   **41.** 39%   **42.** 15%

Exercise 6.3 (page 44)

1. £160   **2.** £36   **3.** 54   **4.** £56.40   **5.** £44   **6.** 30%
7. 18%   **8.** 40%   **9.** £40.50   **10.** £168
11. 52.5%, 53.6%, 45%, 57.1%, 43.3%

Exercise 6.4 (page 45)

1. £54   **2.** £63   **3.** £24   **4.** £76   **5.** £40.28   **6.** £12.32
7. £360   **8.** £80   **9.** £40   **10.** £60   **11.** (i) £32   (ii) £150
12. £420   **13.** £3550   **14.** (i) £69   (ii) £120   **15.** £27.83
16. £62.10, £62.10; both answers are the same. No, she gets £63.

Exercise 6.5 (page 47)

1. 25%	**2.** 30%	**3.** 15%	**4.** 6%	**5.** 30%	**6.** 8%
7. 40%	**8.** $22\frac{1}{2}$%	**9.** £45	**10.** £9.60	**11.** £3150	**12.** £486
13. £200	**14.** £500	**15.** £750	**16.** £130	**17.** £70	**18.** £4
19. 20%	**20.** 25%	**21.** £75	**22.** £55.20		

Exercise 6.6 (page 49)

1. £24	**2.** £42	**3.** £72	**4.** £200	**5.** £63	**6.** £90
7. £374	**8.** £22	**9.** £12.96	**10.** £126.36	**11.** £40.03	**12.** £75.41
13. £175	**14.** £300	**15.** 5 yr	**16.** 6 yr	**17.** 5%	**18.** 6%
19. 8%, £524.88		**20.** First way by £3.06			

Exercise 6.7 (page 51)

1. £466.56   **2.** £726   **3.** £955.06   **4.** £313.60
5. £133.10   **6.** £463.05   **7.** £867.67   **8.** £262.16
9. (i) £4800   (ii) £3840   **10.** £90 000, £81 000, £72 900
11. £18 376   **12.** 140, 196, 274, 384, 538 g (to nearest g); 230 g; after 3.3 days

Exercise 7.1 (page 53)

1. 3:5	**2.** 7:5	**3.** 5:9	**4.** 1:7	**5.** 4:1	**6.** 3:8
7. 2:5	**8.** 20:13	**9.** 5:6	**10.** 1:5	**11.** 9:4	**12.** 8:3

13. 4:3, 3:4   **14.** 3:2, 3:2, 9:4   **15.** 3:7   **16.** 3:2
17. 5:3   **18.** 12:1   **19.** 4:5   **20.** 1:5   **21.** 24:5   **22.** 2:11
23. (i) 6 cm   (ii) 4 cm   (iii) 5.6 cm   (iv) 1 m   (v) 1.6 m   (vi) 45 cm
24. (i) 2 km   (ii) 3.15 km   (iii) 2 cm   (iv) 12 cm   (v) 1.4 cm
25. 27, 18   **26.** 25, 15   **27.** £6, £14   **28.** 0.6 m, 2.4 m
29. £7.50, £1.50   **30.** £2.10, £3.90   **31.** 80p, £1.20, £1.60
32. 24p, 96p, £1.20   **33.** 1.9 kg, 80 g, 20 g   **34.** 50 g, 150 g, 300 g
35. 36 kg   **36.** £18   **37.** (i) 20   (ii) 12   **38.** (i) 1.5   (ii) 9
39. (i) £36   (ii) £25   **40.** (i) 18 cm   (ii) 32 cm
41. 5 cm, 8.5 cm, 11.5 cm   **42.** 21 cm, 28 cm, 31.5 cm
43. (i) £34 000   (ii) £6800, £10 200, £17 000

Exercise 7.2 (page 55)

1. 20 h	**2.** 7 kg	**3.** £81.20	**4.** £54	**5.** 230 km	**6.** 16
7. 3 days	**8.** 270 g	**9.** 146 pages	**10.** 414 steps, 870 m		

Exercise 7.3 (page 56)

1. 370.26 F fr	**2.** $311.04	**3.** 367 336 yen	**4.** 2753.52 S fr
5. 58.57 marks	**6.** 78.41 F fr	**7.** £669.30	**8.** £152.01
9. £54.86	**10.** £11.74	**11.** £19.10	**12.** £6181.94
13. £35.57	**14.** £29.30		

Exercise 7.4 (page 57)

1. (i) 900 km   (ii) 10 km   (iii) 70 km   **2.** (i) 30 min   (ii) 6 min   (iii) 74 min
3. (i) 54 km/h   (ii) 20 km/h   (iii) 7.5 m/s
4. (i) 135 km/h   (ii) 180 km/h   (iii) 37 800 km/h
5. 3 h 40 min   **6.** 20 km   **7.** 116 km   **8.** 20 s   **9.** 54 km/h
10. (i) 7.5 m/s   (ii) 126 km/h   **11.** 4.2 m/s   **12.** 14 min   **13.** 15 min

Exercise 8.1 (page 58)

1. 80 mm, 43 mm, 2.7 mm   **2.** 600 cm, 940 cm, 2.8 cm
3. 5000 m, 6300 m, 47 m   **4.** 4 m, 2.3 m, 76 m
5. 7 m, 0.823 m, 0.064 m   **6.** 2 km, 0.058 km, 3.46 km
7. 4800 g, 75 g, 8.6 g   **8.** 3.2 kg, 0.74 kg, 0.092 kg
9. 0.8 t, 0.034 t, 720 t   **10.** 5600 kg, 370 kg, 0.048 kg

Exercise 8.2 (page 60)

1. (i) 24 cm, 35 cm^2 (ii) 60 cm, 216 cm^2 (iii) 120 m, 875 m^2
   (iv) 8.8 m, 4.68 m^2
2. (i) 24 cm, 36 cm^2 (ii) 36 m, 81 m^2 (iii) 21.6 cm, 29.16 cm^2
3. 12 cm, 40 cm   **4.** 4.8 m, 16.6 m   **5.** 7 cm, 28 cm   **6.** 81 cm^2
7. 28   **8.** 120   **9.** 36 m^2 **10.** 170 cm^2 **11.** 30 cm
12. (i) 100 mm^2 (ii) 300 mm^2 (iii) 570 mm^2
13. (i) 6 cm^2 (ii) 70 cm^2 (iii) 0.85 cm^2
14. (i) 10000 cm^2 (ii) 40000 cm^2 (iii) 83000 cm^2
15. (i) 7 ha   (ii) 4.6 ha   **16.** 48 cm^2 **17.** 1.5 km, 13.5 ha

Exercise 8.3 (page 62)

1. (i) 96 cm^2 (ii) 45 cm^2 **2.** (i) 40 cm^2 (ii) 63 cm^2
3. (i) 42 cm^2 (ii) 18 cm^2 (iii) 25.5 cm^2 **4.** (i) 28 cm^2 (ii) 52 cm^2
5. (i) 57 cm^2 (ii) 102 cm^2 (iii) 68 cm^2 **6.** 8 cm   **7.** 12 cm   **8.** 6 cm

Exercise 8.4 (page 64)

1. (i) 18.2 cm   (ii) 47.8 cm   (iii) 302 cm
2. (i) 26.4 cm^2 (ii) 181 cm^2 (iii) 7240 cm^2
3. (i) 44 cm   (ii) 132 cm   (iii) 22 m
4. (i) 154 cm^2 (ii) 3850 cm^2 (iii) 38.5 cm^2 **5.** 176 cm, 176 m
6. 201 cm, 994   **7.** (i) 28.3 m^2 (ii) 50.3 m^2 (iii) 22.0 m^2
8. (i) 88 m   (ii) 112 m   **9.** 19.7 cm   **10.** 13.5 cm   **11.** 10.2 cm
12. 50.5 cm   **13.** (i) 15.3 cm   (ii) 183 cm^2

Exercise 8.5 (page 66)

1. (i) 11 cm   (ii) 22 cm   (iii) 11 cm   **2.** (i) 38.5 cm^2 (ii) 66 cm^2 (iii) 55 cm^2
3. (i) 4.19 cm   (ii) 12.6 cm^2 **4.** (i) 12.2 cm   (ii) 61.1 cm^2
5. (i) 4.24 cm   (ii) 11.5 cm^2 **6.** (i) 29.9 cm   (ii) 188 cm^2 **7.** 1.57 cm
8. 111 km   **9.** (i) 50.3 cm^2 (ii) 32 cm^2 (iii) 18.3 cm^2; 73.2 cm^2

Exercise 8.6 (page 69)

1. 240 cm^3 **2.** 280 cm^3 **3.** 64 cm^3 **4.** 810 cm^3 **5.** 150 cm^2
6. 94 cm^2 **7.** 222 cm^2 **8.** (i) 8   (ii) 24   (iii) 24   (iv) 8
9. 72   **10.** 84 m^3 **11.** 17   **12.** (i) 54 cm^2 (ii) 4 cm
13. 31, 16 cm^3 **14.** 389 cm^3 **15.** 1010 cm^3 **16.** 2580 cm^3
17. 484 cm^3 **18.** 530 cm^2 **19.** 413 cm^2

Exercise 8.7 (page 70)

1. 6.5 cm   **2.** 4.5 cm   **3.** 15 cm   **4.** 550 kg
5. (i) 2.85 m^2 (ii) 6.84 m^3
6. 135 litres   **7.** (i) 80 m^2 (ii) 1.2 m^3 (iii) 50 cm   **8.** 3.2 cm/min

Exercise 8.8 (page 72)

1. (i) $1540 \, \text{cm}^3$ (ii) $176 \, \text{cm}^3$ 2. (i) $6360 \, \text{cm}^3$ (ii) $2440 \, \text{cm}^3$
3. (i) $440 \, \text{cm}^2$ (ii) $88 \, \text{cm}^2$ 4. $1010 \, \text{cm}^2$ 5. 23 cm by 12 cm
6. $7986 \, \text{cm}^2$ 7. $1100 \, \text{cm}^3$, 1320 g 8. $90\pi \, \text{cm}^3$; 90
9. (i) $201 \, \text{m}^2$ (ii) 176 litres 10. (i) $2163 \, \text{cm}^2$ (ii) $865.2 \, \text{cm}^3$
11. 226 litres 12. (i) $154 \, \text{cm}^2$ (ii) 9 cm
13. 15 cm 14. 6 cm 15. (i) $68\,200 \, \text{cm}^2$ (ii) $924\,000 \, \text{cm}^3$; 6 cm
16. (a) $270\,000 \, \text{cm}^2$ (b) $415\,800 \, \text{cm}^3$ (c) $3850 \, \text{cm}^2$ (d) 108 cm; 83

Exercise 8.9 (page 76)

1. $210 \, \text{cm}^3$ 2. $110 \, \text{cm}^3$ 3. $1800 \, \text{cm}^3$ 4. $1.83 \times 10^6 \, \text{m}^3$
5. $2200 \, \text{cm}^3$ 6. $1232 \, \text{cm}^3$ 7. $9430 \, \text{cm}^3$ 8. $777 \, \text{cm}^3$
9. 75 ml 10. 18 cm, 24 cm, $1700 \, \text{cm}^2$ to 3 s.f.
11. 31.6 cm, 114° 12. (i) $424 \, \text{cm}^2$ (ii) $188 \, \text{cm}^2$ (iii) $236 \, \text{cm}^2$
13. $613 \, \text{cm}^2$ 14. 6.4 l 15. (i) $524 \, \text{cm}^3$ (ii) $1700 \, \text{cm}^3$ (iii) $10\,300 \, \text{cm}^3$
16. (i) $314 \, \text{cm}^2$ (ii) $688 \, \text{cm}^2$ (iii) $2290 \, \text{cm}^2$ 17. $33.5 \, \text{cm}^3$
18. $5.4 \, \text{cm}^2$ 19. $318 \, \text{m}^3$ 20. 4208 litres 21. $160 \, \text{cm}^3$
22. 8 cm

Exercise 9.1 (page 80)

1. £37, £14.80 2. £27; 20.8% 3. £5; £9.60, 12%
4. £82.50 5. £191.10; 3 6. (i) £3840 (ii) 12 (iii) £6560
7. £55.54; 540 8. £47.85
9. (i) £6240 (ii) £2540 (iii) £762 (iv) £14.65 (v) £105.35
10. £15 040; £13 964.80 11. £59.50; 95, 6625
12. £44, £21.78
13. (i) £392 (ii) £42, 12% (iii) 28 months; yes, by £2.25; no, by 75p
14. (i) £133.36 (ii) £84.34 (iii) £420.48 (iv) 46.96%

Exercise 10.1 (page 84)

1. $8a$ 2. $4b$ 3. $12c^2$ 4. 3 5. e^8 6. f^4 7. $21fg$
8. Not possible to simplify 9. n^7k^3 10. p^{10} 11. 0 12. 1
13. c^5 14. 1 15. e^8 16. f^{15} 17. $16g$ 18. 3 19. $3k$
20. Not possible to simplify 21. $27p^6$ 22. $2q^4$ 23. $10tv$
24. w 25. $15a^2b^4$ 26. $7c^4$ 27. $8de^6$ 28. $5f^2$
29. $6a$ 30. 0 31. $c + 5d$ 32. $4f + 2g$ 33. $3h^2$
34. $2k^2$ 35. 1 36. 0 37. $5a + 3$ 38. $b + 1$
39. $21c$ 40. $24de$ 41. $\dfrac{f}{3}$ 42. $\dfrac{g}{2h}$ 43. $\dfrac{3}{5}$
44. $\dfrac{3}{p}$ 45. $6y^4$ 46. $8n$

Exercise 10.2 (page 87)

1. $2, -8, -2, 8$ 2. $-7, 7, 5, -5$ 3. $-11, -5, 11, 5$
4. $13, -13, 5, -5$ 5. $-3, -5, -9$ 6. $-12, 2, 0$
7. $4, -4, 0$ 8. $3, -11, -3$ 9. $3, -3, -7$
10. $2, -4, -2$ 11. $-8, -8, 8$ 12. $-5, -5, 5$

13. $4, -8, 16$ **14.** $-8, 8, 8$ **15.** $-30, 30, -30$
16. $-1, -1$ **17.** $-9, -27$ **18.** $-2a, 0, 0, -2a$
19. $b^2, -b^2, 1, -1$ **20.** $-c^3, -6d, -12f$ **21.** $9g^2, -8h^3, -4k^2, 15m^2$

Exercise 10.3 (page 89)

1. $8, 2, 10, 5, 1, 6, 0, 0, \frac{1}{2}, 2\frac{1}{4}$ **2.** $8, 9, 18, 36, 30, 4, 1, 1, 16, 1$
3. $5, -6, 7, -7, 3, 0, -3, -2, 3, 0, 18, 36, -3, 9, 1, -1, 27, -64, 3, 1$
4. $-12, 2, 8, -8, 32, 64, -3, 9, -4, 36, 12, -64, -16, 108, 324, 0, 0, 0, 12,$
     2
5. (i) $25, 1, 0, 1, 25$ (ii) $-25, -1, 0, -1, -25$ (iii) $8, 1, 0, -1, -8$
     (iv) $-8, -1, 0, 1, 8$
6. $1, 1\frac{1}{2}, 3, -3, -1\frac{1}{2}, -1$ **7.** $1, -1, -3, -5$ **8.** $5, 2, -1, -4$
9. $-6, 6, 10, 6, -6$ **10.** $9, 1, 4, 9, 49$ **11.** $-3, 4, 5, 6, 13$
12. $2, 3, 6, -6, -3, -2$ **13.** $0, 0, 2, 6, 12, 2$ **14.** $5, -3, -3, 5, 21$
15. $-3, 1, 3, 3, 1$ **16.** $-3, 4, 5, 0, -11$ **17.** $33, 3, -3, -33$
18. $4, 2, 4, 20$ **19.** $10, 6, 6, 4, -6$ **20.** $3, 6, -6, -3, 2, 1\frac{1}{2}$
21. $2, 4, 8$ **22.** $-1\frac{1}{2}, 2\frac{1}{4}, -3\frac{3}{8}$ **23.** $12, 18, 24, -12$
24. $4, 36, 9, 4, 36$ **25.** $3, 7, 1, 9, -1, 11$ **26.** $5, 5, 2, 2, -3, -3$

Exercise 10.4 (page 91)

1. $2a + 6$ **2.** $3b - 12$ **3.** $c^2 + c$ **4.** $3d - df$ **5.** $gh + 5h$
6. $k^2 - 6k$ **7.** $3m^2 + 15m$ **8.** $8n^2 - 4np$ **9.** $r^4 + r^2 t$
10. $w^3 x - wx^3$ **11.** $-6 + 2a$ **12.** $-5b^2 + 10b$ **13.** $3c - 4$
14. $4 - 3d$ **15.** $fg - fh + 2f$ **16.** $3k^2 + 6km - 3kn$ **17.** $-2p + 4q - 6$
18. $-r^2 t - rt^2 - rtw$ **19.** $5a + 8$ **20.** $9b - 2$ **21.** $c - 27$
22. $4d + 6$ **23.** $2x - 3$ **24.** -5 **25.** $4 - n$ **26.** $2 - k$
27. $3f^2 - 5f$ **28.** $g^2 + 13g$ **29.** $3h^2 + 2k^2$ **30.** $4m - 2m^2 + 7$

Exercise 10.5 (page 92)

1. $ab + 2a + 3b + 6$ **2.** $cd + c + 5d + 5$ **3.** $e^2 + 8e + 12$
4. $f^2 + 7f + 12$ **5.** $14 + 9g + g^2$ **6.** $18 + 6h + 3m + hm$
7. $n^2 + 10n + 25$ **8.** $4 + 4p + p^2$ **9.** $3ab + 15a + 2b + 10$
10. $2c^2 + 11c + 15$ **11.** $6df + 3d + 8f + 4$ **12.** $15g^2 + 26g + 8$
13. $h^2 + 5hk + 6k^2$ **14.** $2m^2 + 11mp + 12p^2$ **15.** $x^2 + 4xy + 4y^2$
16. $9q^2 + 6qt + t^2$ **17.** $3m$ **18.** $2t^2 + 13t + 17$

Exercise 10.6 (page 92)

1. $ab + 4a - 3b - 12$ **2.** $c^2 + 7c - 18$ **3.** $df - 2d - 5f + 10$
4. $g^2 - 8g + 7$ **5.** $h^2 - 6h - 16$ **6.** $k^2 - 6k - 7$
7. $6 + m - m^2$ **8.** $30 - 11n + n^2$ **9.** $p^2 - 49$
10. $q^2 - 8q + 16$ **11.** $3a^2 + 11a - 4$ **12.** $2b^2 - 11b + 15$
13. $10c^2 - 11c + 3$ **14.** $6d^2 - d - 35$ **15.** $9f^2 - 4$
16. $4g^2 - 20g + 25$ **17.** $16 - 30h + 9h^2$ **18.** $49 - 25k^2$
19. $10m^2 - 7mp + p^2$ **20.** $12x^2 + xy - 6y^2$ **21.** 12
22. $2p^2 - 10p + 6$

Exercise 10.7 (page 94)

1. $a^2 + 6a + 9$ **2.** $b^2 - 8b + 16$ **3.** $36 + 12c + c^2$

4. $25 - 10d + d^2$ **5.** $9f^2 + 6f + 1$ **6.** $4g^2 - 4g + 1$

7. $h^2 + h + \frac{1}{4}$ **8.** $k^2 - \frac{4}{3}k + \frac{4}{9}$ **9.** $m^2 - 9$

10. $n^2 - 25$ **11.** $49 - p^2$ **12.** $16 - q^2$

13. $9r^2 - 1$ **14.** $25t^2 - 1$ **15.** $w^2 - 9x^2$

16. $4y^2 - z^2$ **17.** $4c^2 - 25d^2$ **18.** $9f^2 - 16g^2$

19. $h^6 - 1$ **20.** $k^2 - \frac{4}{9}$ **21.** $n^2 + 6np + 9p^2$

22. $q^2 - 10qr + 25r^2$ **23.** $9t^2 + 30t + 25$ **24.** $16v^2 + 24v + 9$

25. $4w^2 - 20wx + 25x^2$ **26.** $9y^2 - 2yz + \frac{1}{9}z^2$ **27.** $p^4 - 8p^2 + 16$

28. $64 + 16q^3 + q^6$ **29.** $r^2 - 2 + \frac{1}{r^2}$ **30.** $t^2 + \frac{6}{5}t + \frac{9}{25}$ **31.** 961

32. 841   **33.** 2809   **34.** 2304   **35.** 40804   **36.** 39204   **37.** 94.09

38. 392.04   **39.** 16.24   **40.** 1.08   **41.** 24.80   **42.** 15.68   **43.** $20x$

44. $-28y$ **45.** $2n^2 + 18$

Exercise 10.8 (page 95)

1. 3   **2.** 16, 4   **3.** $+49, -7$ **4.** $+100, p + 10$ **5.** $6\frac{1}{4}$ **6.** $2\frac{1}{4}, +1\frac{1}{2}$

7. $30\frac{1}{4}, n + 5\frac{1}{2}$ **8.** $+\frac{1}{4}, q - \frac{1}{2}$ **9.** 25   **10.** 64   **11.** 4   **12.** 25

13. $6\frac{1}{4}$ **14.** $12\frac{1}{4}$

Exercise 10.9 (page 96)

1. $(x + 5)^2 + 3$ **2.** $(x + 4)^2 - 6$ **3.** $(x - 2)^2 + 3$ **4.** $(x - 6)^2 - 16$

5. $(x - 1)^2 - 6$ **6.** $(x + 3)^2 - 12$ **7.** $(x + 2\frac{1}{2})^2 - 4\frac{1}{4}$

8. $(x + 1\frac{1}{2})^2 - 9\frac{1}{4}$ **9.** $(x - 3\frac{1}{2})^2 - 8\frac{1}{4}$ **10.** $(x - 4\frac{1}{2})^2 - 25\frac{1}{4}$

11. $3(x + 5)^2 - 58$ **12.** $2(x + 3)^2 - 13$ **13.** $5(x + 1)^2 - 8$

14. $3(x - 1)^2 - 10$ **15.** $-(x + 3)^2 + 13$ **16.** $-2(x - 4)^2 + 37$

17. Min 8, $x = -3$ **18.** Min 5, $x = 2$ **19.** Min $-7, x = 1$

20. Min $-5, x = -4$ **21.** Max 11, $x = -2$ **22.** Max 28, $x = 5$

23. Min 11, $x = -1$ **24.** Max 60, $x = 5$

Exercise 11.1 (page 98)

1. $\frac{1}{25}, \frac{1}{3}, \frac{1}{8}, \frac{1}{16}$ **2.** $x^{\frac{1}{3}}, x^{\frac{1}{5}}, x^{\frac{1}{4}}, x^{\frac{1}{2}}$ **3.** 3, 5, 2, 3, 2

4. $y^{\frac{2}{3}}, y^{\frac{3}{4}}, y^{\frac{4}{3}}, y^{\frac{2}{5}}, y^{\frac{5}{7}}$ **5.** 8, 4, 8, 32, 81   **6.** $\frac{1}{b^3}, 1, \sqrt[3]{b}, (\sqrt[3]{b})^2$

7. $\sqrt[4]{c}, 1, \frac{1}{c^4}, (\sqrt[4]{c})^3$ **8.** $x^{-3}, x^{\frac{1}{3}}, x^{-5}, x^{\frac{1}{5}}, x^{\frac{2}{5}}$ **9.** 4, 1, $\frac{1}{7}, \frac{1}{8}$, 4

10. $\frac{1}{4}, 1, 1, \frac{1}{4}, 32$ **11.** $\frac{1}{3}, \frac{1}{2}, \frac{4}{5}, 1, \frac{9}{4}$ **12.** $\frac{2}{3}, \frac{2}{3}, \frac{3}{2}, \frac{81}{16}$

13. $\frac{1}{8x^3}, \frac{1}{3y}, 1, \frac{1}{p^8}, \frac{1}{2t^2}$ **14.** (i) 6   (ii) 2   (iii) 2

15. $a^{\frac{3}{4}}, b^{\frac{2}{5}}, c^{2\frac{1}{2}}, d^{\frac{1}{2}}$ **16.** $6f^2, 10g^{\frac{6}{5}}, 3h^{\frac{1}{4}}, 5k^{\frac{1}{2}}$

Exercise 12.1 (page 101)

1. $5(a+b)$ **2.** $c(d-f)$ **3.** $h(g+k)$ **4.** $7(m-p)$ **5.** $2(r-3)$
6. $3(t+3)$ **7.** $2(5+x)$ **8.** $5(3-y)$ **9.** $a(a+3)$ **10.** $b(5-b)$
11. $3(c+1)$ **12.** $5(1-d)$ **13.** $f(e+1)$ **14.** $g(h-1)$
15. $2(3h+4)$ **16.** $3(5+2m)$ **17.** $2a(b+c)$ **18.** $3d(d-e)$
19. $5f(g-2)$ **20.** $2h(h-2)$ **21.** $a(b+c+d)$ **22.** $e(f-g-h)$
23. $k(k-m+3)$ **24.** $5(n-p-q)$ **25.** $3(r+t+2v)$ **26.** $2(x+2y+4)$

Exercise 12.2 (page 101)

1. $(a+3)(a-3)$ **2.** $(b+4)(b-4)$ **3.** $(3+c)(3-c)$
4. $(4+d)(4-d)$ **5.** $(e+1)(e-1)$ **6.** $(1+f)(1-f)$
7. $(3g+4)(3g-4)$ **8.** $(2h+5)(2h-5)$ **9.** $(4k+1)(4k-1)$
10. $(1+6n)(1-6n)$ **11.** $(5p+9)(5p-9)$ **12.** $(10+7t)(10-7t)$

Exercise 12.3 (page 102)

1. $3(x+2)(x-2)$ **2.** $5(y+3)(y-3)$ **3.** $2(n+4)(n-4)$
4. $4(p+1)(p-1)$ **5.** $k(m+3)(m-3)$ **6.** $h(h+g)(h-g)$
7. $r(1+t)(1-t)$ **8.** $4(4+q)(4-q)$ **9.** $(c+\frac{2}{3})(c-\frac{2}{3})$
10. $(d+\frac{1}{2})(d-\frac{1}{2})$ **11.** $(f+1\frac{1}{2})(f-1\frac{1}{2})$ **12.** $(g+2\frac{1}{2})(g-2\frac{1}{2})$
13. $(h+0.3)(h-0.3)$ **14.** $(k+0.5)(k-0.5)$ **15.** $(n^3+7)(n^3-7)$
16. $(p^5+2)(p^5-2)$ **17.** 6600   **18.** 1000   **19.** 12600   **20.** 1840
21. 0.4   **22.** 13.2   **23.** 64   **24.** 4

Exercise 12.4 (page 103)

1. $(a+2)(a+3)$ **2.** $(b+2)(b+4)$ **3.** $(c+1)(c+3)$
4. $(d+1)(d+5)$ **5.** $(e-2)(e-7)$ **6.** $(f-2)(f-9)$
7. $(g-2)(g-6)$ **8.** $(h-3)(h-4)$ **9.** $(k-2)(k+4)$
10. $(m-2)(m+3)$ **11.** $(n-7)(n+2)$ **12.** $(p-5)(p+3)$
13. $(q-6)(q+2)$ **14.** $(r-2)(r+6)$ **15.** $(t-1)(t+12)$
16. $(u-4)(u+3)$ **17.** $(3+w)(5+w)$ **18.** $(2-x)(5-x)$
19. $(2-y)(3+y)$ **20.** $(7-z)(1+z)$ **21.** $(1-n)(1-3n)$
22. $(1-p)(1+5p)$ **23.** $(k+3m)(k+m)$ **24.** $(g-h)(g-7h)$

Exercise 12.5 (page 104)

1. $3x^2+17x+10$, $3x^2+11x+10$, $3x^2+31x+10$, $3x^2+13x+10$;
   (i) $(3x+2)(x+5)$ (ii) $(3x+10)(x+1)$
2. $2x^2-13x+21$, $2x^2-17x+21$, $2x^2-23x+21$, $2x^2-43x+21$;
   (i) $(2x-7)(x-3)$ (ii) $(2x-21)(x-1)$
3. $5x^2+2x-3$, $5x^2-2x-3$, $5x^2+14x-3$, $5x^2-14x-3$;
   (i) $(5x-1)(x+3)$ (ii) $(5x+3)(x-1)$
4. $(3x+5)(x+1)$ **5.** $(3x-5)(x-1)$ **6.** $(3x+1)(x+5)$
7. $(3x-1)(x-5)$ **8.** $(2x+7)(x+1)$ **9.** $(2x-1)(x-7)$
10. $(2x-3)(x-1)$ **11.** $(2x-1)(x-3)$ **12.** $(2x+5)(x-1)$
13. $(2x-5)(x+1)$ **14.** $(2x-1)(x+5)$ **15.** $(2x+1)(x-5)$
16. $(5x+3)(x-1)$ **17.** $(5x-3)(x+1)$ **18.** $(5x+1)(x-3)$
19. $(5x-1)(x+3)$ **20.** $(3a+2)(a-1)$ **21.** $(3b-2)(b-1)$

22. $(3c + 1)(c + 2)$ **23.** $(3d + 1)(d - 2)$ **24.** $(5e + 7)(e + 1)$
25. $(5f - 7)(f + 1)$ **26.** $(7g - 5)(g + 1)$ **27.** $(7h - 1)(h - 5)$
28. $(2k - 5)(k - 2)$ **29.** $(3m + 2)(m - 3)$ **30.** $(5n + 6)(n + 1)$
31. $(3p - 5)(p + 2)$ **32.** $(3a + 1)(2a + 3)$ **33.** $(5b - 3)(2b - 3)$
34. $(2x + 3)(x + 1)$, 23×11 **35.** $(3p + 7)(p + 1)$,
36. $(2q + 1)(2q + 3)$, $3 \times 7 \times 29 \times 67$ $\quad\quad 307 \times 101$

Exercise 12.6 (page 105)

1. $(a + 2)(b + 3)$ **2.** $(c + 5)(d + f)$ **3.** $(g + 3)(g - h)$
4. $(k - 5)(m - 7)$ **5.** $(n - p)(n - 1)$ **6.** $(5 + v)(1 - t)$
7. $(4 + w)(3 + x)$ **8.** $(7 - y)(u + y)$ **9.** $(5 + d)(3 - c)$
10. $(2g + 3)(h - 6)$ **11.** $(k - 7)(2m + 5)$ **12.** $(3p + 4q)(p - r)$
13. $(a - 5c)(2b + 3c)$ **14.** $(d + 3e)(3d - 2f)$ **15.** $(3 - m)(1 - h)$
16. $(1 + n)(2 - k)$ **17.** $(q - 1)(p + 5)$ **18.** $(5w - 2)(3x + 4)$
19. $(7y + 2a)(y - 3b)$ **20.** $(4n - t)(2p + 3)$ **21.** $(a - c)(b + d)$
22. $(x + 5)(y + 2)$ **23.** $(4h + m)(1 + m)$ **24.** $(xy + 2)(5 - x^2)$

Exercise 12.7 (page 106)

1. $3(x + y)$ **2.** $(h + 5)(h - 5)$ **3.** $n(n - p)$
4. $(1 + 3k)(1 - 3k)$ **5.** $(m + 2)(m + 5)$ **6.** $(n - 4)(n + 3)$
7. $(t - 4)(t - 2)$ **8.** $2a(a - 3b)$ **9.** $3(x - 5)(x + 5)$
10. $(1 + 2a)(1 + 5a)$ **11.** $c(d + 3)(d - 3)$ **12.** $(3 - b)(5 + b)$
13. $2(4 + y)(4 - y)$ **14.** $3(x - 3)(x + 1)$ **15.** $3(y - 1)^2$
16. $(2n - 1)(3n + 2)$ **17.** $(5h + 2)(h - 2)$ **18.** $(4a - 5)(2a + 3)$
19. $3b(b + 1)(b - 1)$ **20.** $2(c - 3)(c + 5)$ **21.** $(3x - 5)(x - 2)$
22. $(a - 2)(b + 5)$ **23.** $(3 - 2c)(1 - d)$ **24.** $(4x + 3)(y + 2)$
25. $(k - 4)(m - 3)$ **26.** $(xy - 2)(xy + 5)$ **27.** $(3p + 4q)(p - 2q)$
28. $(a - c)(b + d)$ **29.** $(e - 3f)(e + 2g)$

Exercise 13.1 (page 107)

1. b **2.** c **3.** 3   **4.** 6   **5.** $5h$ **6.** $7k$ **7.** nq
8. rt **9.** $3w$ **10.** $6x$ **11.** $3a$ **12.** bc **13.** d^2 **14.** $3e$
15. gh **16.** $3m$ **17.** pn^2 **18.** $5rt$ **19.** $4vw$ **20.** $6xy^2$ **21.** 15
22. ab **23.** $3c$ **24.** $5g$ **25.** 4   **26.** f **27.** g^2 **28.** $5hk$

29. $12n$ **30.** p^3 **31.** $6rt$ **32.** $10w^3$ **33.** $\dfrac{a}{c}$

34. $\dfrac{5}{7}$ **35.** $\dfrac{e}{f}$ **36.** $\dfrac{g}{h}$ **37.** $\dfrac{k}{m}$ **38.** $\dfrac{p}{3}$

39. $\dfrac{1}{q}$ **40.** $\dfrac{1}{s}$ **41.** $\dfrac{1}{t}$ **42.** $\dfrac{1}{v^3}$ **43.** $\dfrac{2w}{3}$

44. $\dfrac{2}{3x}$ **45.** $\dfrac{b}{a}$ **46.** $\dfrac{2c}{3d}$ **47.** $\dfrac{3f}{4e}$ **48.** 7

49. b **50.** c^2 **51.** $3e$ **52.** $3e$ **53.** $5g$

54. $2h$ **55.** km **56.** $\dfrac{a}{b}$ **57.** $\dfrac{d}{e}$ **58.** $\dfrac{2}{3}$

59. $\dfrac{m+p}{n-q}$ **60.** $\dfrac{2}{3}$ **61.** $\dfrac{c}{d}$ **62.** $\dfrac{3}{f}$ **63.** $\dfrac{3}{7}$

64. 3 **65.** 4 **66.** $\dfrac{x+y}{4}$ **67.** $\dfrac{p-q}{2}$ **68.** $a(c+3)$

69. $5(d-4)$ **70.** eh **71.** $5k$ **72.** $3p+3q$ **73.** $ru+rw$

Exercise 13.2 (page 108)

1. $\dfrac{3}{x}$ **2.** 5 **3.** $\dfrac{1}{a-2}$ **4.** $\dfrac{1}{b-3}$ **5.** $c+5$ **6.** $\dfrac{1}{e-f}$

7. $\dfrac{1}{g+h}$ **8.** $\dfrac{k-1}{3}$ **9.** $a+b$ **10.** $c+3$ **11.** $d+5$

12. $3(e+2)$ **13.** $4(f-3)$ **14.** $g(g+h)$

Exercise 13.3 (page 109)

1. $\dfrac{7a}{10}$ **2.** $\dfrac{4b}{21}$ **3.** $\dfrac{cf-de}{df}$ **4.** $\dfrac{h+g}{gh}$ **5.** $\dfrac{7}{k}$ **6.** $\dfrac{6}{m}$ **7.** $\dfrac{5}{6n}$

8. $\dfrac{2q+3p}{pq}$ **9.** $\dfrac{5v-2r}{rtv}$ **10.** $\dfrac{2w+3}{10x}$ **11.** $\dfrac{ad-be}{bcd}$ **12.** $\dfrac{3f-1}{f^2}$

13. $\dfrac{b+a}{b}$ **14.** $\dfrac{3d-c}{d}$ **15.** $\dfrac{e+gf}{f}$ **16.** $\dfrac{2-h^2}{h}$ **17.** $\dfrac{5c-4}{6c^2}$ **18.** $\dfrac{11}{6a}$

19. $\dfrac{5y^2-2y+1}{y^3}$ **20.** $-\dfrac{1}{6b}$ **21.** $\dfrac{5a+19}{6}$ **22.** $\dfrac{b-5}{4}$ **23.** $\dfrac{8c+30}{15}$

24. $\dfrac{31-5d}{14}$ **25.** $\dfrac{5e-4f}{18}$ **26.** $\dfrac{11-3g}{10}$ **27.** $\dfrac{7x-3}{6}$ **28.** $\dfrac{5x+8}{12}$

29. $\dfrac{4h-21}{24}$ **30.** $\dfrac{11p+14}{30}$ **31.** $\dfrac{a}{3}+\dfrac{b}{2}$ **32.** $\dfrac{d}{e}+\dfrac{d}{c}$ **33.** $\dfrac{3}{g}+\dfrac{f}{g}$

34. $\dfrac{2}{h}+\dfrac{h}{5}$ **35.** $\dfrac{k}{2}-\dfrac{m}{5}$ **36.** $\dfrac{3}{q}-\dfrac{4}{p}$ **37.** $\dfrac{1}{t}-\dfrac{7}{t^2}$ **38.** $\dfrac{3}{2x}-\dfrac{2}{w}$

Exercise 13.4 (page 110)

1. $\dfrac{2a+5}{(a+2)(a+3)}$ **2.** $\dfrac{2b-5}{(b-3)(b-2)}$ **3.** $\dfrac{7}{(3+c)(4-c)}$

4. $\dfrac{5d+7}{(d+2)(d+1)}$ **5.** $\dfrac{3e-5}{(e-1)(e-2)}$ **6.** $\dfrac{2}{(f+3)(f+5)}$

7. $\dfrac{4}{(g-2)(g+2)}$ **8.** $\dfrac{13-2h}{(h-2)(h+1)}$ **9.** $\dfrac{3k+11}{(k-3)(k+2)}$

10. $\dfrac{5m+3}{(m+1)m}$ **11.** $\dfrac{3}{n(n+3)}$ **12.** $\dfrac{6}{(3-y)(3+y)}$

13. $\dfrac{a^2+b^2}{(a-b)(a+b)}$ **14.** $\dfrac{2cd}{(d-5)(d+5)}$ **15.** $\dfrac{2g-1}{g(g-2)(g+1)}$

16. $\dfrac{h}{(h+3)(h+2)}$ **17.** $\dfrac{2}{k(k+1)(k-1)}$ **18.** $\dfrac{2}{m+3}$

19. $\dfrac{1}{n-1}$ **20.** $\dfrac{1}{x}$ **21.** $\dfrac{3y+4}{y^2-4}$

Exercise 13.5 (page 111)

1. $\dfrac{nt}{pw}$ 2. $\dfrac{k}{r}$ 3. $\dfrac{1}{2}$ 4. $\dfrac{c}{b}$ 5. $\dfrac{5e}{df}$ 6. $\dfrac{3}{2}$ 7. $\dfrac{1}{h}$ 8. $2k$

9. $\dfrac{mr}{pq}$ 10. $\dfrac{4t}{5v}$ 11. $\dfrac{2}{3}$ 12. $\dfrac{y}{z}$ 13. $\dfrac{a}{c}$ 14. $\dfrac{d^2}{ef}$ 15. $\dfrac{g}{h}$ 16. $\dfrac{n}{t}$

17. $\dfrac{3}{2}$ 18. $\dfrac{3}{5}$ 19. $\dfrac{2}{3}$ 20. $\dfrac{10}{3}$ 21. $\dfrac{4c}{3}$ 22. $\dfrac{1}{6}$ 23. $\dfrac{4}{9}$ 24. $\dfrac{1}{k}$

25. $\dfrac{m-2}{2}$ 26. $\dfrac{2}{3(n+3)}$ 27. $\dfrac{p+5}{p}$ 28. $\dfrac{r+7}{r}$ 29. $3(h+2)$ 30. $\dfrac{y}{x}$

Exercise 14.1 (page 113)

1. 8 2. 5 3. 35 4. 8 5. -3 6. 15 7. $2\frac{3}{5}$
8. -5 9. 5 10. 2 11. 8 12. 12 13. 2.5 14. $1\frac{1}{3}$
15. -3 16. $-2\frac{1}{3}$ 17. 3 18. 2 19. $3\frac{1}{2}$ 20. -4 21. 3
22. $8\frac{1}{2}$ 23. $4\frac{1}{4}$ 24. $5\frac{3}{4}$ 25. 6 26. $1\frac{1}{2}$ 27. -2 28. $-1\frac{1}{5}$
29. 6 30. $14\frac{1}{2}$ 31. $12\frac{1}{2}$ 32. -4 33. 4 34. $-1\frac{3}{5}$ 35. $\frac{2}{3}$
36. -2 37. -1 38. 15 39. $3\frac{2}{3}$ 40. $3\frac{1}{3}$

Exercise 14.2 (page 115)

1. 12 2. 20 3. 10 4. 16 5. 6 6. 60 7. 24

8. 8 9. 20 10. 2 11. 2 12. -7 13. 6 14. -5

15. 13 16. 5 17. $\dfrac{5}{7}$ 18. $\dfrac{1}{2}$ 19. $5\frac{1}{2}$ 20. 7 21. 5

22. 4 23. $4\dfrac{2}{3}$

Exercise 14.3 (page 116)

1. $7\frac{1}{2}$ 2. $5\frac{3}{5}$ 3. 8 4. $1\frac{2}{3}$ 5. $5\frac{1}{4}$ 6. $-7\frac{1}{2}$ 7. $6\frac{1}{2}$
8. $5\frac{3}{4}$ 9. $5\frac{1}{2}$ 10. $20\frac{1}{2}$ 11. 7 12. $7\frac{1}{2}$ 13. $8\frac{1}{2}$ 14. 7
15. 11 16. -1 17. 4

Exercise 14.4 (page 117)

1. 4 2. 5 3. 16 4. 15 5. 60 6. 72 7. 10 8. 6

Exercise 14.5 (page 120)

1. (i) 8, 9 (ii) $-2, -1, 0, 1$ (iii) 1, 2, 3 (iv) $-4, -3$ (v) 0, 1, 2
2. (i) 1, 2, 3, 4 (ii) $-1, 0, 1, 2$ 3. (i) $-3, -2$ (ii) 1, 2 (iii) $-1, 0, 1, 2$
4. $a > 2$, $b \leq -2$, $-3 < c \leq 1$
6. (i) $x > 7$ (ii) $x < -7$ (iii) $x \leq 3$ (iv) $x \geq 1\frac{1}{3}$ (v) $x > -6$ (vi) $x < -1$
7. (i) $-1 < n < 2$ (ii) $-4 < n < -1$ 8. $-2, -1, 0, 1, 2$
9. $-4 \leq y \leq 4$

10. (i) (2, 1), (3, 2), (3, 1), (4, 3), (4, 2), (4, 1)   (ii) (1, 3), (1, 4), (2, 4)
(iii) (1, 1), (1, 2), (2, 1)   (iv) (3, 4), (4, 3), (4, 4)
11. $-2, -1, 0, 1$ **12.** $-2, -1$

Exercise 15.1 (page 121)

1. (i) 56p   (ii) $c = 100 - kx - ny$ **2.** (i) £80   (ii) $c = x + (n - 1)y$

3. (i) 4h   (ii) $t = \dfrac{d}{s}$ (iii) $3\frac{1}{2}$ h   **4.** (i) £5.90   (ii) $y = x + kn$

5. (i) $y = p - x$ (ii) 7.4 tonnes

6. (i) $L = \dfrac{A}{W}$ (ii) $L = \sqrt{A}$ (iii) $V \approx 4.2\, r^3$

7. (i) $\phi = 180 - 2\theta$ (ii) $\theta = 180 - \dfrac{360}{n}$ **8.** $f = d + 2nx$, 20173

9. $p = \dfrac{ny}{100} - x$, £14, loss of £10   **10.** $A = \pi(r + w)^2 - \pi r^2$, 53

Exercise 15.2 (page 125)

1. $7, x = b - a$ **2.** $\dfrac{7}{9}, x = \dfrac{d}{c}$ **3.** $12, x = fg$ **4.** $11, x = k + h$

5. $\pm 5, x = \pm\sqrt{n}$ **6.** $\pm 5, x = \pm\sqrt{\dfrac{q}{p}}$ **7.** $81, x = r^2$ **8.** $\dfrac{7}{3}, x = \dfrac{t}{u}$

9. $7\frac{1}{2}, x = \dfrac{wy}{v}$ **10.** $2\frac{6}{7}, x = \dfrac{ab}{c}$ **11.** $4\frac{3}{5}, x = \dfrac{f - e}{d}$ **12.** $21, x = (h + m)g$

13. $p = n - q$ **14.** $k = h + m$ **15.** $d = \dfrac{c}{\pi}$ **16.** $d = st$

17. $h = \sqrt{A}$ **18.** $p = k^2$ **19.** $h = \dfrac{V}{lb}$ **20.** $B = 180 - A - C$

21. $h = \dfrac{2A}{b}$ **22.** $r = \sqrt{\dfrac{A}{\pi}}$ **23.** $a = \dfrac{v - u}{t}$ **24.** $d = \dfrac{b^2}{c}$

25. $v = \sqrt{\dfrac{2E}{n}}$ **26.** $R = \dfrac{100I}{PT}$ **27.** $s = \dfrac{v^2 - u^2}{2a}$ **28.** $a = \sqrt{c^2 - b^2}$

29. $r = \sqrt[3]{\dfrac{3v}{4\pi}}$ **30.** $s = (3t)^2$ **31.** $t = \dfrac{d}{s}$ **32.** $b = \dfrac{a}{n^2}$

33. $g = \dfrac{4\pi^2 l}{T^2}$ **34.** $x = \dfrac{2 - 3y}{y}$ **35.** $x = \dfrac{2y + 5}{y - 1}$ **36.** $x = \dfrac{5y - 4}{2y - 3}$

37. $a = \dfrac{2(s - ut)}{t^2}$ **38.** $x = \sqrt{\dfrac{y - 5}{3}}$ **39.** $x = \dfrac{7}{y - 3}$ **40.** $h = \dfrac{S - \pi r^2}{2\pi r}$

41. $v = \dfrac{uf}{u - f}$

Exercise 15.3 (page 126)

1. (i) $V = 9x^2$ (ii) $x = \sqrt{\dfrac{V}{9}}$ (iii) 10   2. (i) 144 (ii) $n = \dfrac{360}{180 - x}$ (iii) 12

3. (i) 52 km (ii) $h = \dfrac{d^2}{13}$ (iii) 325 m

4. (i) $y = 180 - 2x$ (ii) 74 (iii) $x = \dfrac{180 - y}{2}$ (iv) 34

5. (i) 30, -10 (ii) $F = \dfrac{9}{5}C + 32$ (iii) 284, 5 (iv) $-40°$

6. (a) 323 cubic units (b) $R = \sqrt{\left(\dfrac{3V}{2\pi h} + \dfrac{r^2}{2}\right)}$ (c) $R = 2r$

Exercise 16.1 (page 130)

1. 2, 5	2. 17, -3	3. 1, 6	4. -2, 4	5. 3, 2
6. -1, -3	7. 1, 2	8. 5, -2	9. 9, 5	10. 4, 1
11. 2, -1	12. 3, -2	13. 7, 2	14. -2, $3\frac{1}{2}$	15. 5, 4
16. 2, -1	17. 6, 5	18. 3, 0	19. 3, -2	20. -3, 5
21. 5, -2	22. -3, 4	23. -5, 2	24. 2, -3	

Exercise 16.2 (page 131)

1. $3\dfrac{2}{5}, \dfrac{2}{5}$ 2. $3\dfrac{2}{5}, -1\dfrac{3}{10}$ 3. $-2\dfrac{9}{11}, 1\dfrac{8}{11}$ 4. $\dfrac{2}{7}, -1\dfrac{2}{7}$

5. 7, -6 6. 6, -1 7. 6p, 14p   8. 220 seats

9. 19, 13   10. 17, 12   11. 6.3 kg, 4.2 kg   12. 14, 15

13. 10, 20   14. 15, 21

Exercise 17.1 (page 134)

1. 2, 6	2. -5, 8	3. 0, 3	4. -1, 0	5. -3, -2	6. 2, 5
7. 1, 3	8. -7, -1	9. 0, 2	10. -5, 0	11. -1, 3	12. -5, 1

13. -7, 1   14. -1, 2   15. -3, 3   16. -6, 6   17. $-\dfrac{5}{3}, 3\dfrac{1}{2}$

18. $-\dfrac{1}{2}, -\dfrac{2}{5}$ 19. $-2\dfrac{1}{2}, -1$ 20. $\dfrac{1}{3}, 2$ 21. $-\dfrac{1}{2}, \dfrac{1}{3}$ 22. $\dfrac{1}{5}, \dfrac{1}{2}$

23. $-1, \dfrac{2}{5}$ 24. $-\dfrac{1}{2}, 2$ 25. $-\dfrac{1}{2}, -\dfrac{2}{3}$ 26. $1\dfrac{1}{2}, \dfrac{1}{3}$ 27. $-\dfrac{2}{3}, \dfrac{2}{3}$

28. $-2\dfrac{1}{2}, 2\dfrac{1}{2}$ 29, 1, $1\dfrac{1}{3}$ 30. $-\dfrac{1}{3}, -4$ 31. $-2, \dfrac{1}{2}$ 32. $-\dfrac{1}{3}, 3$

33. $-1\dfrac{2}{3}, 0$ 34. $0, \dfrac{2}{3}$ 35. 0, 3   36. -6, 5

Exercise 17.2 (page 135)

1. $x^2 - 5x + 6 = 0$ **2.** $x^2 - 6x + 5 = 0$ **3.** $x^2 + 2x - 15 = 0$
4. $x^2 - 5x - 14 = 0$ **5.** $x^2 + 5x + 4 = 0$ **6.** $x^2 + 9x + 18 = 0$
7. $2x^2 - 5x + 2 = 0$ **8.** $9x^2 - 9x + 2 = 0$ **9.** $2x^2 + x - 6 = 0$
10. $5x^2 - 23x - 10 = 0$ **11.** $9x^2 + 15x + 4 = 0$ **12.** $4x^2 - 4x - 15 = 0$

Exercise 17.3 (page 136)

1. $x^2 - 7x + 12 = 0$ **2.** $x^2 - 9x + 8 = 0$ **3.** $x^2 - 3x - 10 = 0$
4. $x^2 + 2x - 24 = 0$ **5.** $x^2 + 7x + 10 = 0$ **6.** $x^2 + 4x + 3 = 0$
7. $x^2 - 4 = 0$ **8.** $x^2 - 3x = 0$ **9.** $2x^2 - 3x + 1 = 0$
10. $9x^2 - 9x + 2 = 0$ **11.** $3x^2 - 2x - 1 = 0$ **12.** $4x^2 + 8x + 3 = 0$
13. (i) yes   (ii) no   (iii) no   (iv) yes   (v) no   (vi) yes   (vii) no   (viii) yes

Exercise 17.4 (page 138)

1. $1, 5$ **2.** $-4, 2$ **3.** $3, 7$ **4.** $-\dfrac{1}{2}, -\dfrac{2}{3}$ **5.** $-3, \dfrac{1}{2}$ **6.** $-\dfrac{2}{3}, 2\dfrac{1}{2}$

7. $-5.24, -0.76$ **8.** $-4.30, -0.70$ **9.** $-3.45, 1.45$ **10.** $-1.77, -0.57$

11. $-0.28, 1.78$ **12.** $-1.54, -0.26$ **13.** $0.74, 4.77$ **14.** $-2.83, 0.83$

15. $-2.38, 0.63$ **16.** $-1, 2.67$ **17.** $-2.58, 0.58$ **18.** $0.65, 3.85$

Exercise 17.5 (page 139)

1. 3, 8 or $-3, -8$ **2.** 7 or -8 **3.** 5, 6 or $-5, -6$ **4.** 12, 14 or $-12, -14$
5. 8 m   **6.** 5 or 9   **7.** 4 or 7   **8.** 12 cm
9. 8, 15, 17 cm   **10.** (3, 5) and $(-4, -2)$ **11.** (2, 5) and $(-5, -2)$
12. $-3.31, 0.31$

Exercise 17.6 (page 140)

1. $-3, 4$ **2.** $-5, 8$ **3.** $-\dfrac{3}{4}, 2$ **4.** $-2\dfrac{1}{2}, 3$ **5.** $-1, 5$ **6.** $-3, 4$

7. $-4, 7$ **8.** $-\dfrac{2}{3}, 3$ **9.** -4 **10.** 3   **11.** $-3, \dfrac{1}{2}$ **12.** $-1\dfrac{1}{2}, 7$

13. $£\dfrac{30}{x+4}; 6$ **14.** $£\dfrac{60}{x}; 12$ **15.** 12 km/h   **16.** $2\dfrac{1}{2}$ h, $37\dfrac{1}{2}$ km/h

17. 5   **18.** 25

Exercise 17.7 (page 142)

The solutions are given in the form: x_1, y_1 and x_2, y_2.

1. $-2, 3$ and $-3, 2$ **2.** $2, -2$ and $4, 2$ **3.** $-4, -2$ and $2, 4$
4. $\frac{1}{2}, 1$ and $3, \dfrac{2}{3}$ **5.** $-1, 0$ and $4, 5$ **6.** 1, 2 and $1\frac{1}{2}, 1\frac{3}{4}$
7. 1, 5 and 2, 3   **8.** $-2, 1$ and $4, 3$ **9.** 1, 1 and $2, -\frac{1}{2}$
10. 2, 1 and $-1, -3$ **11.** 1, 3 and $-2, 1$ **12.** $-1, -2$ and 3, 4
13. (1, 4) and (3, 6)   **14.** (1, 0) and $(-2, 3)$
15. $10\frac{1}{2}$ cm by 5 cm or 9 cm by $6\frac{1}{2}$ cm

Exercise 18.1 (page 145)

1. $3, -3, -5, -9$ 2. $1, 1, 10, 9\frac{3}{4}$ 3. $6, 0, -3, -9$ 4. $4, 10, -5, 40$

5. $\{1, 4, 7, 10\}$ 6. $\{2, 3\}$ 7. $\left\{-3, -2\frac{1}{2}, -2, -1\frac{1}{2}, -1\right\}$

8. $\left\{\frac{1}{4}, \frac{1}{2}, 1, 2, 4\right\}$ 9. (i) $7, 1$ (ii) $5, -3$ 10. (i) $25, 25$ (ii) $+6, -6$

11. (i) $6, 1\frac{1}{3}$ (ii) $4, 6$

13. (i) $\{x : 0 \leqslant x \leqslant 6\}$ (ii) $\{x : -2 \leqslant x \leqslant 8\}$ (iii) $\{x : -2 \leqslant x \leqslant 7\}$
(iv) $\{x : \frac{1}{3} \leqslant x \leqslant \frac{1}{2}\}$

14. (i) $\{1, 5\}$ (ii) $\{-3, 0, 3\}$ 15. (i) $\{-2, 0, 4\}$ (ii) $\{0, 1, 2, 3\}$

Exercise 18.2 (page 147)

1. (i) $x \to \frac{1}{2}x$ (ii) $x \to x + 5$ (iii) $x \to 4x$ (iv) $x \to \frac{5}{3}x$

2. (i) $5, 8, 14$ (ii) $x \to \frac{1}{3}(x - 2)$ 3. (i) $5, 3, 1$ (ii) $\frac{(5 - x)}{2}$ (iii) $0, 1, 2$

4. (i) $x \to \frac{(x + 3)}{5}$ (ii) $x \to \frac{1}{3}x - 2$ (iii) $x \to 2(x - 7)$

5. (i) $6, 4, 3$ (ii) $x \to \frac{12}{x}$ 6. (i) $7, 5, 3$ (ii) $x \to 7 - x$

7. $f^{-1} : x \to x + 6$, $g^{-1} : x \to 6 - x$, $h^{-1} : x \to \frac{6}{x}$; $-4, 4, 3$

9. (i) $27, -8$ (ii) $\sqrt[3]{x} + 1$ (iii) $4, -1$

10. (i) $5, 5$ (ii) $x \to \pm\sqrt{(x - 1)}$; $+2, -2$ (iii) Has two values

Exercise 18.3 (page 148)

1. $8, 10, 16, 13, 2x + 6, 2x + 3$ 2. $4, 8, 0, 2\frac{2}{3}, \frac{x}{3} - 4, \frac{x - 4}{3}$

3. $9, 7, 1, 49, 10 - x^2, (10 - x)^2$ 4. $3, 2, 10, 0, (2 - x)^2 + 1, 1 - x^2$
5. (i) $7, 10, x + 6$ (ii) $-1, -4, x - 6$ 6. $7, -5, 15, -9$
7. $28, -17, 9x - 8$ 8. $f : x \to x^2, g : x \to x + 3; x \to (x + 3)^2$

9. $f : x \to x + 1, g : x \to \frac{1}{x}, x \to \frac{1}{x} + 1, 1\frac{1}{2}, \frac{1}{3}$

Exercise 18.4 (page 149)

1. (i) $0.4, 12, 0.6$ (ii) $x \to 1 - \frac{18}{x}, 4$ 2. $gfh; x \to 3x^2 - 1$
3. $h : x \to x + 3; g : x \to x^2, f : x \to 2x, x \to 4x^2 + 3, x \to 4x^2 + 24x + 36$
4. (i) $x \to x + 3$ (ii) $x \to 2x - 1$ (iii) $x \to x^2$ (iv) $x \to \frac{6}{x}$

5. (i) $x \to \dfrac{20 - x}{3}$ (ii) $x \to \dfrac{12}{x - 5}$ **6.** (i) ± 3 (ii) $1, -2$

7. $2(x + 2), 2x + 2, \dfrac{4}{x}; 1, -2$ **8.** (i) $\{0, 3, 15\}$ (ii) $\{-1, 0, 2, 3\}$

9. (i) $x \to x + 3$ (ii) $x \to \pm\sqrt{x}$ (iii) $x \to (x - 3)^2$ (iv) $x \to (x - 3)^2 - 3$

(v) $x \to 3 \pm\sqrt{x}$ (vi) $x \to 3 \pm\sqrt{x + 3}; h : x \to x - 6;$

$(hgf)^{-1} : x \to 3 \pm\sqrt{x + 6}; 0.17, 5.83$

Exercise 19.1 (page 151)

1. $(3, 4), (3, -2), (-4, 4), (-1, -4), (4, 0), (0, -3), \left(-3\dfrac{1}{2}, -2\dfrac{1}{2}\right), (0, 0)$

2. Square; $(0, 1)$ **3.** $(3, -3), (-2, -2), (1, -1)$

Exercise 19.2 (page 155)

5. (i) $(0, 3), \dfrac{1}{4}, y = \dfrac{1}{4}x + 3$ (ii) $(0, -5), \dfrac{7}{3}, y = \dfrac{7}{3}x - 5$

(iii) $(0, 10), -\dfrac{4}{3}, y = -\dfrac{4}{3}x + 10$ (iv) $(0, 6), -\dfrac{2}{3}, y = -\dfrac{2}{3}x + 6$

6. (i) 5 (ii) 1 (iii) -2 (iv) -1

9. (i) $y = -\dfrac{3}{7}x + 3, -\dfrac{3}{7}, (0, 3)$ (ii) $y = -2x + 5, -2, (0, 5)$

(iii) $y = 4x - 8, 4, (0, -8)$ (iv) $y = \dfrac{2}{5}x - \dfrac{9}{5}, \dfrac{2}{5}, \left(0, -1\dfrac{4}{5}\right)$

10. (i) 6 (ii) $\dfrac{2}{3}$ (iii) $(-9, 0)$ **11.** (i) 3 (ii) -5 **12.** $1\dfrac{1}{2}, 2\dfrac{1}{2}$

13. $2\dfrac{1}{2}, -1$

Exercise 19.3 (page 159)

1. $\pm 4.47; \pm 2.45, x^2 - 20 = -14, 6$
2. $0.70, 4.30; 0.21, 4.79, x^2 - 5x + 1 = 0; n = -2, 1.3, 3.62$
3. (i) $-2.19, 3.19$ (ii) $-2.70, 3.70$ (iii) $-1.30, 2.30$; no, no
4. (i) $0.77, 5.24$ (ii) $1.59, 4.42$ (iii) $0.17, 5.83$; impossible; -5

Exercise 19.4 (page 161)

1. $(2, 4), (3, 9)$ **2.** $(-1, 1), (3, 9)$ **3.** $3, -2$
4. $-1\dfrac{1}{2}, 2; x^2 - x - 1 = 0; -0.62, 1.62$
5. $5.6, -2.4$ (i) $-4.17, -0.69, 3.48$ (iii) $x^3 - 10x = 0; 0, \pm 3.16$
6. (b) $-0.27, 3.73$ (c) $x^2 - 4x + 1 = 0$
7. (a) The numbers are $-9, 0, 3$. (d) $2x^2 - 5x - 5 = 0, -0.77$.

Exercise 19.5 (page 164)

7. $(4, 2), (4, 3), (5, 2)$
8. $(2, 3), (2, 4), (3, 2), (3, 3), (3, 4)$
9. $(2, 3), (2, 4), (3, 4), (4, 5)$

Exercise 19.6 (page 167)

1. (i) 12.10   (ii) 10 km   (iii) $\frac{1}{2}$ h   (iv) 5 km/h   (v) 4 km/h
3. 11.40; 10.50 and 11.23, $12\frac{1}{2}$ km and $20\frac{3}{4}$ km from P
4. (i) 0.7 m/s²   (ii) 0.9 m/s²   (iii) 4 km/h per s
5. (i) 10 s   (ii) 20 m/s   (iii) 0.75 m/s², 0.5 m/s²   (iv) 1 m/s²   (v) 300 m
6. 0.4 m/s², 0.5 m/s²; 224 m
7. (i) At 15 s and 80 s   (ii) 0.8 m/s²   (iii) 1.2 m/s²   (iv) 1560 m
8. (i) 16   (ii) 0.8 m/s²   (iii) 1.6 m/s²

Exercise 20.1 (page 172)

1. (i) $A \propto r^2, A = kr^2$ (ii) $d \propto v, d = kv$ (iii) $w \propto h^3, w = kh^3$
2. $y \propto x^2, y = 3x^2$ **3.** $y \propto x, y = \frac{1}{2}x$

4.

x	2	4	6	8	10	20
y	3	6	9	12	15	30

5. (i) 6, 14   (ii) 12   **6.** (i) 45, 125   (ii) 2   **7.** 2.4, 3.6
8. 1.6, 10   **9.** (i) 128, 250   (ii) 2   **10.** 3.6, ±7

11. (i) $v = 5h^3, 135$ (ii) $\sqrt[3]{\frac{v}{5}}, 4$ **12.** $y = 3\sqrt{x}, \pm 6; x = \frac{y^2}{9}, 16$

Exercise 20.2 (page 175)

1. (a)(i) divided by 3   (ii) multiplied by 2   (b)(i) 4   (ii) 24
2. y inversely proportional to x (i) 4   (ii) 2
3. y inversely proportional to the square of x (i) $6\frac{1}{4}$ (ii) ±5
4. 8   **5.** $12\frac{1}{2}$ **6.** (i) 8   (ii) 12   **7.** (i) 8   (ii) ±6   **8.** 200, 9.6
10. (i) 9   (ii) 1.44

Exercise 20.3 (page 176)

1. (i) $E = cmv^2$ (ii) $p = \frac{cq}{r}$ (iii) $d = \frac{cm}{x^3}$ **2.** $m = \frac{5}{18}r^2h, 3\frac{1}{3}$

3. $F = \frac{14.4m}{d^2}, 40$ **4.** (i) $y = ax + bx^2$ (ii) $h = \frac{a}{p} + bp$ (iii) $y = a + bn^2$

5. $y = 5x + \frac{12}{x}, 23$ **6.** $S = \frac{1}{2}n + \frac{1}{2}n^2, 28$

Exercise 21.1 (page 178)

1. $\frac{2}{5}$ **2.** $\frac{7}{4}$ **3.** $-\frac{5}{4}$ **4.** $-\frac{3}{5}$

5. The gradients of the tangents are 4 and -2.

6. The gradients at points P, Q and R are 1, 2 and 3.

Exercise 21.2 (page 180)

1. 0–5 s velocity increases; 5–10 s velocity is 6 m/s;
10–15 s velocity decreases; 15–25 s velocity zero;
25–30 s velocity increases

3. (i) 60 m/min, 148 m/min   (ii) 160 m/min, 120 m/min   (iii) 300 m/min at 3 s

4. (i) 10 m/s   (ii) -20 m/s   (iii) 30 m/s

5. (i) 1.1 m/s^2 (ii) 1.5 m/s^2 (iii) at 6 s

6. (i) 2.2 m/s^2 (ii) 1.4 m/s^2 (iii) 6 m/s^2 at $3\frac{1}{2}$ s

Exercise 22.1 (page 184)

1. (i) 4.3 square units   (ii) 4.5 square units

2. 5.3 square units, 5.0 square units

3. 19.3 square units, 22.4 square units   **4.** 28 square units

Exercise 22.2 (page 186)

1. (i) 0.6 m/s^2 (ii) 485 m   **2.** (i) 0.9 m/s^2, 1.7 m/s^2 (ii) $12\frac{1}{6}$ m

3. (i) 2.2 m/s^2 (ii) 46 m   **4.** (i) 120 s   (ii) 80 m/s^2 (iii) 734 km

Exercise 23.1 (page 191)

1. $\begin{pmatrix} 9 & 9 \\ 6 & 6 \end{pmatrix}$ **2.** $\begin{pmatrix} 4 & 7 & 5 \\ 4 & 3 & 5 \end{pmatrix}$ **3.** $\begin{pmatrix} 3 & 3 \\ 2 & 2 \end{pmatrix}$ **4.** $\begin{pmatrix} 0 & 2 & 0 & 1 \\ 0 & 1 & 5 & 4 \end{pmatrix}$

5. $\begin{pmatrix} 3 & 9 \\ 6 & 12 \end{pmatrix}$ **6.** $\begin{pmatrix} 10 & 30 & 5 \\ 15 & 0 & 20 \end{pmatrix}$ **7.** $\begin{pmatrix} 4 & 5 \\ 3 & 2 \end{pmatrix}$ **8.** $\begin{pmatrix} 16 \\ 9 \end{pmatrix}$

9. $\begin{pmatrix} 30 \\ 34 \end{pmatrix}$ **10.** $\begin{pmatrix} 9 \\ 12 \end{pmatrix}$ **11.** $\begin{pmatrix} 23 \\ 19 \end{pmatrix}$ **12.** $\begin{pmatrix} 21 \\ 27 \end{pmatrix}$

13. $\begin{pmatrix} 7 & 18 \\ 6 & 17 \end{pmatrix}$ **14.** $\begin{pmatrix} 12 & 13 \\ 7 & 7 \end{pmatrix}$ **15.** $\begin{pmatrix} 12 & 4 \\ 8 & 7 \end{pmatrix}$ **16.** $\begin{pmatrix} 31 & 33 \\ 14 & 12 \end{pmatrix}$

17. $\begin{pmatrix} 6 & 11 \\ 14 & 24 \end{pmatrix}$ **18.** $\begin{pmatrix} 14 & 9 \\ 0 & 15 \end{pmatrix}$ **19.** $\begin{pmatrix} 3 & 7 \\ 9 & 4 \end{pmatrix}, \begin{pmatrix} 3 & 7 \\ 9 & 4 \end{pmatrix}$; same.

20. $\begin{pmatrix} 5 & 3 \\ 6 & 2 \end{pmatrix}$, rows exchanged; $\begin{pmatrix} 2 & 6 \\ 3 & 5 \end{pmatrix}$, columns exchanged.

21. $\begin{pmatrix} 2 & 0 \\ -6 & 3 \end{pmatrix}$ **22.** $\begin{pmatrix} 9 & -2 & -5 \\ 2 & -7 & 5 \end{pmatrix}$ **23.** $\begin{pmatrix} -2 & -1 \\ -2 & -7 \end{pmatrix}$ **24.** $\begin{pmatrix} 0 & -1 & 1 \\ 0 & -1 & 1 \end{pmatrix}$

25. $\begin{pmatrix} 2 \\ 9 \end{pmatrix}$ **26.** $\begin{pmatrix} -10 \\ -8 \end{pmatrix}$ **27.** $\begin{pmatrix} -7 \\ 7 \end{pmatrix}$ **28.** $\begin{pmatrix} 2 & 2 \\ 2 & 3 \end{pmatrix}$ **29.** $\begin{pmatrix} -2 & 14 \\ -2 & 8 \end{pmatrix}$

30. $\begin{pmatrix} 5 & 6 \\ -11 & -10 \end{pmatrix}$ **31.** $\begin{pmatrix} 8 & 5 \\ 8 & 12 \end{pmatrix}$ **32.** $\begin{pmatrix} 5 & 10 & 5 \\ 9 & 7 & 9 \\ 4 & 2 & 4 \end{pmatrix}$

33. $\begin{pmatrix} 3 & 5 \\ 7 & 1 \end{pmatrix}, \begin{pmatrix} 1 & 1 \\ 1 & 1 \end{pmatrix}, \begin{pmatrix} 6 & 9 \\ 12 & 3 \end{pmatrix}, \begin{pmatrix} 11 & 4 \\ 7 & 8 \end{pmatrix}, \begin{pmatrix} 10 & 5 \\ 6 & 9 \end{pmatrix}$

34. $\begin{pmatrix} 5 & -2 \\ 4 & 3 \end{pmatrix}, \begin{pmatrix} 1 & -2 \\ -4 & 1 \end{pmatrix}, \begin{pmatrix} 10 & 0 \\ 20 & 5 \end{pmatrix}, \begin{pmatrix} -2 & -2 \\ 8 & 2 \end{pmatrix}, \begin{pmatrix} 6 & -4 \\ 12 & -6 \end{pmatrix}$

35. $\begin{pmatrix} 7 & 12 \\ 17 & 2 \end{pmatrix}, \begin{pmatrix} 16 & 9 \\ 12 & 13 \end{pmatrix}, \begin{pmatrix} 7 & 2 \\ 3 & 6 \end{pmatrix}, \begin{pmatrix} 9 & 7 \\ 9 & 7 \end{pmatrix}$ **36.** $4, -3, 2, -3$

37. $2, 6$ **38.** $3, 5, 4, 24$

39. (a) $\begin{pmatrix} 10 & 7 \\ 4 & 7 \end{pmatrix}$, not possible, $\begin{pmatrix} 27 \\ 18 \end{pmatrix}$, not possible, $\begin{pmatrix} 30 & 28 \\ 15 & 17 \end{pmatrix}, \begin{pmatrix} 29 & 27 \\ 16 & 18 \end{pmatrix}$

(b) (i) $\begin{pmatrix} 2 & 3 \\ 2 & 1 \end{pmatrix}$ (ii) $\begin{pmatrix} 6 & 1 \\ 0 & 5 \end{pmatrix}$

40. (a) $\begin{pmatrix} 1 & 3 \\ 1 & -3 \end{pmatrix}, \begin{pmatrix} 7 & 1 \\ -5 & 5 \end{pmatrix}$, not possible, $\begin{pmatrix} -7 \\ 10 \end{pmatrix}, \begin{pmatrix} -6 & -4 \\ 9 & -6 \end{pmatrix}, \begin{pmatrix} -14 & -5 \\ 20 & 2 \end{pmatrix},$

$\begin{pmatrix} 12 & -7 \\ -21 & 19 \end{pmatrix}$ (b) $\begin{pmatrix} -7 & -1 \\ 5 & -5 \end{pmatrix}, \begin{pmatrix} 19 & 7 \\ -11 & 8 \end{pmatrix}$

41. $\begin{pmatrix} 59 \\ 53 \end{pmatrix}$; Sarah spent 59 sols and Tim 53 sols.

42. $\begin{pmatrix} 35 \\ 53 \\ 40 \end{pmatrix}$; A cost £35, B £53, C £40.

43. $\begin{pmatrix} 27 & 40 \\ 27 & 41 \end{pmatrix}$.   (i) Match drawn   (ii) Camptown won by 1 point

Exercise 23.2 (page 196)

1. (i) 2   (ii) 3   (iii) -4 (iv) 1   (v) -1 (vi) 0   (vii) 14   (viii) 0

3. (i) $\begin{pmatrix} 2 & -5 \\ -1 & 3 \end{pmatrix}$ (ii) $\begin{pmatrix} 3 & -2 \\ -10 & 7 \end{pmatrix}$ (iii) $\begin{pmatrix} 9 & -7 \\ -5 & 4 \end{pmatrix}$ (iv) $\begin{pmatrix} -2 & -3 \\ 5 & 7 \end{pmatrix}$

4. $\begin{pmatrix} 2 & -1 \\ -2\frac{1}{2} & 1\frac{1}{2} \end{pmatrix}$ **5.** $\begin{pmatrix} 1 & -\frac{2}{3} \\ -2 & 1\frac{2}{3} \end{pmatrix}$ **6.** $\begin{pmatrix} -1 & 2 \\ 4 & -7 \end{pmatrix}$ **7.** $\begin{pmatrix} -2 & 1\frac{1}{2} \\ 3 & -2 \end{pmatrix}$

8. $\begin{pmatrix} 5 & 3 \\ -7 & -4 \end{pmatrix}$ **9.** $\begin{pmatrix} -\frac{1}{3} & 0 \\ -\frac{5}{6} & -\frac{1}{2} \end{pmatrix}$ **11.** 3   **12.** $12 - 3k, 4$

13. $n^2 - 2n - 15, -3$ and 5

Exercise 24.1 (page 201)

1. (i) $\mathbf{a} + \mathbf{b}$ (ii) $-\mathbf{a} - \mathbf{b}$ (iii) $\mathbf{a} + \mathbf{b} + \mathbf{c}$

2. (i) $\mathbf{d} + \mathbf{e}$ (ii) $-\mathbf{d} + 2\mathbf{e}$ (iii) $-\mathbf{d} + \mathbf{e}$

5. (i) $\begin{pmatrix} 2 \\ 1 \end{pmatrix}$ (ii) $\begin{pmatrix} -2 \\ 2 \end{pmatrix}$ (iii) $\begin{pmatrix} 0 \\ 3 \end{pmatrix}$ (iv) $\begin{pmatrix} 4 \\ -1 \end{pmatrix}$ (v) $\begin{pmatrix} 2 \\ 6 \end{pmatrix}$ (vi) $\begin{pmatrix} 4 \\ 4 \end{pmatrix}$

6. (i) $\begin{pmatrix} 1 \\ 2 \end{pmatrix}$ (ii) $\begin{pmatrix} 5 \\ -6 \end{pmatrix}$ (iii) $\begin{pmatrix} 6 \\ -4 \end{pmatrix}$ (iv) $\begin{pmatrix} -6 \\ 12 \end{pmatrix}$ (v) $\begin{pmatrix} 0 \\ 8 \end{pmatrix}$ (vi) $\begin{pmatrix} -1 \\ 2 \end{pmatrix}$

8. (i) 5   (ii) 13   (iii) 10   (iv) $\sqrt{74}$ **9.** (i) 68.2°   (ii) 26.6°   (iii) 149.0°

10. $2\mathbf{i} + 5\mathbf{j}, 4\mathbf{i} + 10\mathbf{j}, -4\mathbf{i} + 3\mathbf{j}, -2\mathbf{i} + 8\mathbf{j}, 6\mathbf{i} + 2\mathbf{j}$

11. $\begin{pmatrix} 4 \\ 2 \end{pmatrix}, \begin{pmatrix} 5 \\ -1 \end{pmatrix}, \begin{pmatrix} -3 \\ 3 \end{pmatrix}$ **12.** 45° to x-axis   **13.** 5; $-45°$;   (i) 2   (ii) $-1\frac{1}{2}$

14. A $\begin{pmatrix} 1 \\ 2 \end{pmatrix}$, B $\begin{pmatrix} 3 \\ 1 \end{pmatrix}$, C $\begin{pmatrix} 3 \\ 4 \end{pmatrix}$, D $\begin{pmatrix} 0 \\ 4 \end{pmatrix}$, E $\begin{pmatrix} -2 \\ 3 \end{pmatrix}$, F $\begin{pmatrix} 2 \\ -1 \end{pmatrix}$, G $\begin{pmatrix} -2 \\ -2 \end{pmatrix}$

15. $\begin{pmatrix} 2 \\ -1 \end{pmatrix}, \begin{pmatrix} 6 \\ -3 \end{pmatrix}$; 3

16. (i) $\begin{pmatrix} 5 \\ 7 \end{pmatrix}$ (ii) $\begin{pmatrix} -2 \\ 6 \end{pmatrix}$ (iii) $\begin{pmatrix} 6 \\ 12 \end{pmatrix}$ (iv) $\begin{pmatrix} 2 \\ 3 \end{pmatrix}$ (v) $\begin{pmatrix} 6 \\ -3 \end{pmatrix}$ (vi) $\begin{pmatrix} 8 \\ 0 \end{pmatrix}$

17. (i) 8, 1   (ii) 3, -4 (iii) $-3, -3$ (iv) $-5, 2$

Exercise 24.2 (page 205)

1. $\begin{pmatrix} 6 \\ 9 \end{pmatrix}, \begin{pmatrix} 1 \\ 1\frac{1}{2} \end{pmatrix}, \begin{pmatrix} -10 \\ -15 \end{pmatrix}$; $3a, \frac{1}{2}a, -5a$ **2.** $\begin{pmatrix} -3 \\ 4 \end{pmatrix}, \begin{pmatrix} -3 \\ 4 \end{pmatrix}$; collinear

3. (i) Parallel and HK = 5 LM   (ii) Collinear and PQ = 2 QR

5. (i) $2\mathbf{p}$ (ii) $2\mathbf{q}$ (iii) $2\mathbf{q} - 2\mathbf{p}$ (iv) $\mathbf{q} - \mathbf{p}$; parallel and DE = 2 BC

6. (i) $\mathbf{b} - \mathbf{a}$ (ii) $\frac{1}{2}\mathbf{b} - \frac{1}{2}\mathbf{a}$ (iii) $\frac{1}{2}\mathbf{a} + \frac{1}{2}\mathbf{b}$ (iv) $2\mathbf{a}$ (v) $\mathbf{b} - 2\mathbf{a}$

(vi) $\frac{1}{3}\mathbf{b} - \frac{2}{3}\mathbf{a}$ (vii) $\frac{2}{3}\mathbf{b} + \frac{2}{3}\mathbf{a}$; collinear and PT = 3 TV

7. (i) $\mathbf{b} - \mathbf{a}, \frac{1}{2}\mathbf{b} - \frac{1}{2}\mathbf{a}, \frac{1}{2}\mathbf{a} + \frac{1}{2}\mathbf{b}$ (ii) $\frac{1}{3}\mathbf{a} + \frac{1}{3}\mathbf{b}, \frac{1}{3}\mathbf{b} - \frac{2}{3}\mathbf{a}$ (iii) $\frac{1}{2}\mathbf{b} - \mathbf{a}, k = \frac{3}{2}$

8. (a) (i) $\mathbf{b} - \mathbf{a}$ (ii) $\frac{1}{2}\mathbf{b} - \frac{1}{2}\mathbf{a}$ (iii) $\frac{1}{2}\mathbf{a} + \frac{1}{2}\mathbf{b}$

(b) $\frac{1}{2}\mathbf{a} + \frac{1}{2}\mathbf{b} - \mathbf{c}$ (c) (i) $\frac{1}{3}\mathbf{a} + \frac{1}{3}\mathbf{b} + \frac{1}{3}\mathbf{c}, \frac{2}{3}\mathbf{b}$ (ii) On OB so that OP = $\frac{2}{3}$ OB

9. $5\mathbf{q} - 3\mathbf{p}, 2\mathbf{r} - 5\mathbf{q}, 3\mathbf{p} - 2\mathbf{r}$; $\mathbf{q} - \mathbf{p}, \frac{3}{2}\mathbf{q} - \frac{3}{2}\mathbf{p}$; PQ:QR = 2:3

10. (i) $\mathbf{a} - \mathbf{b}$ (ii) $3\mathbf{a} + 2\mathbf{b}$ (iv) $\frac{3}{5}, \frac{1}{5}$ (v) $\frac{2}{3}$

Exercise 25.1 (page 210)

2. (iii) $\begin{pmatrix} -2 \\ -3 \end{pmatrix}$ **3.** (i) $(-2, 4)$ (ii) $(2, -2)$ (iii) $(4, 2)$ (iv) $(-4, -2)$

4. (i) $(-3, 3)$ (ii) $(2, -4)$ (iii) $(5, 3)$ **5.** (iii) $\begin{pmatrix} -2 \\ -2 \end{pmatrix}$

6. Rotation 90° anticlockwise about O
7. (i) $(-2, 3)$ (ii) $(2, -3)$ (iii) $(2, 5)$ (iv) $(-4, 3)$ (v) $(3, 2)$
(vi) $(-3, -2)$
8. (i) $(-1, 3)$ (ii) $(5, 3)$ (iii) $(-3, 3)$ **9.** Reflection in x-axis
10. $x + y = 7, 7$ **11.** (ii) $(5, 4)$ (iii) $\begin{pmatrix} -1 \\ -3 \end{pmatrix}$
12. (i) Rotation through 180° about $(3, 0)$ (ii) Anticlockwise rotation through
90° about $(1, 2)$
13. (i) $(5, 5)$ (ii) no

Exercise 25.2 (page 217)

2. (iii) Centre C, scale factor $\dfrac{2}{3}$

4. $(12, 0)$, 2

5. Translation $\begin{pmatrix} 3 \\ 2 \end{pmatrix}$

6. (i) 10 cm, 4 cm (ii) D, $\dfrac{5}{2}$ (iii) 7 cm, 10.5 cm (iv) $-\dfrac{3}{2}$

Exercise 26.1 (page 222)

1. $(5, -2)$, $(-5, -2)$, $(2, -5)$, $(2, 5)$
2. $(-4, 2)$, $(-1, 5)$, $(-6, 4)$; anticlockwise rotation through 90° about O
3. $(2, 3)$, $(4, 5)$, $(2, 6)$
4. $(-1, -1)$, $(-1, -3)$, $(-2, -3)$; reflection in the line $y = -x$

5. $(0, 0)$. $(3, 0)$, $(3, 2)$, $(0, 2)$; stretch in x direction, magnitude $1\frac{1}{2}$; $\begin{pmatrix} 3 & 0 \\ 0 & 1 \end{pmatrix}$
6. $(5, 10)$, $(11, 2)$, $(7, -1)$; $y = 2x$
7. (i) $(3, 3)$, $(9, 3)$, $(9, 6)$, $(3, 6)$; enlargement, scale factor 3, centre O
(ii) Enlargements, centre O, scale factors 5 and $\dfrac{1}{2}$
8. $(7, 6)$, $(0, 1)$, $(-8, -5)$; original points obtained
9. Enlargement, centre O, scale factor 2. $\begin{pmatrix} 0 & 1 \\ 1 & 0 \end{pmatrix}$

Exercise 26.2 (page 224)

1. (i) $(-1, 3)$, $(4, 1)$; $\begin{pmatrix} -2 & 2 \\ 4 & -1 \end{pmatrix}$ (ii) $\begin{pmatrix} -4 & 1 \\ 2 & 1 \end{pmatrix}$; $(-1, 5)$; no

2. (i) $(9, -4)$, $(18, -21)$; $\begin{pmatrix} -6 & 8 \\ 9 & -10 \end{pmatrix}$ (ii) $\begin{pmatrix} -10 & 12 \\ 6 & -6 \end{pmatrix}$, $(26, -12)$, no

3. Reflection in $y = -x$; reflection in y-axis; clockwise rotation through 90°
about O: $\begin{pmatrix} 0 & 1 \\ -1 & 0 \end{pmatrix}$

4. (i) $\begin{pmatrix} -1 & 0 \\ 0 & -1 \end{pmatrix}$, $\begin{pmatrix} 0 & 1 \\ -1 & 0 \end{pmatrix}$. Anticlockwise rotations through 90°, 180° and 270°
about O

4. (ii) $\begin{pmatrix} 9 & 0 \\ 0 & 9 \end{pmatrix}, \begin{pmatrix} 27 & 0 \\ 0 & 27 \end{pmatrix}$; enlargements, centre O, scale factors 3, 9 and 27

5. $(2, 0), (2, 1)$; stretch in x-direction, magnitude 2; $(0, 2), (1, 2)$; $\begin{pmatrix} 0 & 1 \\ 2 & 0 \end{pmatrix}$

Exercise 26.3 (page 225)

1. $(7, 2), (29, 12)$; $\begin{pmatrix} 2 & -5 \\ -1 & 3 \end{pmatrix}$ **2.** $(7, 6), (-9, -8)$; $\begin{pmatrix} -1 & 1\frac{1}{2} \\ -2 & 2\frac{1}{2} \end{pmatrix}$

3. $(4, 1)$; $\begin{pmatrix} 2 & -5 \\ -3 & 8 \end{pmatrix}$ **4.** $(13, 9)$; $\begin{pmatrix} -\frac{1}{2} & 1\frac{1}{2} \\ -1 & 2 \end{pmatrix}$ **5.** $(-5, 2)$

6. $(0, 0), (12, 6), (16, 8), (4, 2)$; 0; singular
7. $(10, 15), (-2, -3), (4, 6)$
8. $(1, 0), (3, 3), (1, 3), (-1, 0)$; 6 square units; $3:1$

9. (i) $\begin{pmatrix} -1 & \frac{3}{4} \\ -1 & \frac{1}{2} \end{pmatrix}$ (ii) $(3, -1)$ (iii) $\begin{pmatrix} -8 & 6 \\ -8 & 4 \end{pmatrix}$, $k = 8$

(iv) $\begin{pmatrix} 8 & 0 \\ 0 & 8 \end{pmatrix}$; enlargement, centre O, scale factor 8

10. $63°$; 5; $\sqrt{5}$; $\frac{1}{5}\begin{pmatrix} 1 & 2 \\ -2 & 1 \end{pmatrix}$

11. (b) $\begin{pmatrix} 0 & 8 & 2 & -6 \\ 0 & 6 & 14 & 8 \end{pmatrix}$ (d) 10; 100

(e) Apply det **T** to original area   (f) $\begin{pmatrix} 0.4 & 0.3 \\ -0.3 & 0.4 \end{pmatrix}$; $(1, -2)$

Exercise 27.1 (page 229)

1. (i) $23°, 84°, 67°$ (ii) $108°, 144°, 98°$ (iii) $305°, 230°, 195°$
2. (i) $110°$ (ii) $47°$ (iii) $60°$ **3.** (i) $80°$ (ii) $55°$ (iii) $60°$ (iv) $20°$
4. (i) $130°$ (ii) $70°$ (iii) $90°$ (iv) $12°$
5. (i) $70°$ (ii) $10°$ (iii) $55°$ (iv) $18°$
6. (i) $100°$ (ii) $120°$ (iii) $115°$ (iv) $50°$
7. (i) $130°, 50°, 130°$ (ii) $64°, 116°, 64°$ (iii) $45°$ (iv) $50°$
8. (i) $90°$ (ii) $120°$ (iii) $300°$ **9.** (i) $30°$ (ii) $210°$ (iii) $10°$
10. (i) $15°$ (ii) $55°$ (iii) $720°$ **11.** $72°$

Exercise 27.2 (page 232)

1. $50°, 70°$ **2.** $125°, 38°$ **3.** $76°, 104°$ **4.** $80°, 55°, 45°$
5. $75°, 42°, 63°$ **6.** $50°, 30°, 80°$ **7.** $50°$ **8.** $30°$ **9.** $75°, 105°$
10. (i) g, m (ii) g, h (iii) h, m (iv) $115°$

Exercise 27.3 (page 236)

1. (i) 55° (ii) 48° (iii) 50°   **2.** (i) 125° (ii) 82° (iii) 75°
3. (i) No (ii) no (iii) yes (iv) no
4. (i) Isosceles (ii) 52°, 76° (iii) 70° (iv) 42° (v) 60°; equilateral
5. 25°   **6.** 80°, 50°   **7.** 40°, 40°, 20°   **8.** 70°, 70°, 40°
9. 70°, 130°, 50°   **10.** 20°, 40°, 40°, 100°   **11.** 30°, 64°, 86°, 86°
12. 40°, 40°, 50°   **13.** 30°, 85°, 65°   **14.** 20°   **15.** 96°
16. 15°; 37°, 102°, 41°

Exercise 27.4 (page 239)

1. 540°; 106°   **2.** 720°; 130°   **3.** (i) 1800° (ii) 2700°
4. (i) 45°, 135° (ii) 72°, 108° (iii) 30°, 150°

5. (i) 162° (ii) 177° (iii) $128\frac{4}{7}°$ **6.** (i) 6 (ii) 20 (iii) 36°

7. (i) Yes, 9 (ii) no (iii) no (iv) yes, 8 (v) yes, 36
8. (i) 50° (ii) 102°   **9.** 30   **10.** 44   **11.** 150°   **12.** 36°
13. 36°, 108°, 36°; 36°, 72°, 72°   **14.** 30°, 90°, 60°

Exercise 28.1 (page 242)

1. 15, 12   **2.** 20 cm, 8 cm   **3.** 4 cm, 8 cm   **4.** 5.6 cm, 7.2 cm
5. 27 cm, 13.5 cm   **6.** 4.5, 7.5   **7.** 6.4 cm, 2.4 cm, 9.6 cm   **8.** 5 cm
9. ADB and CEB   **10.** (i) HED, 6 cm (ii) EHK, 3.6 cm, 7.5 cm, 3 cm
11. 6.4 cm, 14.4 cm   **12.** 23 m   **13.** 1.8 m

Exercise 28.2 (page 246)

1. (i) 1:4 (ii) 1:4 (iii) 1:4 (iv) 6 cm² (v) 96 cm² (vi) 1:16
2. 72 m²   **3.** (i) 80 cm² (ii) 5 cm²   **4.** 4, 112 cm²   **5.** 81 cm²
6. 9 cm   **7.** (i) 2:3 (ii) 2:3; yes; 4:9; 45 kg
8. (i) 9:25, 3:5 (ii) 1 cm to 180 m

Exercise 28.3 (page 247)

1. 128 cm³   **2.** 32 cm³   **3.** 16 cm³   **4.** 160 cm³
5. (i) 3:2 (ii) 216 cm², 96 cm², 9:4 (iii) 216 cm³, 64 cm³, 27:8
6. (i) 2:3 (ii) 4:9 (iii) 8:27   **7.** (i) 4:5 (ii) 64:125

8. (i) 5:4 (ii) 125:64 (iii) $1562\frac{1}{2}$ ml

9. (i) 3:4 (ii) 27:64 (iii) 9:16; 1920 g

Exercise 28.4 (page 249)

1. EDF and JGH; DF and GH   **2.** STR and ZYX; Ŝ and Ẑ
3. ABC and GHJ   **4.** ABC, DEF and KLM
5. NPQ, RST and XYZ

Exercise 29.1 (page 253)

8. H, N, X, Z   **10.** 4, 5, 3; third one.   **14.** $y = x, x + y = 6$

Exercise 29.2 (page 257)

1. 55°, 70°   **3.** ii, iv, vii
5. (i) Isos. trap.   (ii) rhom., rect., squ.   (iii) rhom., squ.
6. (i) Kite   (ii) rhom., rect., squ.   (iii) rect., squ.
7. (i) Isos. trap., rect., squ.   (ii) kite, rhom., squ.   (iii) isos. trap., kite
8. 90°, 60°; 45°, 105°, 90°   **9.** 45°, $67\frac{1}{2}°$, $67\frac{1}{2}°$
11. GH = HJ = JK = KG; GM = HM = JM = KM. GJ = HK; angles 45° and 90°.

Exercise 30.1 (page 258)

1. 9.43 cm   **2.** 11.2 cm   **3.** 13 cm   **4** 9.1 cm   **5.** 6.71 cm
6. 8 cm   **7.** 13.3 cm   **8.** 3.9 cm   **9.** (i) 10 cm   (ii) 14.2 cm
10. (i) 12 cm   (ii) 11.5 cm   **11.** 5.20 m   **12.** 18.0 n.m.   **13.** 20 cm
14. 14.1 cm   **15.** 10   **16.** 10.8 cm   **17.** 5, 8.60, 4.56 units
18. 4 cm   **19.** 12 cm, 180 cm²   **20.** (i) 12 cm   (ii) 108 cm²
21. 9.75 cm   **22.** (i) 20 cm   (ii) 25 cm   **23.** 2.24 m; 3.16 m

Exercise 30.2 (page 261)

1. (i) 5 m   (ii) 15 m   (iii) 50 cm   (iv) 25 cm
2. (i) 13 cm   (ii) 26 cm   (iii) 52 m   (iv) 1300 m
3. (i) 4 cm   (ii) 3 cm   (iii) 6 m   (iv) 12 m   **5.** $\hat{V}$ **8.** 20 cm, 15 cm
9. 30
10. (i) Obtuse   (ii) right angle   (iii) 3 acute   (iv) obtuse   (v) 3 acute
     (vi) right angle

Exercise 31.1 (page 265)

1. 9.21 cm   **2.** 5.80 cm   **3.** 6.83 cm   **4.** 2.47 cm   **5.** 10.3 cm
6. 4.37 cm   **7.** $\frac{g}{h}, \frac{f}{g}$ **8.** $\frac{g}{h}, \frac{f}{h}, \frac{g}{f}$ **9.** 2.30 cm   **10.** 5.00 cm
11. 38.7 cm   **12.** 15.2 cm   **13.** 8.16 cm   **14.** 7.06 cm   **15.** 10.9 cm
16. 1.26 cm

Exercise 31.2 (page 267)

1. 55.2°   **2.** 36.9°   **3.** 30°   **4.** 58.0°   **5.** 39.7°
6. 48.2°   **7.** 29.1°   **8.** 61.0°   **9.** 36.9°   **10.** 51.2°
11. 41.4°   **12.** 56.6°   **13.** 68.2°   **14.** 39.5°

Exercise 31.3 (page 269)

1. 65.4°   **2.** 3.09 cm, 6.18 cm   **3.** 51.3°, 77.4°   **4.** 5.75 cm
5. 8.16 cm   **6.** 68.7°, 68.7°, 42.6°   **7.** 73.7°   **8.** 72°, 11.8 cm
9. 11.3 cm   **10.** 10.5 cm   **11.** 29.1 cm   **12.** 16.4 cm   **13.** 23.4 cm
14. 16.0 cm

Exercise 31.4 (page 270)

1. $\frac{4}{5}, \frac{3}{4}$ **2.** $\frac{4}{5}, \frac{3}{5}$ **3.** $\frac{8}{17}, \frac{8}{15}$ **4.** $\frac{3}{\sqrt{13}}, \frac{9}{13}$

5. $\dfrac{\sqrt{40}}{7}, \dfrac{40}{49}$ **6.** $\dfrac{\sqrt{21}}{2}, \dfrac{21}{25}$ **7.** $\dfrac{1}{2}$ **8.** $\dfrac{\sqrt{3}}{2}, \dfrac{1}{\sqrt{3}}$ **9.** $1, \dfrac{1}{\sqrt{2}}, \dfrac{1}{\sqrt{2}}$

Exercise 31.5 (page 273)

1. 9.90 m   **2.** 35.5°   **3.** 29.7 m   **4.** 80.0 m   **5.** 031.6°
6. (i) 8.32 km N   (ii) 39.1 km W   **7.** 50.5 m   **8.** 1300 m   **9.** 425 m
10. (i) 5.62 km   (ii) 13.9 km   **11.** 397 m, 261 m, 136 m   **12.** 391 m
13. (i) 12.2 m   (ii) 35.7 m   (iii) 23.5 m   **14.** (i) 9.83 cm   (ii) 6.88 cm
15. 143 m, 275 m, 59°, 149°, 235 m   **16.** 350 m, 127 m, 21°
17. 10.1 m, 21.6 m, 25°   **18.** (i) 107 m, 87 m, 138 m   (ii) 43°

Exercise 31.6 (page 276)

1. Negative, positive, negative   **2.** 1, 0, −1, 0, −1, 0
3. Increases from 0 to 1, then decreases to −1 and then increases to 0
4. Both 0.4226 to 4 d.p.   **5.** Both 0.2588 to 4 d.p.
6. cos 110° = −cos 70°, cos 250° = −cos 70°, cos 290° = cos 70°;
tan 110° = −tan 70°, tan 250° = tan 70°, tan 290° = −tan 70°.
7. Both 0.5446 to 4 d.p.   **8.** 0.8387 and −0.8387 to 4 d.p.
9. 0.6494 and −0.6494 to 4 d.p.
10. 30° and 150°; 210° and 330°   (i) line symmetry about $x = 90$
(ii) line symmetry about $x = 270$ (iii) point symmetry about $(180, 0)$;
$\theta = 20°$, $\phi = 65°$
11. 60°, 300°, 120° and 240°; line symmetry about $x = 180$; 335° and 200°; point
symmetry about $(90, 0)$; 155°
12. 58°, 238°; 122°, 302°; 11.43, 28.64, −19.08; approaches infinity
13. 120° and 240°

Exercise 31.7 (page 278)

1. (i) 160°   (ii) 100°   (iii) 130°   (iv) 148°
2. (i) −cos 70°   (ii) sin 50°   (iii) −tan 15°   (iv) sin 27°   (v) −cos 33°
(vi) −tan 81°
3. (i) 66°   (ii) 114°   (iii) 38°   (iv) 142°
4. (i) 130°   (ii) 155°   (iii) 158°   (iv) 106°
5. (i) 62°, 118°   (ii) 33°, 147°
6. (i) 30°   (ii) 150°   (iii) 124°   (iv) 56°
7. (i) 56°   (ii) 124°   (iii) 64° or 116°   (iv) 33°   (v) 147°

Exercise 32.1 (page 281)

1. 3 cm   **2.** 13 cm   **3.** 8 cm   **4.** 24 cm   **5.** 8.60 cm   **6.** 6.32 cm
7. 13.9 cm   **8.** 6 cm, 45°, 90°   **9.** 0.7, 44.4°, 88.8°   **10.** 13.1 cm
11. (i) 14 cm   (ii) 2 cm   **12.** 10 cm   **13.** (i) 2 cm   (ii) 16 cm

Exercise 32.2 (page 283)

1. 2 cm   **2.** 6 cm   **3.** 7.5 cm   **4.** 8.4 cm   **5.** 15 cm   **6.** 12 cm
7. 5 cm   **8.** 7 cm   **9.** 4 cm   **10.** 10 cm   **11.** 5 cm
12. 7.8 cm   **13.** 4 cm   **14.** 3 cm   **15.** Equal; 8, 10, 5 cm
16. 13, 9, 6, 12 cm   **17.** 2.4 cm, 12 cm   **18.** $8\dfrac{1}{4}$ cm, $13\dfrac{1}{2}$ cm

19. (i) 6 cm, 6 cm   (ii) both $\sqrt{xy + y^2}$ **20.** $10 - x$, 1 cm

Exercise 32.3 (page 288)

1. 140°   **2.** 54°   **3.** 230°   **4.** 155°   **5.** 100°, 260°, 130°
6. 134°   **7.** 200°, 160°, 80°   **8.** 120°, 30°, 30°   **9.** 30°, 70°, 30°
10. 28°, 65°, 87°   **11.** 24°, 76°, 80°   **15.** 90°, 52°
17. 58°, 116°, 32°   **18.** 136°, 68°   **19.** 110°, 140°, 55°
20. 50°, 60°, 50°   **21.** 56°, 34°, 34°, 17°, 17°

Exercise 32.4 (page 291)

1. 70°   **2.** 102°   **3.** 45°, 115°   **4.** 72°   **5.** 56°, 44°   **6.** 50°, 50°
8. 124°, 28°   **9.** 70°, 110° parallel   **10.** 90°, 70°, 110°, 20°, 50°
11. 70°, 110°, 35°

Exercise 32.5 (page 293)

1. 64°   **2.** 34°   **3.** 10 cm, 11 cm, 13 cm   **4.** 61°, 57°, 62°, 61°, 57°
5. 150°   **6.** 54°, 40°, 86°   **7.** 8 cm, 36.9°   **8.** 15 cm, 56.1°
9. 8.49 cm, 79.0°   **10.** 8.75 cm, 92.8°   **11.** 11.9 cm, 15.6 cm
12. 16.4 cm, 18.2 cm

Exercise 32.6 (page 295)

1. 49°, 83°   **2.** 68°, 35°   **3.** 50°   **4.** 110°, 40°
5. 90°, 40°, 40°   **6.** 30°   **7.** 80°, 48°, 52°   **8.** 65°, 65°, 50°
9. 52°, 52°, 76°   **10.** 54°, 54°   **11.** 62°   **12.** 46°, 88°

Exercise 33.1 (page 299)

10. 5.47 cm   **11.** 5 cm, 4.36 cm, 143.4°, 96.6°
12. 10 cm, 5.85 cm, 136.9°, 73.1°   **13.** 9.95 cm, 3.12 cm
14. 3.44 cm   **15.** 50 m, 54 m; 240°

Exercise 33.2 (page 302)

2. (4, 3), (−4, 3), (−4, −3), (4, −3)   **3.** It is mid-point of ML
4. 3.74 cm   **6.** 3 cm
7. (i) Horizontal straight line, 10 cm above ground
      (ii) Circle, centre C, radius 30 cm
      (iii) Quarter of a circle
8. 5.3 cm
9. (a) Perpendicular bisector of XY
      (b) Two straight lines parallel to XY and 5 cm from it
      (c) Major arc of circle
10. 12.3 cm

Exercise 34.1 (page 305)

1. 8.36 cm   **2.** 4.94 cm   **3.** 5.38 cm   **4.** 16.9 cm
5. 65°, 7.15 cm   **6.** 47°, 28.8 cm   **7.** 8.54 cm   **8.** 9.09 cm
9. 0.3750, 22.0°   **10.** 24.8°   **11.** 22.8°   **12.** 50.0°, 24.0°

Exercise 34.2 (page 307)

1. 8.34 cm   **2.** 13.5 cm   **3.** 13.9 cm   **4.** 3.91 cm   **5.** 12.2 cm
6. 17.2 cm
7. 5.62 cm   **8.** 20.7 cm   **9.** $p^2 = q^2 + r^2 - 2qr \cos P$, 9.14 cm
10. $q^2 = p^2 + r^2 - 2pr \cos Q$, 8.75 cm   **11.** 73.4°   **12.** 55.8°   **13.** 61.1°
14. 68.7°   **15.** 44.6°   **16.** 60.3°   **17.** 101.5°   **18.** 109.5°
19. 127.2°   **20.** 107.2°   **21.** 40.5°   **22.** 81.7°

Exercise 34.3 (page 308)

1. 48°, 259 m, 220 m   **2.** 60°, 8.68 km, 7.08 km   **3.** 36.6 km
4. 5.57 km   **5.** 33.9 m   **6.** 055.1°, 134.7°   **7.** 60.6°, 119.4°
8. 1240 m   **9.** 104.5°, 165.5°   **10.** (a) 26.8 nautical miles   (b) 26.5 m
11. (i) 3.66 nm   (ii) 3.54 nm   (iii) 14.05 h

Exercise 34.4 (page 309)

1. 44.9 cm^2 **2.** 38.8 cm^2 **3.** 72°, 17.1 cm^2, 85.6 cm^2
4. 260 cm^2 **5.** 18.8 cm^2, 37.6 cm^2, $xy \sin \theta$

Exercise 35.1 (page 311)

5. (i) 9°   (ii) 5, 12   (iii) 63°, 90°   (iv) 54°, 6
6. 48°, 60°, 48°, 36°, 36°, 132°   **7.** 72°, 88°, 64°, 136°
8. 24°, 51°, 96°, 69°, 120°   **9.** 24°, 30°, 24°, 66°, 54°, 42°, 120°

Exercise 36.1 (page 314)

1. (i) 6   (ii) 12   (iii) 3   (iv) 1.4   **2.** (i) 16   (ii) 13   (iii) 15
3. (i) 19   (ii) 16   (iii) 24   **4.** 55000   **5.** 570   **6.** 76, 14
7. 16   **8.** £89.88   **9.** 22.1 yr   **10.** 47.4 kg   **11.** 175 cm
12. 13   **13.** (i) 6   (ii) 19   (iii) 15   (iv) 0.5
14. (i) 6   (ii) 13   (iii) 2.5   (iv) 0.4
15. 65 kg   **16.** 155.5 cm   **17.** (i) 93 km   (ii) 88 km
18. (i) 26   (ii) 40   **19.** (i) 9   (ii) 5   **20.** (i) 3   (ii) 3
21. (i) 7   (ii) 4   **22.** 8, 7   **23.** 2, 1.9   **24.** 63
25. (i) 76   (ii) 264   **26.** 202   **27.** (i) 31   (ii) 97.5   **28.** 1980

Exercise 37.1 (page 320)

1.

No. of goals	0	1	2	3	4	5	6
Frequency	4	9	8	4	3	0	2

1; 5

2.

No. of pairs	4	5	6	7	8	9	10
Frequency	4	8	11	7	6	3	1

Mode is 6

3. (i) 4, 5   (ii) 16, 15   **4.** 1, 2, 4 years; 1–3   **5.** 1–2 kg
6. 3, 3, 6, 6 cm

Exercise 37.2 (page 323)

1. (i) £5.20   (ii) 13p   **2.** (i) 60   (ii) 3   **3.** 1, 1.8
4. 1, 2, 2.26   **5.** 7, 6, 6.2
6. 17.95 min   **7.** 3.8 kg   **8.** 3.45 km   **9.** (i) 0   (ii) 2   (iii) 2.4

Exercise 38.1 (page 325)

1. (a) 12; 11; 13, 8, 5   (b) 8; 8; 9, 5, 4   (c) 27; 14; $23\frac{1}{2}$, $6\frac{1}{2}$, 17
2. (a) 9, 6, 11, 5   (b) 6, 4, 8, 4   (c) 8, 5, 9, 4
3. (i) 5.9 and 9.8   (ii) 2.4 and 4.2
4. $10\frac{1}{2}$ and $4\frac{1}{2}$ degrees Celcius

Exercise 38.2 (page 327)

1. 47, 24; 24; 16   **2.** 507 g, 12 g; 15%
3. 18 km, 12 km; 13 km, $21\frac{1}{2}$ km   **4.** $11.27\frac{1}{2}$, 46, 48
5. 890 g, 900 g; 95 g, 102 g; 36%, 42%

Exercise 39.1 (page 330)

1. (i) $\frac{1}{6}$ (ii) $\frac{1}{2}$ (iii) $\frac{1}{3}$ (iv) 0   **2.** (i) $\frac{5}{9}$ (ii) $\frac{4}{9}$ (iii) $\frac{1}{3}$ (iv) $\frac{4}{9}$

3. (i) $\frac{2}{5}$ (ii) $\frac{9}{25}$ (iii) $\frac{6}{25}$ **4.** (i) $\frac{4}{11}$ (ii) $\frac{7}{11}$ (iii) $\frac{3}{11}$ (iv) 0

5. (i) $\frac{1}{2}$ (ii) $\frac{3}{8}$ (iii) $\frac{5}{8}$ (iv) $\frac{1}{8}$ (v) $\frac{7}{8}$ (iv) 0

6. (i) $\frac{1}{4}$ (ii) $\frac{3}{4}$ (iii) $\frac{1}{13}$ (iv) $\frac{12}{13}$ (v) $\frac{1}{52}$ (vi) $\frac{1}{2}$ (vii) $\frac{2}{13}$ (viii) $\frac{3}{13}$

7. (i) 20   (ii) 60   (iii) 40   **8.** 45   **9.** 50   **10.** 250, 200, 150
11. 6 red, 14 green   **12.** (i) $\frac{1}{8}$ (ii) $\frac{1}{4}$ (iii) $\frac{1}{16}$ (iv) 0

13. (i) $\frac{3}{16}$ (ii) 0   (iii) $\frac{5}{16}$ (iv) $\frac{1}{4}$ **14.** (i) $\frac{1}{4}$ (ii) $\frac{1}{2}$ (iii) $\frac{1}{4}$

15. (i) $\frac{1}{6}$ (ii) $\frac{1}{12}$ (iii) $\frac{8}{9}$ (iv) $\frac{5}{12}$ **16.** $\frac{1}{72}$

Exercise 39.2 (page 333)

1. (i) $\frac{1}{2}$ (ii) $\frac{1}{6}$ (iii) $\frac{2}{3}$ **2.** $\frac{8}{15}$ **3.** (i) $\frac{1}{4}$ (ii) $\frac{1}{2}$ (iii) $\frac{1}{8}$

4. (i) $\frac{1}{13}$ (ii) $\frac{2}{13}$ (iii) $\frac{3}{13}$ **5.** $\frac{1}{24}$ **6.** (i) $\frac{1}{4}$ (ii) $\frac{1}{13}$ (iii) $\frac{4}{13}$;
not mutually exclusive, since there is an ace of diamonds

7. (i) 0.4   (ii) 0.3   (iii) 0.7   **8.** $\dfrac{1}{36}$

9. (i) $\dfrac{4}{9}$ (ii) $\dfrac{1}{3}$; not mutually exclusive; $\dfrac{2}{3}$

10. (i) $\dfrac{1}{36}$ (ii) $\dfrac{5}{36}$ (iii) $\dfrac{5}{36}$ (iv) $\dfrac{5}{18}$ **11.** (i) $\dfrac{1}{16}$ (ii) $\dfrac{1}{16}$ (iii) $\dfrac{1}{8}$

12. (i) $\dfrac{1}{36}$ (ii) $\dfrac{1}{36}$ (iii) $\dfrac{1}{6}$ **13.** (i) $\dfrac{1}{7}$ (ii) $\dfrac{1}{49}$ (iii) $\dfrac{1}{7}$

14. (i) 0.04   (ii) 0.32   (iii) 0.36

15. (i) (a) $\dfrac{17}{36}$ (b) $\dfrac{1}{8}$ (c) $\dfrac{95}{144}$ (d) $\dfrac{7}{72}$ (ii) 49

Exercise 39.3 (page 336)

1. $\dfrac{5}{36}, \dfrac{5}{36}, \dfrac{25}{36}$ (i) $\dfrac{1}{36}$ (ii) $\dfrac{5}{18}$ (iii) $\dfrac{25}{36}$ **2.** (i) $\dfrac{1}{16}$ (ii) $\dfrac{3}{8}$ (iii) $\dfrac{9}{16}$

3. (i) $\dfrac{1}{25}$ (ii) $\dfrac{8}{25}$ (iii) $\dfrac{16}{25}$ **4.** (i) $\dfrac{25}{64}$ (ii) $\dfrac{15}{32}$ (iii) $\dfrac{9}{64}$

5. $\dfrac{5}{14}, \dfrac{15}{28}, \dfrac{3}{28}$ **6.** (i) $\dfrac{7}{15}$ (ii) $\dfrac{7}{15}$ (iii) $\dfrac{1}{15}$

7. (i) $\dfrac{4}{25}$ (ii) $\dfrac{12}{25}$ (iii) $\dfrac{9}{25}$ **8.** (i) $\dfrac{1}{15}$ (ii) $\dfrac{8}{15}$ (iii) $\dfrac{2}{5}$

9. (i) $\dfrac{1}{64}$ (ii) $\dfrac{9}{64}$ (iii) $\dfrac{27}{64}$ (iv) $\dfrac{27}{64}$

10. (a) $\dfrac{13}{30}$ (b) $\dfrac{26}{145}$ (c) $\dfrac{48}{145}$ (d) $\dfrac{97}{145}$

Index

acceleration   167
acute angle   228
allied angles   232
alternate angles   231
alternate segment   294
angle of depression   271
angle of elevation   271
approximations   30, 33, 36
arc length   65
are   60

bar chart   310
bearings   272
bilateral symmetry   251
binomial products   91

capacity   69
chord   280
circle area   63
circle circumference   63
column matrix   189
common factor   17, 100
common multiple   17
complementary angles   228
complement of a set   7
composite functions   148
compound interest   50
cone   75
congruent triangles   248
coordinates   151
corresponding angles   231
cosine graph   278
cosine of an angle   263, 276
cosine rule   306
cost price   46
cross multiplying   116
cube   34, 68
cube root   34
cuboid volume   68
cumulative frequency   326
cylinder surface area   72
cylinder volume   72
cyclic quadrilateral   290

deciles   327
decimal places   31
denominator   20

dependent events   336
depreciation   50
determinant   194
directed numbers   85
direct proportion   54, 170
direct variation   170
domain   143
distance–time graphs   165, 179

empty set   2
enlargement   214, 221
equal sets   1
equations   112, 128, 133, 152
equilateral triangle   235
equivalent fractions   20, 107
even numbers   16

factors   16, 100
foreign exchange   55
formulae   121
frequency   317
frustum   77
functions   143

gradient of straight line   153, 177
gradient of curve   177
graphs   150

hectare   60
highest common factor   17
histogram   319
hyperbola   174
hypotenuse   235, 268

identities   93
identity matrix   189
image   143, 208
improper fraction   20
independent events   332
indices   84, 97
inequalities   118, 163
integers   16
intercept   153
interest   48
intersection of graphs   160
intersection of sets   3
inverse functions   146

inverse matrix   194
inverse proportion   54, 173
inverse variation   173
irrational numbers   16
isosceles triangle   235, 268

joint variation   176

kite   255
knot   56

linear equations   112, 128, 152
linear inequalities   118
litre   69
loci   300
lowest common multiple   17

mapping   143, 225
matrices   188, 220
maximum value   96
mean   313, 321
median   313, 320
minimum value   96
mixed number   20
mode   113
modulus of vector   199
multiples   17
mutually exclusive events   332

natural numbers   16
nautical mile   56
null matrix   189
null set   2
numerator   20

obtuse angle   228
odd numbers   16
ogive   326
order of matrix   189

parabola   157
parallel lines   231
parallelogram   61, 256
percentages   41
percentiles   327
perfect cubes   34
perfect squares   34, 94
pi (π)   63
pie chart   310
point symmetry   252
polygon   238
position vector   200
power   34
prime numbers   16
principal   48
prism   68

probability   329
profit   46
proper fraction   20
proper subset   2
proportion   54
pyramid volume   74
Pythagoras' theorem   258, 270

quadrilateral   239, 255
quadratic equation   133, 158
quartiles   325

range of function   143
range of set of values   325
ratio   52
rational numbers   16
reciprocal   36
rectangle   59, 256
recurring decimals   32
reflection   208, 221
reflex angle   228
regular polygon   239
rhombus   256
roots   133
rotation   210, 221
rotational symmetry   252
row matrix   189

scalars   197
scale factor   214
scalene triangle   235
scale of a map   53
sector area   65
segments   287
selling price   46
set   1
significant figures   30
similar figures   241
simple interest   48
simultaneous equations   128, 142, 155
sine graph   277
sine of angle   263, 276
sine rule   304
singular matrix   195
speed   56, 166
sphere   76
square   34, 59, 256
square root   34, 39
standard form   37
straight line graph   152
stretch   216, 221
subset   2
supplementary angles   228
surd   40
symmetric difference of sets   11
symmetry   251

tangent graph   278
tangent of angle   263, 276
tangent to circle   292
tonne   58
translations   208
transversal   231
trapezium   62, 255
travel graphs   165, 179
tree diagrams   335
triangle, types of   235
triangle area   61, 309
trinomial   102

union of sets   4

unit base vectors   200
unit matrix   189, 194
universal set   7

variation   170
vectors   197
velocity   166
velocity–time graph   166, 180, 185
Venn diagram   3, 12
vertically opposite angles   228
volume   68

zero index   97
zero matrix   189, 194